Springer
Berlin
Heidelberg
New York
Barcelona
Hong Kong
London
Milan
Paris
Singapore
Tokyo

Progress in Colloid and Polymer Science

Editors: F. Kremer, Leipzig and G. Lagaly, Kiel

Volume 113 · 1999

Analytical Ultracentrifugation V

Volume Editor:
H. Cölfen

Springer

O
QD
1
K809
v. 113

The series Progress in Colloid and Polymer Science is also available electronically (ISSN 1437-8027)

- Access to tables of contents and abstracts is *free* for everybody.
- Scientists affiliated with departments/institutes subscribing to Progress in Colloid and Polymer Science
  as a whole also have full access to all papers in PDF form. Point your librarian to the LINK access registration form
  at http://link.springer.de/series/pcps/reg-form.htm

ISSN 0340-255X
ISBN 3-540-66175-1
Berlin, Heidelberg, New York

The use of general descriptive names, registered names, trademarks, etc. in this publication does not imply, even in the absence of specific statement, that such names are exempt from the relevant protective laws and regulations and therefore free for general use.

© Springer-Verlag
Berlin Heidelberg 1999

Typesetting: SPS, Madras, India

Cover: Estudio Calamar, F. Steinen-Broo, Pau/Girona, Spain

Printing: Druckhaus Beltz, Hemsbach, Germany

SPIN: 10701577

Printed on acid-free paper
Printed in Germany

Progr Colloid Polym Sci 113: V–VI (1999)
© Springer-Verlag 1999

PREFACE

The series of symposia on analytical ultracentrifugation taking place every 2 years at universities or other institutions in Germany has developed from a German users group meeting to an international conference. The more than 20-years tradition of this meeting was continued on March 25–26, 1999 with the 11th symposium organized by the Max Planck Institute of Colloids and Interfaces at the University of Potsdam.

More than 80 participants from leading groups in 12 countries around the world as well as a large number of new users of analytical ultracentrifugation came together for lively discussions, exchange of ideas and to establish new contacts. In 33 oral presentations and more than 30 posters, a survey of recent developments and results was given, reflecting the wide potential of analytical ultracentrifugation for biophysics/biochemistry, polymer and colloid chemistry as well as hydrodynamics, featuring

- Technical and methodological innovations
- Innovations in data analysis
- Hydrodynamics/Modeling
- Synthetic polymers, colloids and supramolecular systems
- Biological systems
- Interacting systems and assemblies

as well as comparisons of the results of analytical ultracentrifugation with related techniques.

Since the introduction of the new XL-A/I ultracentrifuge generation in 1991 by Beckman Instruments, analytical ultracentrifugation has seen a considerable renaissance about 70 years after the introduction of this technique by Svedberg. Due to the easy data capture, analysis and instrument handling, the popularity of analytical ultracentrifugation is increasing again, reflected in the great interest shown and in the number of new users attending this meeting. Catalysed by the common availibility of cheap and powerful computer resources many innovations in data analysis were presented as well as new measuring methods and technical developments. A large number of the contributions concerned the classical application field of biological systems with a focus on complicated or interacting systems. Also, several contributions were related to the combination of results from analytical ultracentrifugation with those from other techniques, for example, to characterize solution structures of macromolecules. It is remarkable that the number of contributions other than from the classical fields of biophysics and hydrodynamics has increased, indicating that analytical ultracentrifugation might find broader application in polymer and colloid chemistry than has been the case in the last few decades.

Thirty-three contributions were selected for publication in this special volume of *Progress in Colloid and Polymer Science*. It is hoped that this comprehensive collection of the most recent developments in the field of analytical ultracentrifugation will be helpful for all scientists who use this fascinating and powerful analytical technique.

The 11th Symposium on Analytical Ultracentrifugation was generously sponsored by BASF AG (Ludwigshafen), Bayer AG (Leverkusen), Beckman-Coulter GmbH (Munich), the Deutsche Forschungsgemeinschaft (DFG), the Max Planck Society, Roche Diagnostics GmbH (Penzberg) and Schering AG (Berlin). Warm thanks for help during the symposium and its preparation go to Annette Pape, Antje Völkel and Bettina Zilske.

May 1999, Helmut Cölfen

CONTENTS

Progr Colloid Polym Sci (1999) 113:1–9
© Springer-Verlag 1999

TECHNICAL AND METHODOLOGICAL INNOVATIONS

W. Mächtle

The installation of an eight-cell Schlieren optics multiplexer in a Beckman Optima XLI/XL analytical ultracentrifuge used to measure steep refractive index gradients

W. Mächtle
BASF Aktiengesellschaft
Polymer Research Laboratory
D-67056 Ludwigshafen, Germany
e-mail: walter.maechtle@basf-ag.de
Tel.: +49-621-60 48176
Fax: +49-621-60 92281

Abstract Because the Beckman Company was unable to deliver any further drives for our Model E analytical ultracentrifuge (AUC), we were forced to install a modified Model E Schlieren optics in a Beckman Optima XL in order to transform it from a preparative ultracentrifuge into an analytical one. We need to have Schlieren optics available, because Schlieren optics are more universal as interference optics (especially for complex samples). It is only possible to resolve very steep refractive index gradients in the AUC cell and to see the different colours of samples of mixed coloured with white-light Schlieren optics. The configuration of the Schlieren optics is described in detail in the first part of this paper. We used an analytical 8-cell XLI-Ti rotor, a flash lamp as the light source (which can be moved by remote control on a sliding carriage) and a 70 mm reflex film camera instead of the Model E photographic plate. To trigger the flash lamp, we installed a photoelectric cell under the rotor and a small line mirror at the base of the rotor. An electrical circuit (completely independent of the Beckman XL circuits) controlled by a 10 MHz clock allows us to flash any of the eight cells independently and to superimpose any combinations of cells. The modification of the XL itself was only a minor one. We merely drilled holes for the optical path into the heat sink and installed the condensing lens in the rotor chamber. The light source, collimating lens, phase plate, camera lens, cylindrical lens and photographic camera are situated outside the chamber. This modified XL is completely equivalent to the old Model E with Schlieren optics, a flash lamp and an 8-cell multiplexer. Some practical examples of measurements made with this system are described in the second part of this paper. These include steep density gradients in a chemically heterogeneous 11-latex mixture, a rapid steep dynamical H_2O/D_2O density gradient run, synthetic boundary runs of complex coloured core/shell nanoparticles and flash superimposition of different cells during a sedimentation run.

Key words Analytical ultracentrifuge · Schlieren optics · Polymer characterization · Colloidal nanoparticles

Introduction

The analytical ultracentrifuge (AUC) is a versatile instrument for characterizing macromolecules and polyelectrolytes, and it is especially effective for characterizing colloidal particle systems. Because the AUC allows high-resolution fractionation of macromolecules and nanoparticles in terms of both size and density (i.e., chemical composition), it is possible to measure distributions of molar mass, particle size and chemical heterogeneity. There has been a tendency to use the AUC more in recent years because of the launch of the new modern Beckman Optima XLI AUC on the market [1].

In our industrial research laboratory, we test over 3000 samples every year. We use three of the famous old Beckman Model E AUCs, which we have modernized [2–5], two new Beckman Optima XLI [6] and two 8-cell AUC particle sizers [7–9], which we have developed ourselves. This means that four different AUC detectors are available for measuring the concentration of our fractionated samples in the AUC measuring cell: a Schlieren optics detector, an interference optics detector, an 180–800 nm UV scanner and a light scattering/turbidity detector. We have to use the Schlieren optics detector for taking the first "fingerprint" of complex samples that are simultaneously composed of oligomers, macromolecules and (mostly turbid) nanoparticles. This detector is fitted to the Model E, but it is not fitted to the Optima XLI (I stands for interference optics). It therefore came as a shock to us when the Beckman-Coulter Company (Palo Alto, Calif., USA) wrote to us in September 1998 to inform us that they were unable to deliver drives for our three Model Es, and they were unwilling to supply Schlieren optics detectors for their own Optima XLI AUC. This meant that we would have had to give up our Model Es within the next 6–12 months when the present drives wear out, and after that we would have no Schlieren optics in our laboratory. Schlieren optics are essential for us, because they have the following advantages over interference optics:

1. Schlieren optics are more universal than interference optics, especially for complex composed samples.
2. It is only possible to resolve steep refractive index gradients within an AUC cell with Schlieren optics (e.g. in steep density gradients, at very high concentrations, if a Johnston-Ogston effect is present, especially in polyelectrolytes).
3. It is only possible to see different colours in samples of mixed colour with white-light Schlieren optics.
4. Schlieren optics are needed in order to superimpose pictures of different cells taken with flash light.
5. It is only possible to work with the simplest AUC cell, that is the single-sector cell, with Schlieren optics. This cell is cheap, easy to disassemble, to clean and to tighten up, even at maximum rotor speed.

Figure 1 shows the main features of the modern Schlieren optics fitted to our Model E [2–4], and which we wanted to install in our Optima XLI AUC: a triggerable light source (laser or flash lamp), a triggering device, a Schlieren slit (a real one or a virtual one, formed by two cylindrical lenses), a collimating lens in front of and a condensing lens behind the measuring cell, an eight-cell rotor, a phase plate, a

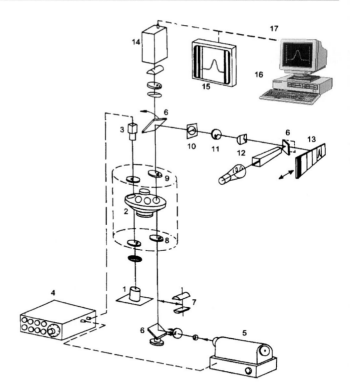

Fig. 1 Optical set-up in our modified Model E AUC [4]: *1* lamp, *2* rotor, *3* photodiode, *4* triggering device, *5* flash lamp or triggerable laser, *6* deflecting mirror, *7* Schlieren slit (real for flash lamp or virtual for laser, formed by two cylindrical lenses), *8* collimating lens for flash lamp or flat window for laser, *9* condensing lens, *10* phase plate, *11* camera lens, *12* cylindrical lens, *13* photo plate camera, *14* digital video camera, *15* live TV monitor, *16* image processing PC, *17* PC monitor with processed digitalized Schlieren curve

camera lens, a cylindrical lens and a photographic plate at the end of the optical light path. We can use a digital TV camera in conjunction with an optical image processing system instead of the photographic plate to analyze Schlieren images, but this is not essential because very complex samples yield complex Schlieren pictures which cannot be analyzed simply and automatically. It is important to have a real photograph in cases such as these.

As Beckman were unable or unwilling to supply the Schlieren optics for their own Optima XLI, we decided to develop them ourselves in the BASF mechanical and electronic workshops. We embarked on a crash course which only took four months, and we achieved our aims simply and cheaply. In order to save time, we decided not to design a completely new compact Schlieren optics for the Optima XLI. Instead, we removed the Schlieren optics from an approved original Model E (with only minor modifications) and installed them in an Optima XL, which is the basic, preparative UC version of the Optima XLI with no interference optics detector. In addition to the XL, we also purchased an analytical XLI eight-cell rotor (Beckman Type AN-50 Ti), which both together cost $70 000.

The physical set-up of this new Optima XL-SO with an eight-cell Schlieren optics (SO) multiplexer is described below, and the performance of this new XL-SO device is demonstrated subsequently with the aid of several different practical examples.

Physical and optical set-up of the XL-SO

Figure 2 shows a cross-section of our XL-SO device and the optical light path. The drawing is true to scale. The numerical values of the focal lengths of the different lenses, the distances of the different optical elements, optical lever arm, calibration constants, etc., have been omitted, because they are exactly the same as in the original Model E Schlieren optics [10]. The arrangement of the optical elements within the optical path in Fig. 2 is virtually the same as in Fig. 1, with, from left to right: (white light) flash lamp with Schlieren slit and green light filter (546 mm) in front of it, collimating lens, 90° glass prism, vacuum-sealed window in the heat sink at the bottom of the vacuum chamber, measuring cell within the rotor, condensing lens, vacuum-sealed window in the top cover of the vacuum chamber, phase plate, camera lens, cylindrical lens, 90° mirror and reflex camera with viewer at the end.

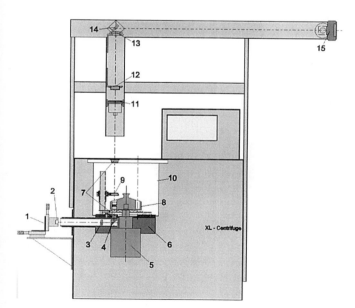

Fig. 2 Physical set-up and optical light path in the XL-SO AUC: *1* flash lamp, *2* Schlieren slit, *3* collimating lens, *4* 90° glass prism, *5* XL drive, *6* heat sink, *7* vacuum-sealed window, *8* eight-cell rotor, *9* condensing lens, *10* vacuum chamber, *11* phase plate, *12* camera lens, *13* cylindrical lens, *14* deflecting mirror, *15* 70-mm film reflex camera with no objective

Our modifications to the original Optima XL centrifuge shown in Fig. 2 were minimal:

(1) We installed an aluminum profile rack for the optical tower above it.
(2) We added a triangular support for the light source outside on the left.
(3) We drilled a horizontal hole into the heat sink and inserted a tube carrying the condensing lens and the 90-degree glass prism.
(4) We drilled two vertical holes into the heat sink and the vacuum chamber cover in order to insert two vacuum-sealed flat-glass windows.
(5) We installed a small support tower for the condensing lens inside the vacuum chamber.

After this general view of our set-up we go into more details. Thus our Schlieren light source, a triggerable flash lamp, is mounted on an adjustable x-y-z-stage, shown in Fig. 2 (on the left side, items 1 and 2). The light source is the same as described in [2]: a Cathodeon C 82007 straight line configuration xenon flash lamp supplied by LOT/Oriel (Darmstadt, Germany) (rise time 5 μs, pulse width 5–50 μs, flash sequence 0–100 Hz, flash energy 1 J, ca. 300 flashes per photo).

We use a Cathodeon-designed power supply type C 512 to run this flash lamp. A small electrical motor is fitted on top of the x-y-z-stage. This is used to continuously move the flash light vertically in the z-direction. This often has to be done to detect very steep density gradients (= refractive index gradients) inside the AUC cell.

The ends of two adjustable tubes jutting out of the centrifuge housing can be seen on the right of the lamp housing in Fig. 2. Their other ends fit into the horizontal 50-mm hole drilled into the heat sink. The 90° glass prism is attached to the end of the tube with the larger diameter, and the condensing lens is fitted to the end of the inner tube with the smaller diameter.

Figure 3 shows our modifications to the heat sink at the bottom of the vacuum chamber. We drilled three holes in it: a large, horizontal 50-mm hole for the tubes with collimating lens and 90° glass prism, a second, vertical 25-mm hole for a vacuum-sealed flat glass window and third a small, vertical threaded 12-mm hole for a vacuum-sealed electrical connector between a photoelectric cell inside the vacuum chamber and our electronic triggering device outside.

Figure 4A shows the open vacuum chamber assembled without the rotor (i.e. the white safety vessel above the heat sink has been fitted and fixed with nine screws). The vacuum-sealed bottom window and the support tower for the condensing lens, with a slit diaphragm above the lens, are situated left of the central rotor drive shaft. This lens is mounted on a black lever, which hinges backwards and forwards through 90° in order to allow the rotor to be inserted. The base of this support

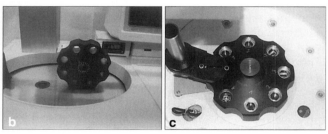

Fig. 3 Photograph of the modified XL heat sink with three additional holes (50, 25 and 12 mm) which we drilled

Fig. 4 A Photograph of the open XL-SO vacuum chamber with drive shaft, adjacent photoelectric cell, vacuum-sealed window, support tower with hinged 90° lever, bearing the condensing lens with a slit diaphragm above it. **B** Photograph of the analytical 8-cell rotor with a polished line mirror at its base, situated on top of the vacuum chamber cover plate with a vacuum-sealed window. **C** Photograph of the open vacuum chamber with the rotor introduced and the lever, bearing the condensing lens with a slit diaphragm above it, in measuring position

tower is fixed with four screws at the bottom, using four of the original threaded holes in the heat sink.

A tiny but very important modification of the original Optima XL is the small black spot near the drive shaft that can be seen in Fig. 4A. We mounted a photoelectric cell onto this drive shaft with a simple plastic ribbon. Some of the electrical wiring which connects the photoelectric cell with the vacuum-sealed connector can be seen on the left underneath the white safety vessel. The photoelectric cell, a "photoreflector (LED-PTr) P 5589", was supplied by Hamamatsu Photonics Deutschland, Herrsching. It consists of a permanently emitting LED (λ = 900 nm) and a fast photo diode (rise and fall time 20 μs) to detect the reflected light.

Figure 4B shows a small polished line mirror that we fitted to the base of our rotor (in the 10 o'clock position near the rotor centre). The LED light reflected by this line mirror yields one sharp pulse per rotor revolution. This pulse is fed into our electronic device incorporating a 10 MHz clock to create the electrical signals that are required to trigger our flash lamp (the principle is described in [3, 4]). For safety reasons it is very important that the triggering of the flash lamp is completely independent of the Beckman Optima XL electronics. The upper vacuum-sealed flat glass window

which we introduced in the cover plate of the vacuum chamber can also be seen in Fig. 4B, to the left of the rotor. The lower end of the optical tower, a square-profile 100-mm aluminum tube, is visible above the window.

Figure 4C shows the rotor installed in the vacuum chamber. The lever bearing the condensing lens with the slit diaphragm above it is shown in the measuring position (it is shown in the retracted position in Fig. 4A). Again, the lower end of the square-profile optical tower is visible above the condensing lens.

Figure 2 shows inside the vertical portion of the optical tower above the vacuum chamber, the positions of the phase plate, the camera lens, the cylindrical lens and the 90° mirror. These four optical elements are all on adjustable stages and they are all original components of the Model E Schlieren optics. We fitted an electrically driven goniometer to the original phase plate to change the Philpot angle precisely. It can also be seen from Fig. 2 that, besides the 90° mirror, the horizontal part of the optical tower only contains a motor-driven 70-mm film Mamiya 640 reflex camera with no objective. We mounted the old viewer of our Model E above the mirror inside the reflex camera. This made it possible for us to take 70-mm film Schlieren photographs and to see the Schlieren pictures live during an AUC run.

Examples of measurements made with the XL-SO

The following examples demonstrate the high Schlieren optical quality, the versatility and the runnability of our new XL-SO device, especially as an eight-cell multiplexer.

We took a very complex colloidal sample for the first examples. The characterization of samples of this type is a speciality of the AUC (see [11]), and virtually no other method can compete. We prepared this colloidal sample by mixing 11 ethylhexyl acrylate-methyl acrylate (EHA/MA) copolymer latices which had all been polymerized separately by standard emulsion polymerization. The percentage w/w of each latex in the sample was the same. All 11 components had nearly the same particle diameter of 200 nm, but the particle densities ρ were different because the compositions of the 11 different copolymers were different [p(MA) = 0/10/20/30/40/50/60/70/80/90/100% w/w]. These particle densities (calculated theoretically from the homopolymer densities and verified via Paar density balance measurements) were $\rho = 0.980$ (P-EHA)/1.000/1.021/1.043/1.066/1.089/1.114/1.140/1.167/1.196/1.225 (P-MA) g/cm^3.

This complex mixture of 11 latices was added to an 88 H$_2$O/12 metrizamide (% w/w) density gradient (DG) mixture, total latex concentration $c = 1$ g/l. The DG mixture was measured separately in our old Model E

and our new XL-SO under the same conditions: 3-mm single sector cell, $-2°$ wedge window, 30 000 rpm, equilibrium running time 27 h. Figure 5 shows the results in the form of two Schlieren photographs taken separately on 70-mm film in the two different AUCs with the same exposure time of 3 s, using a 100 ASA black and white 70-mm Ilford film.

These two Schlieren photographs were identical, which led us to conclude that our new Optima XL-SO and our old Model E are completely equivalent. The quality of the Schlieren optics are the same. It can be seen from the two DG Schlieren photographs in Fig. 5 that the same six latex components nos. 3, 4, 5, 6, 7 and 8 (out of the total 11 components) are present at the same rotor radius positions inside the AUC measuring cells in the form of small turbidity bands.

Both of the Schlieren photographs in Fig. 5 are superimpositions of the single-sector cell in the middle and the counterbalance radius reference cell outside (the two radius reference holes appear on the left and on the right). The superimposition is done automatically by alternately flashing the measuring cell and the reference cell (see [2–4]).

Figure 6 shows a second, very important example of measurements taken with our new XL-SO, a simultaneous eight-cell run. The Schlieren photo of cell no. 1, the counter balance radius reference cell, is shown in Fig. 6 at the top. The Schlieren photographs of seven single-sector cells (electronically linear stretched or compressed in the horizontal rotor radius direction to fit roughly to the exponential ρ-density axis of the photographs at the bottom of Fig. 6) are shown below. Each of the single-sector measuring cells again contained our 11-latex mixture, but this time in seven different (methanol/H$_2$O)/metrizamide DG mixtures, expressed

Fig. 5 Schlieren optics comparison of Model E and Optima XL-SO. Two 70-mm film Schlieren photographs of the same 88 H$_2$O/12 metrizamide density gradients of an 11-latex mixture, showing only the latex components number 3, 4, 5, 6, 7 and 8 as narrow turbidity bands. The Schlieren photographs are identical

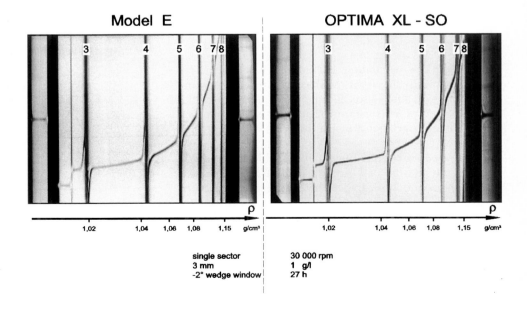

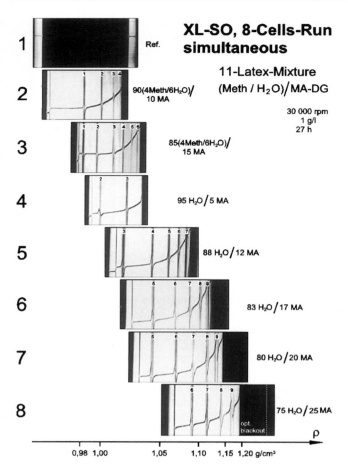

XL-SO, 8-Cells-Run simultaneous

11-Latex-Mixture
(Meth / H₂O)/MA-DG

1 — Ref.

2 — 90(4Meth/6H₂O)/ 10 MA

30 000 rpm
1 g/l
27 h

3 — 85(4Meth/6H₂O)/ 15 MA

4 — 95 H₂O / 5 MA

5 — 88 H₂O / 12 MA

6 — 83 H₂O / 17 MA

7 — 80 H₂O / 20 MA

8 — 75 H₂O / 25 MA — opt. blackout

ρ

0,98 1,00   1,05   1,10   1,15 1,20 g/cm³

Fig. 6 Eight Schlieren photographs of eight different cells measured simultaneously in one XL-SO run: one counterbalance radius reference cell and seven measuring cells. The same 11-latex mixture is measured in seven different (methanol/H₂O)/metrizamide density gradients. In the different gradients, different latex components out of the total 11 components are visible as narrow turbidity bands. In cells 6, 7 and 8, with the steepest density gradients with 17, 20 and 25 wt% metrizamide, an optical blackout is seen near the cell bottom (on the right)

in % w/w. The percentage of the heavy sugar metrizamide (without the light methanol) progressively increased from cell no. 4 to cell no. 8, and p(MA) = 5/12/17/20/25% w/w. This meant that the DGs (and thus the refractive index gradients inside the cells) became steeper and steeper.

As can be seen from Fig. 6, all the different components in the mixture of 11 latices had different densities, as was expected. We can see the four narrow turbidity bands of the latex components nos. 1, 2, 3 and 4 with the lowest particle densities in cell 2, and components nos. 6, 7, 8 and 9 with the high densities in cell 8. Surprisingly, we cannot see the two latex components nos. 10 and 11 with the highest particle densities in cell 8, because there is an optical blackout near the bottom of the cell (on the right), which makes components nos. 10 and 11

invisible. There is also a minor optical blackout in cell 7 and cell 6, with the next steepest DGs.

The situation in respect of the steepest, 25% w/w MA, DG is shown in closer detail at the bottom photograph of Fig. 7. The optical blackout near the cell bottom can be seen more clearly (this DG is steeper than in Fig. 6 because of the longer running time of 118 h, and latex component no. 5 also migrated into the DG). The reason for this optical blackout is that the refractive index continuously increases radially inside the cell, and the cell acts as a deflecting light prism for the parallel light beams reaching it. The deflection of the beams is so strong near the cell bottom that the cell causes them to hit the walls of the optical tower tube instead of the photographic plate.

It can be seen from Fig. 7 at the top that we run the same DG in our Optima XLI as in our XL-SO (at the

XLI - 75 H₂O/25 MA - DG of 11 - Latex - Mixture

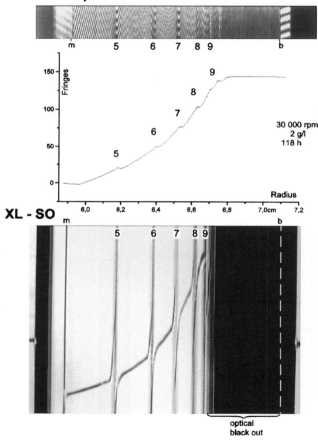

XL - SO

Fig. 7 Comparison of a XLI interference optics TV "photograph" (*above*) and a XL-SO Schlieren photograph (*below*) of the same 75 H₂O/25 metrizamide density gradient of an 11-latex mixture, showing only the latex components numbers 5, 6, 7, 8 and 9. The components numbered 10 and 11 are not visible in both photographs, because they are lying in the optical blackout area near the cell bottom

bottom) under the same conditions, but with a 3-mm double-sector cell instead a 3-mm single-sector cell. The result obtained with the XLI is naturally the same. An optical blackout can be seen near the cell bottom on the "photograph" of the interference fringes taken with the XLI-TV camera. It can be seen even more clearly in the Fourier transformation analysis of this TV photograph shown in the diagram in the middle of Fig. 7, in which the fringe number is plotted against the rotor radius. The steepest interference fringes near the cell bottom cannot be resolved. Instead, a horizontal line appears in this area, forming an incorrect "plateau". We are therefore unable to see the latex components nos. 10 and 11 in our XLI within this cell bottom area, nor with our XL-SO. We have to accept this in our XLI, but not in our XL-SO.

From Fig. 8, it can be seen that our XL-SO offers a very simple solution to the problem. If we move our flash lamp vertically with the small electrical motor from the standard position (lower part in Fig. 8) to a deeper position (upper part in Fig. 8), the optical blackout near the cell bottom disappears and is shifted to the meniscus area within the AUC cell. The bottom area is now clear, and we can now detect latex components nos. 10 and 11 as narrow turbidity bands with the highest particle densities.

Figure 9 illustrates another example of a measurement made with our XL-SO which is difficult to do with an XLI. It is a rapid special synthetic boundary run. We use a 12-mm 2° single-sector synthetic boundary cell (0.50 ml capacity) with a centrifugal force valve inside and a small 0.10-ml storage vessel for the supernatant liquid inside the cell. We call this a rapid dynamic H_2O/D_2O density gradient run. It was introduced by Lange [12] and later modified by us [13]. The sample to be measured was the same 11-latex mixture, but this time dispersed, $c = 3$ g/l, in the heavy D_2O with a density of $\rho(D_2O) = 1.10$ g/cm^3. The upper Schlieren photo in Fig. 9 shows that, at the beginning of the run, the cell is half filled with the 11-latex/D_2O dispersion. At a rotor speed of 10 000 rpm, the valve opens and pure light H_2O from the storage vessel with $\rho(H_2O) = 1.00$ g/cm^3 is superimposed on the heavy D_2O. A rapid, steep, dynamic density gradient is created in the vicinity of the turbid H_2O/D_2O interface after a running time of only 6 min. It is visible as a negative Gaussian Schlieren line in the lower Schlieren photographs in Fig. 9. The integral of this line yields the radial ρ-density axis under the lowest Schlieren photograph. In the 6 min photograph, the first seven latex components out of our total of 11 components had been separated, i.e. fractionated, as narrow turbidity bands. In the 10 min photo, we can only see the five components nos. 2, 3, 4, 5 and 6.

All of the Schlieren photographs referred to up to now were taken with monochromatic light, with a green interference filter ($\lambda = 546$ nm) in front of the white light flash lamp, using a black and white film. The next example employing the XL-SO in Fig. 10 was performed without this green filter, i.e. with white light and a colour film. We use this method in order to be able to see colours in coloured samples, which is sometimes very valuable from the analytical point of view.

Figure 10 shows a standard synthetic boundary sedimentation run of a coloured sample. The sample consisted of complex nanoparticles (food dye) with a diameter of about 100 nm and a core/shell structure,

Fig. 8 Two XL-SO Schlieren photographs of the same density gradient as in Fig. 7. The photograph *below*, with the flash lamp in the normal position, is the same as in Fig. 7, with an optical blackout near the cell bottom. The photograph above is done with the flash light in a position deeper in the *z*-direction. Now the optical blackout is shifted to the meniscus area of the cell and latex components numbers 10 and 11 are visible as narrow turbidity bands

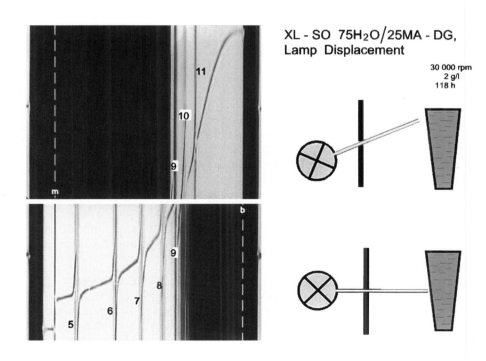

XL - SO 75H$_2$O/25MA - DG, Lamp Displacement

30 000 rpm
2 g/l
118 h

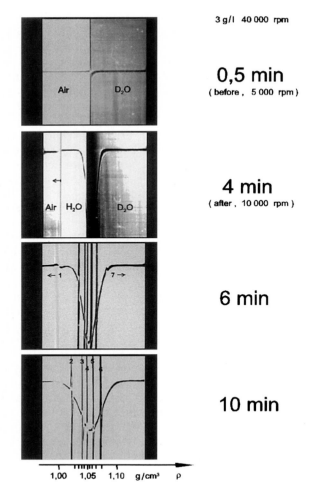

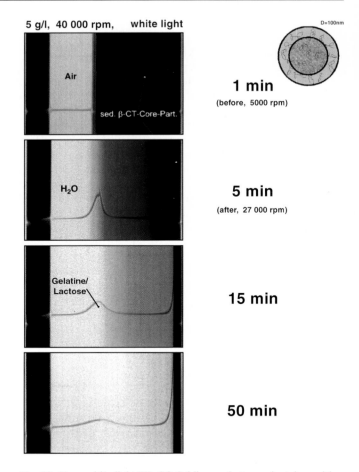

Fig. 9 Four XL-SO Schlieren photographs taken during a special synthetic boundary run, a dynamical H_2O/D_2O density gradient run, where during the run upon a dispersion of the 11-latex mixture in heavy D_2O light H_2O is layered. Within this rapid, steep density gradient the 11-latex mixture is fractionated into the different components. Numbers 2, 3, 4, 5 and 6 are visible in the *lowest* photograph

Fig. 10 Four white-light XL-SO Schlieren photographs taken with a colour film during a synthetic boundary run of a coloured sample of core/shell nanoparticles with a β-carotene core and a gelatine/lactose shell. The yellow/red coloured shadow *right* of the gelatine/lactose Schlieren peak stems from associations between gelatine/lactose and dissolved β-carotene molecules

composed of a β-carotene core and a shell of gelatine and lactose. When we redispersed this powder in water, $c = 5$ g/l, and did a synthetic boundary sedimentation run, the turbid β-carotene core particles completely sedimented in the first 5 min and were visible on the cell bottom as a black turbidity band. We were then able to see into the supernatant aqueous serum, which was clear but coloured yellow/red. Firstly, we can see the Schlieren peak of monomolecular dissolved gelatine and lactose molecules of the particle shell. Secondly, we can see a very slowly sedimenting yellow/red shadow to the right of the peak. This shadow stems most likely from associates between gelatine or lactose with single β-carotene molecules, which are dissolved out of the core.

The final example, shown in Fig. 11, demonstrates the possibility of flash light superimposition in our XL-SO device. Three conventional black and white Schlieren

photographs with monochromatic light, i.e. with the green filter in the optical path, are shown on the left. These three photographs of three different single cells were taken after a running time 240 min during a sedimentation run of almost monodisperse polystyrene macromolecules (molar mass $M = 233\,000$ g/mol), dissolved in toluene with $c = 3$, 4 and 5 g/l. A photograph of these three cells at nearly the same running time created by superimposing different flashes onto the same 70-mm film is shown on the right. This type of superimposition is very valuable in some differential analytical techniques.

Conclusion

This paper describes how we were able to install an 8-cell Schlieren optics multiplexer into an Optima XL, the basic preparative version of the Optima XLI-AUC. We

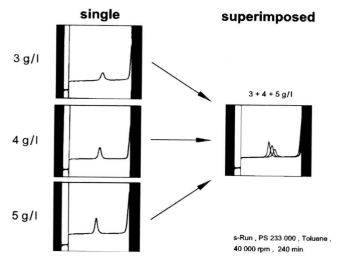

single   **superimposed**

3 g/l

4 g/l

5 g/l

3 + 4 + 5 g/l

s-Run , PS 233 000 , Toluene ,
40 000 rpm , 240 min

Fig. 11 XL-SO Schlieren photographs of three single cells and a superimposition of them by light flashes taken at nearly the same running time, 240 min, during a sedimentation run

were able to find a quick, simple and inexpensive solution to our problem. Our new XL-SO device is equivalent to the famous old Model E, so we can stop using our three Model Es, now that the manufacturer is unable to deliver any further drives. It would not be very complicated to install Schlieren optics in every Optima XLI by the same method. We eventually plan to do this, but the next step is to install a digital TV camera with optical image processing in our XL-SO as in [4]. Automatic image processing works only for simple dissolved samples with a defined, uninterrupted Schlie-

ren line across the complete photograph. Until the arrival of more highly sophisticated and powerful new image processing programs, we need to have some means of taking simple Schlieren photographs of complex colloidal samples in order to evaluate them "by hand".

The same is true of the Beckman XLI image processing program. The whole XLI interference optics needs to be substantially improved, because there are a lot of problems with baseline noise, jitter and the correct triggering of the XLI laser diode when the rotor speed rate is adjusted suddenly and when the rotor speed approaches the maximum. New possibilities are offered by the incorporation of an additional optical light path into the Optima XL/XLIs and, especially, by the introduction of a trigger unit, based on a simple photoelectric cell and a polished line mirror at the base of the rotor, that operates independently of the Beckman electronics (important for safety reasons). It would easily be possible to install new detectors in the AUC, such as a fluorescence detector [14], a better UV scanner than the one that is currently fitted to the Optima XLA, and a better interference optics detector. If we were simply to replace our XL-SO flash light source with a laser with beam-broadening optics and to use double-sector cells, our Schlieren optics would be transformed into interference optics – the optics fitted to the old Model E!

Acknowledgements The author would like to thank M. Stadler for his great contributions for the quick realization of the mechanical XL-SO set-up, A. Lehmann and P. Münsterer for building the electronic triggering unit and Dr. P. Rossmanith and M. Kaiser for helpful discussions.

References

1. Cölfen H (1998) Polym News 23:152–162
2. Mächtle W, Klodwig U (1976) Makromol Chem 177:1607–1612
3. Mächtle W, Klodwig U (1979) Makromol Chem 180:2507–2511
4. Klodwig U, Mächtle W (1989) Colloid Polym Sci 267:1117–1126
5. Mächtle W (1991) Prog Colloid Polym Sci 86:111–118
6. Rossmanith P, Mächtle W (1997) Prog Colloid Polym Sci 107:159–165
7. Mächtle W (1988) Angew Makromol Chem 162:35–52
8. Mächtle W (1992) Makromol Chem Macromol Symp 61:131–142
9. Mächtle W (1999) Biophys J 76:1080–1091
10. Beckman (1974) Instruction manual E-IM-3. Spinco Division, Beckman Instruments, Palo Alto, Calif
11. Mächtle W (1992) In: Harding SE, Rowe AJ, Horton JC (eds) Analytical ultracentrifugation in biochemistry and polymer science. Royal Society of Chemistry, Cambridge, pp 147–175
12. Lange H (1980) Colloid Polym Sci 258:1077–1085
13. Mächtle W (1984) Colloid Polym Sci 262:270–282
14. Schmidt B, Rappold W, Rosenbaum V, Fischer R, Riesner D (1990) Colloid Polym Sci 268:45–55

Progr Colloid Polym Sci (1999) 113 : 10–13
© Springer-Verlag 1999

TECHNICAL AND METHODOLOGICAL INNOVATIONS

D. Kisters
W. Borchard

New facilities to improve the properties of Schlieren optics

D. Kisters · W. Borchard (✉)
Institut für Physicalische
und Theoretische Chemie
Angewandte Physikalische Chemie
Gerhard-Mercator-Universität
Lotharstrasse 1
D-47057 Duisburg, Germany

Abstract A general problem when using a Schlieren optical system in an Analytical Ultracentrifuge is the modulation of the beam of parallel light, which passes the sector-shaped cell during a measurement. In this paper, two new facilities to modulate the laser beam are described. In the first set-up presented, a laser diode is used as the source of light. This diode can be modulated with a multiplexer up to at least 50 000 rpm in a simple way. The second set-up is realized with a Model E by the use of a fast shutter camera. The TTL signal, which is generated by the multiplexer in dependence on the cell position, controls the shutter of this camera. By means of both new set-ups the light intensity, which is available for the measurements, is raised and the calibration procedure is easier in comparison to the equipment developed earlier.

Key words Fast-shutter camera · Laser diode · Multiplexing · Schlieren optics · Analytical ultracentrifugation

Introduction

The detection of the gradient of concentration of the polymer component in thermoreversible gels is difficult in comparison to polymer solutions. Owing to the turbidity of the gels, it is not always possible to detect the gradient of concentration of the polymer component in the gel during a run. Therefore it is very important to raise the available light intensity.

The Analytical Ultracentrifuge (Beckman Model E) at the department of Applied Physical Chemistry in Duisburg is equipped with a laser, an acousto-optical modulator (AOM) and a multiplexer, which was constructed in the department at Duisburg [1, 2].

The old set-up (AOM)

Figure 1 shows the working principle of the multiplexer, which was developed in our group at the university. At the rotor fork a reflection light barrier is mounted, which generates a DC signal every time the reflection field on the rotor code ring passes the barrier. The impulse former converts this signal into a rectangular signal, which is used as the reference signal at 0° of the rotor angle. By means of the multiplexer it is possible to generate two TTL signals with adjustable angles according to the reference signal. A TTL signal is a rectangular DC signal with an amplitude of 5 V and a variable width. These two TTL signals are used to control a high-frequency driver, which produces a radiofrequency signal. A TeO_2 crystal is used as the AOM. The radiofrequency signal induces local changes in the density of the crystal and therefore changes in the index of refraction of the crystal [3].

The result of these changes is that every time a measuring cell passes the optical path, the index of refraction of the crystal is changed in such a way that the laser beam is refracted and passes the apertures. For all other times the light is not refracted and is faded out by means of the apertures [3, 4].

Owing to the fact that for each rotation two adjustable TTL signals are generated, it is possible to measure two different cells per revolution, e.g. a

Fig. 1 Schematic representation of the old set-up

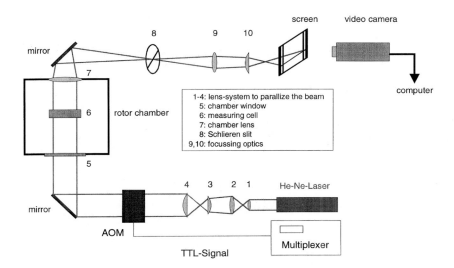

1-4: lens-system to parallize the beam
5: chamber window
6: measuring cell
7: chamber lens
8: Schlieren slit
9,10: focussing optics

measuring cell and the counterbalance cell or other combinations in a multihole rotor.

A disadvantage of this experimental set-up is the fact that about 95% of the incident light is dissipated by the AOM and that four lenses are needed to make the beam parallel. Therefore we attempted to achieve modulation without the use of an AOM.

The new set-ups

In Fig. 2 the two new set-ups, which have been realized, are presented schematically. Both set-ups work without the AOM and therefore the light intensity, which is available for the measurements, is raised considerably. Another advantage of both new set-ups is the fact that

only two cylindrical lenses are needed to form a band of parallel light. Thus the calibration procedure becomes easier.

Modulation of the laser diode

The multiplexer is connected with the power supply of the laser diode. This power supply was also constructed in our department. The laser diode (Philips, 100 mW, 655 nm) is powered down automatically to 80% of the initial power. The reason for this is a raised service life of the diode. The TTL signals modulate the power supply directly, resulting in the modulation of the beam. The time of a single laser pulse is adjustable by means of the change of the width of the TTL signal, which can be varied using the multiplexer. By means of this technique

Fig. 2 Schematic representation of the two new set-ups (see text): **a** modulation of the laser diode; **b** modulation of the camera

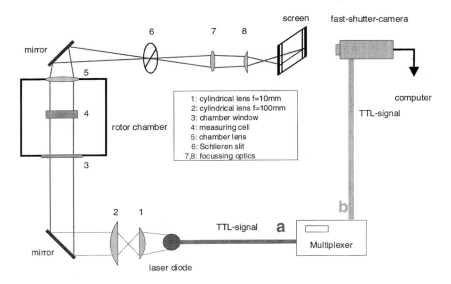

1: cylindrical lens f=10mm
2: cylindrical lens f=100mm
3: chamber window
4: measuring cell
5: chamber lens
6: Schlieren slit
7,8: focussing optics

it is also possible to measure two cells for each revolution of the rotor.

Modulation of the camera

In the last few years, fast-shutter cameras have become cheap. A fast-shutter camera (e.g. FlashCam, PCO Computer Optics) can also be modulated directly by means of the TTL signal, e.g. every time the cell passes the optical path, the shutter of the camera will be opened and a Schlieren picture is received. The minimal exposure time of this type of camera is 1 μs, e.g. it is possible to take Schlieren pictures of 1° cells up to rotational speeds of at least 70 000 rpm. The exposure time is adjustable from 1 μs up to 1 ms in a simple way and therefore it is possible to measure at nearly all conceivable rotor speeds.

In contrast to the modulation of the laser diode, only a single TTL signal can be used. Thus only one cell can be measured for one revolution of the rotor.

A useful function of this camera is the fact that 10 pictures of a cell for 10 rotations can be integrated. Furthermore, the camera has a centronics interface. Therefore a direct connection to a computer is possible.

Figure 3 shows two Schlieren pictures of a water filled cell (sector angle 4°), received by means of the fast shutter camera at an exposure time of 10 μs. This exposure time is equivalent to a rotational speed of about 70 000 rpm for a 4° sector shaped cell. The left picture was received by means of the integration of 10 pictures whereas the right one was received as a single picture. By comparison it becomes clear that the integration function of the camera is very useful to improve the sharpness of the Schlieren line and the contrast of the pictures.

In Fig. 4 a Schlieren picture of a 1° water filled sector shaped cell, received with an exposure time of 2 μs and

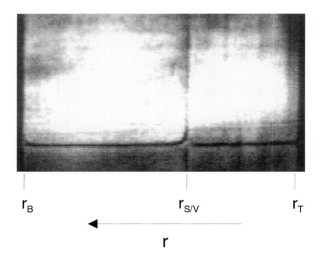

Fig. 4 Schlieren picture of a 1° sector-shaped cell, filled with water, at an exposure time of 2 μs, received by means of the fast-shutter camera (Schlieren angle = 80°); r_B = cell bottom, $r_{S/V}$ = meniscus solvent (water)/vapor, r_T = top of the cell, r = distance from the axis of rotation

integration of 10 pictures, is presented. This exposure time is equivalent to a rotational speed of about 70 000 rpm for a 1° sector shaped cell. The reason for the relatively low sharpness of the Schlieren line in Figs. 3 and 4 is the fact that the frame-grabber card, which is used to digitize the pictures, works only with a resolution of 640 × 480 pixel, whereas the CCD chip in the camera works with a higher resolution of 756 × 580 pixel.

Conclusions

In the present work, two new facilities of the modulation have been realized, which are necessary for measurements by means of an optical system in analytical

Fig. 3 Schlieren pictures of a 4° sector-shaped cell, filled with water, at an exposure time of 10 μs, received by means of the fast-shutter camera (Schlieren angle = 80°) (see text); r_B = cell bottom, $r_{S/V}$ = meniscus solvent (water)/vapor, r_T = top of the cell, r = distance from the axis of rotation

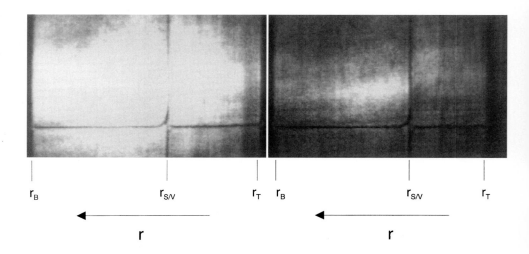

ultracentrifugation. In the first case the light source, a laser diode, is modulated directly by means of the TTL signal, generated by the multiplexer. The power supply was constructed in our department.

The second facility is the use of a fast-shutter camera. The TTL signal controls directly the shutter of this camera.

The advantage of both new set-ups is the fact that the light intensity, which is available for the measurements, is raised and the calibration procedure of the optical system is much easier in comparison to the old set-up. The fast-shutter camera might be a very suitable tool as a component of the detection optics in analytical ultracentrifugation, because the camera is very easy to handle and can be installed into every existing modulation system. Only a TTL signal must be generated in dependence of the actual cell position. An advantage of this type of camera is the integration function. By means of this function it is possible to integrate 10 pictures of a distinct cell. The result is an improvement of the contrast and sharpness of the resulting Schlieren picture. The Schlieren pictures in Figs. 3 and 4 show clearly that it is possible to use this type of camera to perform measurements with 1° sector shaped cells up to at least 70 000 rpm.

Another useful function using the fast-shutter camera is an adjustable delay between the TTL signal and the receiving of the picture, from 1 μs to 1 ms in steps of 1 μs. This function allows the user to measure cells at every position in a multihole rotor if it is not possible to vary the position of the TTL signal by means of a multiplexer.

A general disadvantage of the set-ups we have realized is the fact that the frame-grabber card, used to digitize the pictures, works only with a resolution of 640×480 pixel. In the future we want to replace the old frame-grabber card with a new one to improve the sharpness of the Schlieren pictures. Furthermore, it might be possible to replace the multiplexer by a computer A/D card. For this case, the control of the measurements by use of a computer would become more comfortable.

Acknowledgements We thank the Deutsche Forschungsgemeinschaft for financial support. Furthermore, we thank O. Kramer for the development of the power supply for the laser diode.

References

1. Cölfen H, Borchard (1994) Prog Colloid Polym Sci 272:90–101
2. Cölfen H, Borchard W (1994) Anal Biochem 219:321–334
3. Holtus G (1990) Dissertation, Duisburg
4. Cölfen H (1994) Dissertation, Verlag Köster, Berlin

Progr Colloid Polym Sci (1999) 113 : 14–22
© Springer-Verlag 1999

P. Lavrenko
V. Lavrenko
V. Tsvetkov

Shift interferometry in analytical ultracentrifugation of polymer solutions

P. Lavrenko (✉) · V. Tsvetkov
Institute of Macromolecular Compounds
Russian Academy of Sciences
199004 St. Petersburg, Russia
e-mail: lavrenko@mail.macro.ru

V. Lavrenko
University of Massachusetts
Department of Computer Science
Amherst, MA 01003, USA

Abstract The polarized-light interference method is applied to visualize a distribution of macromolecule
concentrations based on displacements in an ultracentrifugal cell. The
optical system based on the Lebedev
interferometer is presented. The
sensitivity and precision of the system are compared with those of the
widely used Schlieren optical system
and Rayleigh interference optics
when applied to synthetic polymer
dilute solutions. Some questions of
methodology are discussed.

Key words Polarizing interference
optics · Sensitivity · Distributions

Introduction

Analytical ultracentrifugation studies of many synthetic
polymers in organic solvents have been performed [1, 2].
To obtain reliable experimental data, it is very important
to minimize the concentration effects distorting the
shape of the sedimentation curves. For this purpose, the
concentration of the solutions to be investigated should
be as low as possible, and, naturally, this increases the
requirements imposed on the sensitivity of the instrument, i.e., its ability to record small changes in solute
concentration, c, and in the concentration gradient,
$\partial c/\partial x$.

The local changes in c and $\partial c/\partial x$ are usually recorded
by refractometric methods [3] in which the dependence
of the refractive index of the solution, n, on its
concentration is utilized:

$$\partial n/\partial x = (\mathrm{d}n/\mathrm{d}c)\; \partial c/\partial x \; , \tag{1}$$

where $\mathrm{d}n/\mathrm{d}c$ is the specific refractive increment of the
polymer–solvent system. Almost every commercial an-
alytical ultracentrifuge (AUC) is equipped with the well-
known Schlieren optical system that provides direct
viewing.

The sensitivity of the refractometric methods may be
increased by using various interferometric schemes. All
the schemes that have been applied to polymer solutions
may be subdivided into two families. To the first one

belong the schemes where two light beams (which
further form the interference pattern) are shifted from
one another in the direction normal to the direction (x)
of the variable refractive index, $n(x)$, [solute concentration $c(x)$] as illustrated in Fig. 1a. The contour of the
resulting fringe (Fig. 1c) corresponds to the radial
distribution of $n(x)$. To the second family belong the
schemes where these beams are separated in the
direction x of the variable refractive index (Fig. 1b).
The contour of the fringe (Fig. 1c) here has a form
intermediate between the distributions of $n(x)$ and
$\partial n/\partial x(x)$ depending on the relation between the shift
value and the width of the concentration boundary [3].

Hence, the schemes with Jamen [4, 5] and Rayleigh
interference optics belong to the first family, and those
with Lebedev, Bringdahl, and Beutelspacher [3, 6]
interference optics belong to the second one. All of
them are characterized with the great sensitivity that
exceedes that of the refractometric Schlieren systems by
an order of magnitude.

A considerable enhancement in the quality of the
Rayleigh interference patterns was achieved with a
pulsed laser source [7, 8]. However, the application of
Rayleigh interference optics in sedimentation velocity
studies is significantly limited by disturbing patterns,
when applied in the high-speed regime, due to optical
noise (inhomogeneities) which arises in the cell glass
along the deformation of the cell centerpiece in very

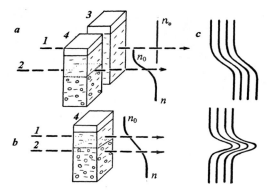

Fig. 1 Schematic path of light beams in the shift interferometers of **a** first and **b** second families as defined in the text, **c** presents the resulting interference patterns. *1* and *2* mean two systems of the light beams shifted from one another; *3* means the reference sector with a solvent, and *4* is the operating one; n and n_0 are refractive indexes of the solution and the solvent, respectively

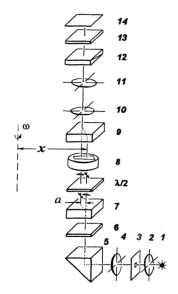

Fig. 2 Scheme of a polarizing interference attachment developed for an analytical ultracentrifuge of the MOM series 3180. x is the coordinate of the point in the cell as counted from the axis of rotation. Light source (*1*) (mercury lamp); lenses (*2, 4, 10, 11*); point diaphragm (*3*); prism with total internal reflection (*5*); polaroids (*6, 13*); spars with a twinning of a (*7, 9*); half-wave plate ($\lambda/2$); cell (*8*); wedge (*12*); photographic film (*14*)

strong centrifugal fields; therefore, it is used primarily for low-speed sedimentation equilibrium experiments [7].

Another great potential of Beutelspacher and Lebedev optics in the applicability to the AUC can be explained by the smaller shift of the interfering beams, and, hence, by the lower (more optimum) sensitivity of these systems to optical noise from the optical elements that arises in strong gravitational fields.

Lebedev interference optics

Among the schemes under discussion, a polarizing interferometer of the Lebedev type [9] has certain advantages; therefore, polarizing–interference attachments based on the Lebedev interferometer were developed, designed, constructed, and used in combination with AUCs from the Hungarian Optical Works MOM of various series such as G-110, G-120, 3130, 3170, and 3180 [10, 11].

Principal optical scheme

The optical scheme of the instrument is shown in Fig. 2. The light beam from the source (1) (mercury lamp) passes through a monochromatizing device, and is focused by a lens (2) on the point diaphragm (3). With the aid of another lens (4) it travels further as a parallel beam. The beam is polarized with a polaroid (6) (with optical axis oriented at an angle of 45° to the radial direction) and separated into two beams by a birefringent plate (7) made, for example, from Iceland spar. Consequently, the interfering beams in the cell are mutually shifted from one another by the distance a (spar twinning), the direction of the shift being normal

to the plane of the solution–solvent interface (e.g., the twinning a value is approximately equal to $0.11 \times \Delta$ for Iceland spar, and $0.0059 \times \Delta$ for quartz, with Δ being the light path in the plate). After passing a half-wave plate ($\lambda/2$) (mica plate rotating the planes of polarization of both beams by 90°) and a cell (8), the two light beams are combined into one again by the second spar plate (9) equivalent to the first (7). A quartz wedge (12) (Babinet compensator) and a polaroid (13), crossed with the first (6), make it possible to observe the interference pattern of the two polarized beams. A telescopic system of lenses (10, 11) transforms one parallel light beam into another one, varying the optical magnification of the pattern. With the aid of that system, the interference pattern is projected onto a photographic film (14). The system also projects the middle of the cell (8) onto the film 14 [the plane (14) is located in the focal plane of the lens (11)]. The edge of the wedge (12), and the axes of spar the plates (7, 9) are mutually oriented in the radial direction.

Image of the cell

In the absence of optical inhomogeneities in the liquid filling the cell (8), the interference pattern on the screen (14) has alternating light and dark fringes, whose interval is dependent on the wedge angle of the compensator. The fringes recorded on photographic film are parallel to the radial direction, x.

The pattern formed is shown in Fig. 3, where images of the empty cell and of the counterbalance cell are compared with an experimental pattern [10]. It is easy to see that all the reflexes here are doubled with the same shift of a. This originates from the twinning plate (9) that is located between the cell and the observer and forms a doubled image of any point of the cell (8) in the plane (14). If the axis of rotation were in the field of observation, we would also see its doubled image. So, we refer to one half of the reflexes as the "left" image, and to the other half as the "right" one. Naturally, the radial distance of all left reflexes is counted from the left image of the axis of rotation, and the right reflexes are related to the right image of the axis. An example is presented in Fig. 3c. Since the standard interval in the counterbalance cell is 1.6 cm, we calculate the coefficient of magnification in a radial direction by $K_m = (x_4 - x_1)/1.6 \equiv (x_4' - x_1')/1.6$, the meniscus position by $x_{men} = 5.7 + (x_2 - x_1)/K_m \equiv 5.7 + (x_2' - x_1')/K_m$, and the boundary position by $x_{max} = 5.7 + (x_3 - x_1)/K_m - a/2 \equiv 5.7 + (x_3 - x_1')/K_m + a/2$, where all the radial distances are in centimeters.

The interference picture is obviously formed in the x range where the left image of the cell coincides only with the right one.

Application to experiments with a double-sector cell

Lebedev interference optics (in contrast to Rayleigh ones) do not require the use of only a double-sector cell. Moreover, these optics are not successful when applied with the double-sector cell as illustrated in Fig. 4a. The interference pattern is not legible here because it results from the superposition of two patterns. One of them is formed by light beams that went through a sector with solution, and the other is formed by light that went through the solvent. The incompatibility of these two patterns may be due to different optical noise in the two directions.

Nevertheless, the optics may be used successfully with the double-sector synthetic-boundary cell when a reference sector is shadowed completely (Fig. 4b) [10].

Interference pattern

During a sedimentation run, the fringes are curved in the direction normal to x (Fig. 3c). If there are optical inhomogeneities in the cell due to changes in the solution concentration, c, in the direction x, then, in accordance with Eq. (1), the optical difference, δ, in the paths of the two interfering beams 1 and 2, as expressed in wavelengths, is given by

$$\delta = \frac{h(n_1 - n_2)}{\lambda} = \beta \int_{x-a}^{x} (\partial c/\partial x)\mathrm{d}x = \beta[c(x) - c(x - a)] \ ,$$

$$(2)$$

$$\beta = (h/\lambda)(\mathrm{d}n/\mathrm{d}c) \ , \tag{3}$$

where h is a cell light path, λ the wavelength, x the rotor radius of the point in the cell, $\partial c/\partial x$ the concentration gradient at the point x, and $c(x)$ and $c(x-a)$ the solute concentrations at two points of the cell with the rotor radiuses x and $x-a$, respectively.

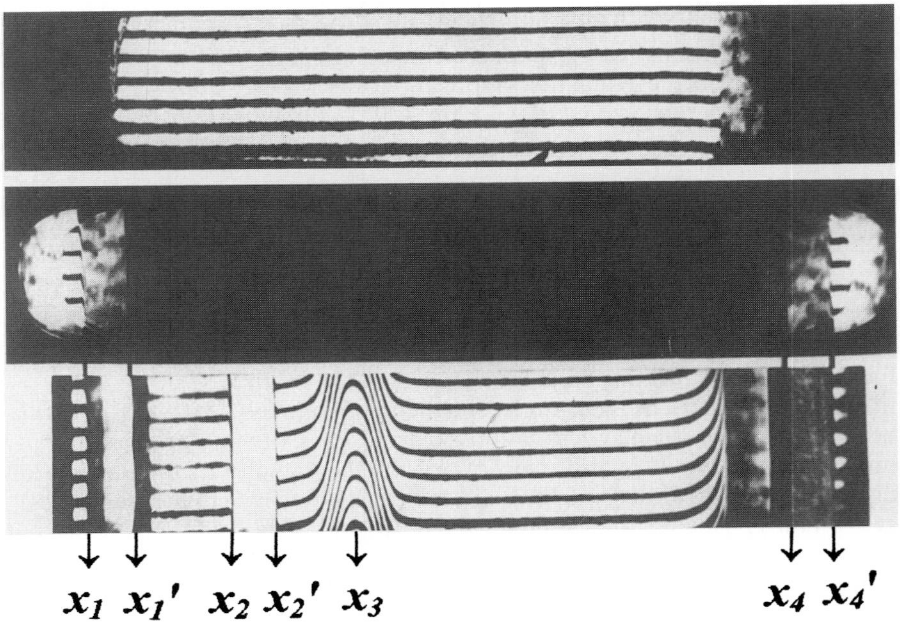

Fig. 3 Image of **a** an empty cell and **b** a counterbalance cell in comparison with **c** the experimental pattern. **a** and **b** are obtained with a stationary rotor, and **c** is the picture of a current sedimentation run. x_1, x_1' and x_4, x_4' are the double reflexes of the counterbalance cell, x_2, x_2' mark the double image of the meniscus, x_3 is the position of the sedimenting boundary

Fig. 4 Interference pattern obtained with the double-sector synthetic-boundary cell when **a** both the sectors are open for light, and **b** when a reference sector is shadowed

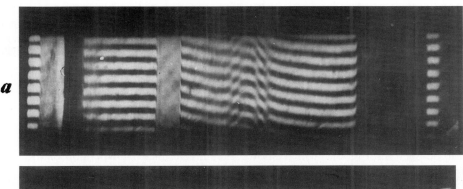

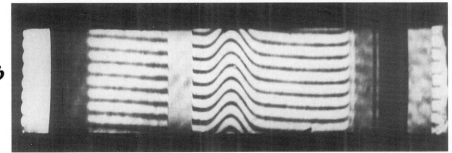

The optical retardation, δ, is revealed in the nonuniform shift of fringes in the direction parallel to the liquid interface (normal to the x-axis) and in their corresponding curvature. In value, δ is clearly equal to the fringe-shift number. An example of the pattern is given in Fig. 3c. The shape of the fringes makes it possible to determine the concentration distribution function of the rotor radius x.

If the rotor radius is counted from the left image of the axis, and two points of the cell with radiuses x and $x - ma$ (where m is a real number) are considered, then according to Eq. (2) the relation between the concentrations $c(x)$, and $c(x - ma)$ at these points is given by

$$c(x) = c(x - ma) + \frac{1}{\beta}\sum_{i=0}^{m-1} \delta(x - ia), \quad m = 1, 2, 3, \ldots ,$$
(4)

where $\delta(x - ia)$ are the ordinates of the points on the fringe with the abscissas $x, x - a, \ldots, x - (m-1)a$, respectively.

In turn, if the rotor radius is counted from the right image of the axis, then, instead of Eq. (2), we have

$$\delta(x) = \beta[c(x + a) - c(x)]$$
(5)

and

$$c(x) = c(x + ma) - \frac{1}{\beta}\sum_{i=0}^{m-1} \delta(x + ia), \quad m = 1, 2, 3, \ldots ,$$
(6)

where $\delta(x + ia)$ are the ordinates of the points on the fringe with the abscissas $x, x + a, \ldots, x + (m-1)a$, respectively.

Equations (4) and (6) allow the concentration distribution function $c = c(x)$ to be calculated using the experimental $\delta(x)$ profile.

In this case, it is convenient to use Eq. (4), beginning the measurements in the range of the neat solvent. For this purpose, the base point with the abscissa $x - ma$ is chosen in the region of the solvent and, correspondingly, $c(x - ma)$ is assumed to be zero. The value of m is found by measuring the radial distance between the point considered and the base point. The values of $\delta(x + ia)$ are determined from the contour of the fringe.

It is more convenient to apply Eq. (6) when the concentrations in the region of the interface are compared to the loading solute concentration, c_{o}. In this case, the base point with the abscissa $x + ma$ is chosen in the region of the initial solution (relatively far from the interface), where $c(x + ma) = c_{\text{o}}$.

After plotting the integral function $c = c(x)$, the gradient distribution function, $\partial c/\partial x(x)$, is found by graphic differentiation of the $c(x)$ curve. In the other method, the gradient profile $\partial c/\partial x(x)$ is found by

$$\partial c/\partial x(x) = \partial c/\partial x(x - ma) + \frac{1}{\beta}\sum_{i=0}^{m-1} \partial\delta(x - ia)/\partial x,$$

$$m = 1, 2, 3, \ldots ,$$
(7)

if the rotor radius is counted from the left image of the axis, or by

$$\partial c/\partial x(x) = \partial c/\partial x(x + ma) - \frac{1}{\beta}\sum_{i=0}^{m-1} \partial\delta(x + ia)/\partial x,$$

$$m = 1, 2, 3, \ldots ,$$
(8)

if the rotor radius is counted from the right image of the axis.

Visual distributions of $c(x)$ and $\partial c/\partial x(x)$

If the visible width of the sedimentation boundary (the region in which $\partial c/\partial x \neq 0$) is less than the a value, then the experimental $\delta(x)$ profile corresponds to the concentration distribution function, $c(x)$, in the cell. This can be seen in Fig. 5a, which shows the interference pattern obtained during a sedimentation run experiment. Spars with a value for twinning of 0.10 cm were used. According to Eq. (7), the right branch of the curve corresponds to the expression $\delta(x) = \beta[c(p) - c(x)]$ because the counts are carried out here from the plateau region where the concentration is $c(p)$.

If the sedimentation boundary is much broader than a, then the experimental $\delta(x)$ profile qualitatively resembles the gradient distribution function $\partial c/\partial x(x)$ but does not coincide with it quantitatively (because of the finite a value). This can be seen in Fig. 5b, which shows the interference patterns obtained for the same

solutions as in Fig. 5a but using spars with smaller twinning, $a = 0.021$ cm.

The method described in the previous section should be used for the quantitative treatment of experimental interference curves at any spar twinning. Equations (4)–(8) are applied here and, as a result, the concentration profile, $c(x)$, and the gradient curve, $\partial c/\partial x(x)$, are obtained.

Dispersion of the $\partial c/\partial x(x)$ distribution

The second central moment (dispersion), $\overline{\sigma^2}$, of the $\partial c/\partial x(x)$ distribution is defined by

$$\overline{\sigma^2} = m_2/m_0 - (m_1/m_0)^2 \; , \tag{9}$$

where m_i is the moment of the ith order with respect to an arbitrary point on the abscissa (x) lying beyond the gradient curve $\partial c/\partial x(x)$:

$$m_i = \int_0^\infty x^i (\partial c/\partial x)\mathrm{d}x \; . \tag{10}$$

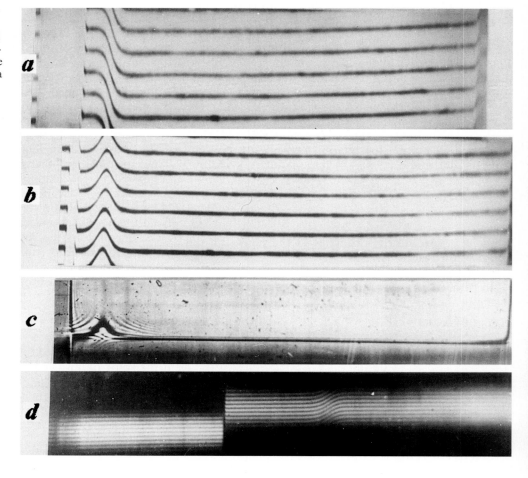

Fig. 5a, b Fringes of Lebedev optics obtained with a spar twinning of **a** $a = 0.103$ cm and **b** $a = 0.020$ cm, **c** Schlieren pattern, and **d** a Rayleigh one. The patterns were obtained during a sedimentation run experiment with a polyacrylate solution in heptane (concentration 1.4 g/l, refractive increment ($\mathrm{d}n/\mathrm{d}c)_{546} = 0.10$ ml/g)

Evidently, m_0 is equal to the $\partial c/\partial x(x)$ curve area, whereas m_1/m_0 gives the radius value of its center (centroid).

From the interference pattern, the dispersion, $\overline{\sigma^2}$, is found by

$$\overline{\sigma^2} = (a^2/2^3)\{\mathrm{argerf}[a(2\pi\overline{\Delta^2})^{-1/2}]\}^{-2} \ , \tag{11}$$

where $\overline{\Delta^2}$ is the dispersion of the experimental $\delta(x)$ profile and argerf means an argument of the probability integral $\Phi(z)$:

$$\Phi(z) = \frac{2}{\sqrt{\pi}} \int_0^z \exp(-y^2)\mathrm{d}y \ .$$

In turn, $\overline{\Delta^2}$ is calculated by $\overline{\Delta^2} = (j_2/j_0) - (j_1/j_0)^2$, with moments of the $\delta(x)$ curve given as

$$j_i = \int_0^\infty x^i - \delta(x)\mathrm{d}x \ . \tag{12}$$

Gaussian approximation

Let the gradient distribution function $\partial c/\partial x(x)$ be of a Gaussian type:

$$\partial c/\partial x(x) = [\Delta c/(2\pi)^{1/2}\sigma] \exp(-x^2/2\overline{\sigma^2}) \ , \tag{13}$$

where σ is the standard deviation, $\sigma = (\overline{\sigma^2})^{1/2}$. The area, m_0, and maximum ordinate, H, of this distribution curve are related to $\overline{\sigma^2}$ by

$$\overline{\sigma^2} = (m_0/H)^2/2\pi \ . \tag{14}$$

This permits the determination of the dispersion, $\overline{\sigma^2}$, from the values of m_0 and H (the so-called "height–area" method).

The substitution of Eq. (13) into Eq. (2) and the introduction of the variable $y = x/(2^{1/2}\sigma)$ readily transforms Eq. (2) into

$$\delta = (\beta\Delta c/2\varphi) \ , \tag{15}$$

where

$$\varphi = \Phi\left(\frac{x + a/2}{\sigma\sqrt{2}}\right) - \Phi\left(\frac{x - a/2}{\sigma\sqrt{2}}\right) \ . \tag{16}$$

At the start moment ($t = 0$, $\sigma = 0$), the $\delta(x)$ function and, correspondingly, the fringe contour have the form of a rectangle with a base equal to a, a height $\delta_\mathrm{m}^\mathrm{o}$ and an area Q given by

$$\delta_\mathrm{m}^\mathrm{o} = \Delta n(h/a) = \beta\Delta c \tag{17}$$

$$Q = \delta_\mathrm{m}^\mathrm{o}a = \beta a\Delta c \ . \tag{18}$$

A sedimentation run leads to a change in the fringe shape, whereas the area Q remains unchanged. In all the

equations, the area under the curve is expressed in terms of unit length multiplied by the number of interference orders. To express the Q value in the usual units of area, it should be multiplied by the distance between the fringes.

If a spar twinning a value is sufficiently high, and the initial boundary is relatively sharp, it is possible to carry out measurements in a relatively short experimental time when $\sigma \leq a/2$. Under these conditions, we may use the total area, Q, under the curve and its "external" part, ΔQ, lying in the range of abscissa x which satisfies the conditions $-\infty \leq 2x/a \leq -1$ and $1 \leq 2x/a \leq \infty$. It follows from Eqs. (15) and (16) at $\sigma \leq a/2$ to within 2% that

$$\sigma = 1.25\Delta Q/Q \ . \tag{19}$$

From the time moment when $\sigma > a/4$, according to Eqs. (15)–(18), the following equation can be used

$$\delta_\mathrm{m}/Q = (1/a)\Phi(a/\sigma 8^{1/2}) \ , \tag{20}$$

where δ_m is the maximum ordinate (height) of the fringe curve; hence, σ is calculated here by using the experimental δ_m and Q values.

Accuracy and resolving power

The sensitivity of Lebedev interference optics in application to sedimentation runs may be compared with those of the schlieren optical system. The scheme in Fig. 6 shows the origins of the errors, Δx_1, and Δx_2, in the experimental determination of the boundary radial position, and, hence, of the sedimentation coefficient, s, that follow from a finite optical width of the interference band and from a width of the boundary, respectively. Let us define an optical width d of the fringe as the distance between the equal-density lines with intensity

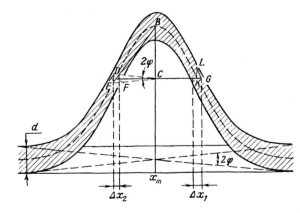

Fig. 6 Schematic origins of uncertainties in the determination of the sedimentation boundary position, x_m; Δx_1 and Δx_2 are uncertainties in the x_m determination caused by the width of the curve, d, and by the boundary width, respectively. See the text for more details

equal to half the maximum blackness of the fringe registered. The position of a fringe contour is detected with an accuracy d/k. We analyze the contour of the unimodal sedimentation curve with a maximum ordinate H and an area Q as obtained at time t and at an angle speed ω of the rotor rotation. The relative error, $\Delta s/s$, in the determination of the sedimentation coefficient can be estimated by [12]

$$\Delta s/s = (\Delta s/s)_0 + (\Delta s/s)_s + (\Delta s/s)_D + (\Delta s/s)_{PI} \ , \qquad (21)$$

where $(\Delta s/s)_{0,s,D,PI}$ is the error that arises from disturbance of the boundary due to a start width (0), the inhomogeneity of a sample (s), diffusion effects (D), and the additional width of the interference curve (PI), with a being the twinning value:

$$(\Delta s/s)_0 = (4/3k\omega^2 s)(d/Q)(\Delta_0^2/t)(1 + k\Delta_o) \qquad (22)$$

$$(\Delta s/s)_s = (48\omega^2/k)(d/Q)(\overline{\sigma_s^2}/s)t(1 + 6k\omega^2\sigma_s t) \qquad (23)$$

$$(\Delta s/s)_D = (8/3k\omega^2)(d/Q)(D/s)[1 + k(2Dt)^{1/2}] \qquad (24)$$

$$(\Delta s/s)_{PI} = (1/3k\omega^2 s)(d/Q)(a^2/t)(1 + ka) \ . \qquad (25)$$

Here s and D are the sedimentation and diffusion coefficients, respectively, σ_s is standard deviation of the differential sedimentation coefficient distribution function, and Δ_0^2 is the start dispersion of the boundary in an experiment with a synthetic-boundary cell. The uncertainty in the s determination thus evaluated obviously depends on the sensitivity of the registration optical system (through the d/Q parameter) and quality of the pattern. Figure 5 demonstrates that fringes in the Lebedev pattern (Fig. 5a, b) are optically more uniform, they can be registered with higher accuracy than those of the Schlieren pattern (Fig. 5c), and the d/Q value here is 3–5 times lower.

The resolving power of the sedimentation run method is determined by the minimum difference in s, $\Delta s = s_1 - s_2$, that leads to the formation of the bimodal gradient profile. It is evaluated by [13]

$$(\Delta s/s)_{res} = (1/\omega)(D/s)^{1/2} \ . \qquad (26)$$

Errors in the s determination, originating from the quality of the optical system and estimated by Eq. (21), are comparable in value with the resolving power, $(\Delta s/s)_{res}$. Hence, these errors reduce the resolving power of the AUC, and, with interference optics, they are several times smaller.

Inhomogeneity parameters from low-speed sedimentation experiments

At low frequency of the AUC rotor rotation, spreading of the concentration boundary (usually formed in the synthetic-boundary cell) with time is mainly due to the diffusion phenomenon. When the polymer is monodisperse with respect to molecular weight, the curve $\partial c/\partial x(x)$ corresponds to the Gaussian function (Eq. 13); however, if the polymer is polydisperse, then the $\partial c/\partial x(x)$ curves prove to be non-Gaussian. The dispersion, $\overline{\sigma^2}$, of the $\partial c/\partial x(x)$ profile as determined by the height–area method differs from that obtained by the method of moments, and their relation reflects the inhomogeneity of the sample [14]. This can be successfully detected in low-speed experiments as illustrated below.

Suppose we have an ensemble of hard particles, spherical in form, with different radiuses r and with a weight r distribution of the Kraemer–Lansing type [15]:

$$(1/c_0)(dc/dr) = (\gamma/\beta\pi^{1/2})\exp\{-[\ln^2(\gamma r)]\beta^2\} \ , \qquad (27)$$

with $\beta = [2\ln(r_w/r_n)]^{1/2}$ and $\gamma = (r_w/r)^{-1}\exp(\beta^2/4)$. Here c_0 is the concentration of the initial solution, β and γ are the parameters of the distribution, r_w and r_n are the weight- and number-average values of r. The diffusion coefficient of these particles in dilute solution is defined by $D = kT/6\pi\eta_0 r$ in accord with the Stokes–Einstein law. Here k is the Boltzmann constant, T the absolute temperature, and η_0 the solvent viscosity. Computer calculations yield the relationship between the r_w/r_n and D_w/D_A parameters that is given in Fig. 7. The results are applied below to the experimental data obtained for fullerene (C_{60}) in solution in 1-methyl-2-pyrrolidone. A great aggregation of the C_{60} molecules was recently observed in this solvent.

A contour of the experimental fringe, $\delta(x)$, and its change with time as obtained for this system in a low-

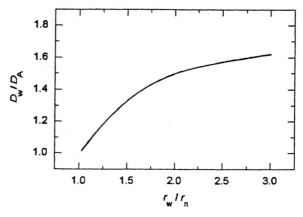

Fig. 7 The relationship between D_w/D_A and r_w/r_n as calculated for a Kraemer–Lansing distribution. D is the diffusion coefficient determined either by the method of moments (D_w) or by the height–area method (D_A). r_w (r_n) is the weight- (number-) average value of the radius of a sphere

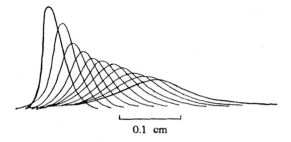

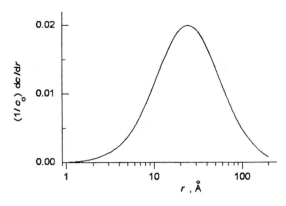

Fig. 8 Contour of the fringe $\delta(x)$ and its change with time as observed in a solution of C_{60} in 1-methyl-2-pyrrolidone at 22 °C at times (from *left* to *right*) 6, 10, 15, 20, 25, 30, 35, 40, 50, 60, and 70 min, respectively, after the start. The frequency of the rotor rotation was 8,000 rpm. The solute concentration was 0.44 g/l. For convenience, every second curve is shifted to the right by 0.016 cm

Fig. 10 The linear dimension distribution function of the Kraemer–Lansing type for the particles responsible for the diffusion phenomenon observed in a solution of C_{60} in 1-methyl-2-pyrrolidone. The inhomogeneity parameter is $r_w/r_n = 2$

speed experiment is shown in Fig. 8. The curves are symmetric in form, and the experimental time is sufficiently long for a significant change in the curve form that makes it possible to calculate the diffusion coefficient D. The curves were treated with both the "height–area" method using Eq. (20) and with the method of moments using Eq. (12). The results are presented in Fig. 9, where the $\overline{\sigma^2}$ values are plotted against time. The experimental points form a linear function; hence, the diffusion coefficient is calculated using $D = (1/2)\partial\overline{\sigma^2}/\partial t$. The $\overline{\sigma^2}$ values obtained with the method of moments (points 1′ in Fig. 9) are obviously higher, and the corresponding D_w value is 1.56 times the one of D_A. This result corresponds to the $r_w/r_n = 2$ value which follows from the $D_w/D_A–r_w/r_n$ relationship given in Fig. 7. The model distribution of the particles (aggregates) responsible for the diffusion phenomenon in a solution of fullerene C_{60} in 1-methyl-2-pyrrolidone is shown in Fig. 10 with respect to r (the radius of the hydrodynamically equivalent hard sphere).

Equilibrium analytical ultracentrifugation

Polarizing interferometry is a useful tool for the study of not only velocity sedimentation but also sedimentation equilibrium in all its stages [16]. The intersection of the $\delta(x)$ curve with the meniscus ($x = x_{\text{men}}$) allows the determination of c in the region of the meniscus, whereas the value of $\partial c/\partial x$ is determined directly from the slope of the curve $\delta(x)$ in this region. The same is true for the bottom region and for any other cell zones. This justifies the use of the equilibrium centrifugation method as well as the approaches to the equilibrium state (Archibald method).

Conclusions

The main advantage of the polarizing interferometer as presented here is its high sensitivity compared to the well-known Schlieren optics supplied with almost all commercial AUCs. This can be seen in Fig. 5, which shows the sedimentation curves obtained simultaneously using a MOM AUC with the aid of three optical systems: Lebedev interference optics, the Schlieren optical system, and the Rayleigh one. At solute concentrations of about 0.5 g/l, the Schlieren optical system is no longer productive, whereas with the aid of a polarizing interferometer we can reliably measure not only the radial position of the sedimentation boundary, but also analyze the shape of the fringe contour characterizing the molecular-weight distribution of the polymer. In comparison with Rayleigh optics, the Lebedev interferometer seems to be more applicable in high-speed experiments. Additionally, the Lebedev system contains no astigmatic optical elements and, therefore, provides an experimenter with a more adequate image of a process in the cell.

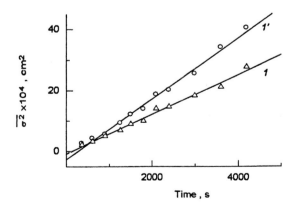

Fig. 9 Dispersion, $\overline{\sigma^2}$, of the $\partial c/\partial x$ distribution versus time as obtained for C_{60} in 1-methyl-2-pyrrolidone at 22 °C from the curves presented in Fig. 8 and treated by the method of moments (*1′*) and by the height-area method (*1*)

References

1. Klämer PEO, Ende HA (1989) In: Brandrup J, Immergut EH (eds) Polymer handbook, 3rd edn Wiley-Interscience, New York, p 1378
2. Lavrenko PN (1977) In: Kabanov VA (ed) Encyclopedia of polymers. Sovetskaya Encyclopedia, Moscow, 3:394 (in Russian)
3. Frenkel SY, Lavrenko PN (1983) In: Ioffe BV (ed) Refractive methods in chemistry. Khimia, Leningrad, p 352 (in Russian)
4. (a) Scheibling G (1950) J Chim Phys 47:688 (b) Scheibling G (1951) J Chim Phys 48:559
5. Beams JW, Snidow N, Robenson A, Dixon HM (1954) Rev Sci Instrum 25:295
6. Beutelspacher H (1966) German Patent DE 1222283
7. Harding SE, Rowe AJ, Horton JC (eds) (1992) Analytical ultracentrifugation in biochemistry and polymer science. Royal Society of Chemistry, Cambridge, pp 629
8. Flossdorf J, Schillig H, Schindler K-P (1978) Makromol Chem 179:1617
9. Lebedev AA (1931) Tr Gos Opt Inst Leningr 5:53
10. Tsvetkov VN, Skazka VS, Lavrenko PN (1971) Polym Sci USSR Ser A 13:2530
11. Lavrenko PN (1971) Polarizing interferometry in sedimentation analysis of polymers. Ph D Thesis Institute of macromolecular compounds, Leningrad
12. Lavrenko PN (1978) Polym Sci USSR Ser A 20:1149
13. Bodmann von O, Kranz D, Mutzbauer H (1965) Makromol Chem 87:282
14. Nefedov PP, Lavrenko PN (1979) Transport Methods in analytical chemistry of polymers. Khimia, Leningrad, pp 232 (in Russian)
15. Kraemer EO, Lansing WD (1934) Nature 133:870
16. Tsvetkov VN (1989) Rigid-chain polymers. Consultants Bureau, New York, pp 499

Progr Colloid Polym Sci (1999) 113 : 23–28
© Springer-Verlag 1999

L. Börger
H. Cölfen

Investigation of the efficiencies of stabilizers for nanoparticles by synthetic boundary crystallization ultracentrifugation

L. Börger · H. Cölfen (✉)
Max Planck Institute of Colloids
and Interfaces
Colloid Chemistry, Am Mühlenberg
D-14424 Potsdam, Germany
e-mail: helmut.coelfen@mpikg-golm.
mpg.de
Tel.: +49-331-5679513
Fax: +49-331-567-9502

Abstract A recently established synthetic boundary crystallization ultracentrifugation method was applied to stabilized inorganic nanoparticles. As a model system stabilized CdS showing the quantum size effect was chosen allowing both independent access to particle size via UV-vis absorption measurements and the elaboration of the influence of different stabilizer molecules on the CdS particle size distributions. The results were also compared to those derived from light scattering. It was found that the analytical ultracentrifugation method is sensitive to very small changes in the particle size distributions in the angström range, yields particle sizes down to the size range of a critical crystal nucleus and allows the calculation of the thickness of the stabilizer shell if the overall diameter is known from light scattering. Thus, synthetic boundary crystallization ultracentrifugation is suitable to study the general influence of additives on crystallization processes.

Key words Analytical ultracentrifugation · Cadmium sulfide · Nanoparticles · Synthetic boundary · Stabilizer efficiency

Introduction

Crystallization is an important process in colloid chemistry and materials science in general. The controlled nucleation and growth of inorganic materials under the influence of organic additives resulting in hybrid materials is a topic of recent investigations in the field of biomineralization and biomimetics [1, 2]. Despite the increasing importance of such processes it is remarkable that they are still not well understood. To obtain information about the crystallization process, the particle size distribution of the colloidal system has to be determined by a reliable method. There exist different models concerning the growth mechanism and the expected size distribution of colloidal particles. The particle size distribution can either be represented by a continuous curve with one maximum [3] or can possess more than one maximum indicating favoured particle sizes [4]. The particles can grow by diffusion-forced growth of single particles or by aggregation of smaller clusters which can result in certain preferred agglomeration numbers ("magic numbers") [5]. Magic numbers are also found if the particles grow shell by shell. Therefore, by analysing the particle size distribution it should be possible to distinguish between the different mechanisms.

Different organic additives also have an influence on the crystallization of inorganic systems thus affecting the particle size distribution. This process is of enormous technical relevance. By studying the differences introduced, it should be possible to evaluate concepts of stabilizer efficiency and colloidal stabilization in general. Analytical ultracentrifugation (AUC) has been applied to particle size measurements since the early days of colloid science [6, 7] and it has proved to work well in the relevant particle size range of a few nanometres especially for highly complex colloids [8].

In this work we applied a recently established method called synthetic boundary crystallization ultracentrifugation [9]. In synthetic boundary cells a reactant containing solution is layered onto a more dense solution containing the other reactant when the centrifuge is speeded up so that a very sharp and defined reaction boundary is formed. Due to the high centrifugal field, even the smallest colloids formed sediment out of the reaction boundary and so their growth is quenched. Simultaneously, the particle size distribution can be determined via sedimentation. Thus, the whole method makes use of the stopped growth once the particles leave the sharp reaction boundary paired with the observation of their movement in the centrifugal field.

Synthetic boundary crystallization ultracentrifugation fulfils several requirements for the investigation of nucleation processes:

1. The method allows the determination of particle size distributions with robust statistical significance.
2. In many inorganic systems, nucleation and growth are very fast processes with a time scale of the order of microseconds. Quenching in an external reactor would not give access to the primary particles; therefore, the experiment has to be run in a fractionating analytical device with simultaneous preservation of the different stages of particle growth by quenching.
3. Particles as small as about 1 nm are detectable.
4. The experiments can be run under exactly reproducible conditions as crystallization reactions are very sensitive to changes in reaction parameters.

Despite the fact that synthetic boundary cells have been applied for various investigations in the field of biopolymers [10, 11], to our knowledge it is the first time that synthetic boundary cells have been used to carry out chemical reactions between two reactants which are themselves not able to sediment considerably but only their reaction product is.

Theory

Synthetic boundary crystallization experiments

As the theory of synthetic boundary crystallization experiments has been described comprehensively [9] we just briefly summarize the main results here. Overlaying of the reaction solutions leads to a sharp boundary at which nucleation takes place as indicated in Scheme 1. The newly formed particles can either stay in the reaction boundary and grow further or can diffuse or sediment out of the reaction boundary resulting in a quenching of the reaction. The particles can then sediment further or diffuse back into the reaction boundary and continue their growth. As the grown particles are then again subjected to further or quenched growth particles of different sizes are generated.

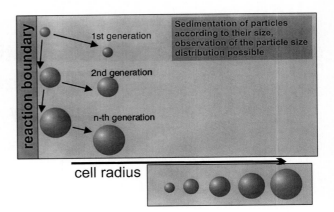

Scheme 1 Schematic representation of synthetic boundary crystallization ultracentrifugation

It has been shown [9] both by theoretical investigations and by wavelength experiments on semiconductor nanoparticles that diffusion is the predominant process in the early stage of the experiment, whereas with increasing experimental time sedimentation becomes the prevailing process. Therefore the particles can grow in the way described by diffusing back into the reaction boundary and the fractionation order of the particles is according to Scheme 1, with the biggest particles sedimenting first.

The experimentally accessible sedimentation coefficient distribution (s distribution) is broadened due to diffusion of the particles. The diffusion broadening is significant as the particle size is only a few nanometres. The s distribution can be corrected by subtracting the broadening due to a mean diffusion coefficient calculated from the mean particle size experimentally obtained by ultracentrifugation [12]; however, this approach only makes sense if the particle density is known, which is the case for polymer latices but usually not for organic/inorganic hybrid colloids. Another way for the diffusion correction is the extrapolation of the s distribution at different experimental times to infinite time [13–17]. Applying one of these corrections gives the diffusion-corrected s distribution which can be converted to a particle size distribution assuming the validity of Stokes' law:

$$d_P = \sqrt{\frac{18\eta s}{\rho_p - \rho_s}} \, , \tag{1}$$

where d_P is the particle diameter, η is the solvent viscosity, s is the sedimentation coefficient, ρ_p is the average particle density and ρ_s is the solvent density.

Quantum size effect

As colloidal semiconductor materials belong to a state of matter between molecules and solids they show size-dependence of physical and chemical properties [18]. Of special interest is the dependence of the energy of the bandgap between the conduction and valence bands on the particle size reflected in a blue shift of the UV-vis absorption onset with decreasing particle diameter. Therefore by evaluating the energy according to the absorption onset one has independent access to the particle size [19].

Particle morphology

For core–shell particles the volume of core and shell can be calculated via Eq. (2) if the total volume of the particle, the particle density, the core density and the shell density are known.

$$\frac{V_{core}}{V_P} = \frac{\rho_p - \rho_{shell}}{\rho_{core} - \rho_{shell}} , \qquad (2)$$

where V_{core} is the core volume, V_P is the particle volume, ρ_{shell} is the density of the shell and ρ_{core} is the density of the core.

The core volume gives access to the number of unit cells the particle contains as the volume of the unit cell is known for all kinds of inorganic minerals. Assuming spherical geometry the core diameter and shell thickness are calculated easily from the corresponding volumes but other geometries also do not present a problem.

Experimental

Methods

A Beckman Optima XL-I analytical ultracentrifuge (Beckman Instruments, Spinco Division, Palo Alto) with scanning absorption optics (200–800 nm, accuracy ±2 nm) and an on-line Rayleigh interferometer was used for all measurements. The ultracentrifuge software corrects the measured light intensity, I, for the initial intensity, I_0, to give the absorbance, $-\log I/I_0$.

In sedimentation velocity experiments the absorbance is measured as a function of the radius at different fixed wavelengths. Continuous radial scans with a step size of 50 μm were performed at different scan intervals and a rotational speed of 40 000 rpm. All experiments were performed at 25 °C.

In this work we applied a home-made 12-mm charcoal-filled Epon double-sector synthetic boundary cell of the Vinograd type [9, 11] as this special cell type is not commercially available anymore.

This cell utilizes the centrifugal force to form a reaction boundary by layering Na$_2$S solution (ρ_s = 0.9974 g/ml, η = 0.00904 P) in a reservoir onto a more dense CdCl$_2$/thiol solution (ρ_s = 0.998 g/ml, η = 0.0091 P) in one sector of a double-sector centrepiece via a thin capillary. The two solutions build a sharp boundary at which the reaction takes place. Although the diffusion rate of the ions is very high, the fast reaction to an insoluble semiconductor CdS prevents extensive diffusion broadening of the boundary.

For the UV-vis and dynamic light scattering (DLS) experiments the samples were prepared in the ultracentrifuge under synthetic boundary crystallization conditions. The run was stopped immediately after the overlaying was observed to ensure that the colloids formed were comparable to the systems analysed in the sedimentation velocity experiments. The solution was then cleaned by dialysing against pure water.

The UV-vis spectra were recorded with a UVIKON 940/941 dual-beam grating spectrophotometer (Kontron Instruments) with a wavelength range of 190–900 nm. The solutions were measured in a quartz glass cell (Suprasil; Hellma, Mülheim/Germany) with a pathlength of 1 cm at 25 °C.

The DLS experiments were performed with a laboratory-built goniometer with temperature control (±0.05 K), an attached single-photon detector (ALV/SO-SIPD) and a multiple-tau digital correlator (ALV5000/FAST) from the ALV (Langen/Germany). The light source was an Innova (Coherent; Dieburg/Germany) 300 argon-ion laser operating at 488.0 nm in single-frequency mode with a power of approximately 500 mW. The light-scattering cell (Hellma, Mülheim/Germany) was filled with 1 ml sample and put into the thermostated sample holder. The correlation function was recorded immediately afterwards in periods of 120 s. The resulting correlation function was transformed [20] into an intensity-weighted size distribution.

Materials

The synthesis of the colloidal CdS particles was performed in the presence of thiolactic acid (Aldrich, 95%), thioglycerine (Fluka, p.a.) or L-cysteine (Fluka, ≥99%) [21] as stabilizing agents.

The CdS particles were prepared in the synthetic boundary cell at 25 °C by layering 15 μl of a 10 mM aqueous solution of Na$_2$S (Fluka, p.a.) (pH = 10.5) onto 265 μl of a 5 mM CdCl$_2$ (Fluka, p.a.) solution containing 0.5 mM thiol. The CdS particles (TG1), also examined by UV-vis and DLS measurements, were prepared analogously to Ref. [22] with thioglycerine as stabilizer. The pH values of the CdCl$_2$ solutions were adjusted to different values with an accuracy of 0.1 pH units using a Metrohm 716 DMS Titrino (Metrohm, Switzerland). The chemicals were used as obtained from the supplier.

Results and discussion

Particle characterization

To achieve a complete characterization of the hybrid colloids obtained it is necessary to know the particle size and the particle density. The combination of UV-vis and DLS data on the one hand, and synthetic boundary sedimentation data on the other yields these parameters leading to information about the structure of the particles. Thiol-stabilized CdS particles are known to show core–shell morphology [22], with a core consisting of pure CdS and a shell consisting of stabilizer (Fig. 1).

CdS particles (TG1) with a relatively narrow particle size distribution are crystallized in synthetic boundary crystallization experiments in the AUC in the presence of thioglycerine. The sedimentation velocity experiments yield a sedimentation coefficient distribution with a weight average sedimentation coefficient of 3.6 ± 0.9 S. Assuming the bulk density 4.83 g/cm^3 of CdS [23] one would obtain an apparent mean particle size of 1.2 nm. (ρ_s = 1.0098 g/cm^3, η = 0.00917 P). This density is certainly too high as the stabilizer shell has to be taken

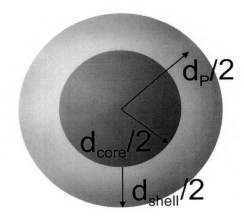

Fig. 1 Core–shell particle

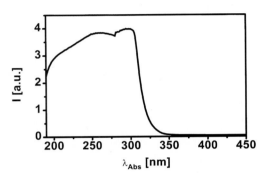

Fig. 2 UV-vis absorption data of CdS-TG1 in a wavelength range of 190–900 nm

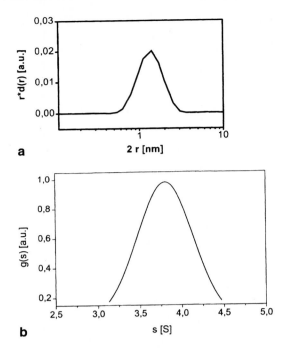

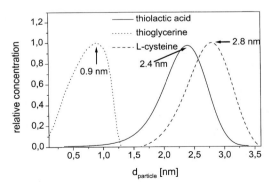

Fig. 3 a Particle size distribution of CdS-TG1 system obtained by dynamic light scattering with a 500 mW argon-ion laser at 488 nm. **b** Sedimentation coefficient distribution of CdS-TG1 system obtained by synthetic boundary crystallization at 25 °C, 40 000 rpm

It is shown by these results that the combination of synthetic boundary crystallization ultracentrifugation with DLS und UV-vis absorption measurements allows a comprehensive characterization of CdS nanoparticle systems.

Comparison of stabilizer influence

In a next step we studied the effect of chemically quite similar stabilizers with almost equal densities on the particle size distribution. In Fig. 4 the AUC particle size

into account. The particle sizes obtained from UV-vis (Fig. 2) and DLS (Fig. 3a) measurements give the same mean diameter of 1.6 nm. The size distribution obtained by DLS is relatively broad, which is a usual phenomenon accompanying relaxation techniques such as DLS. As just the average particle size is used for further calculations this is no restriction here. For comparison the sedimentation coefficient distribution obtained from the AUC measurement and corrected in the way described is given (Fig. 3b).

Assuming that the total diameter is well-determined by these density independent measurements, it is possible to calculate the average particle density, ρ_p, via Eq. (1) by combination of the weight-average sedimentation coefficient with the particle size obtained from the intensity-weighted DLS size distribution (z-average) [24]. The mean particle density of the TG1-CdS system is derived to be 3.2 g/cm^3. Assuming that the density of the core is that of bulk CdS and that the density of the shell is that of the pure stabilizer (1.2 g/cm^3) it is possible to calculate the diameter of the core and the shell (Fig. 1). The contribution of the shell is 0.5 nm in diameter, which is in good accordance with literature values 0.4 nm [22]. The core contains 56 unit cells of CdS (112 atoms of Cd and S, respectively).

Fig. 4 Comparison of approximately diffusion-corrected particle size distributions obtained by synthetic boundary crystallization at 25 °C and 40 000 rpm with different stabilizers

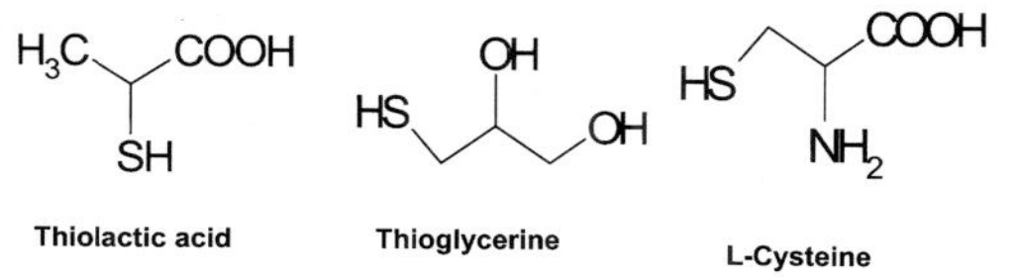

Fig. 5 Chemical structure of the stabilizers applied

distributions obtained with different thiol stabilizers are compared to visualize the differences induced by chemically similar stabilizers. The distributions are diffusion-corrected according to a modified van Holde–Weischet algorithm [16]. For a first approach we assume the particle density to be 3.2 g/cm^3 as derived for the thioglycerine case. This implies that the contributions of the different stabilizers to the particle density are approximately equal. As the stabilizer densities are similar within 0.05 g/ml, this is a reasonable assumption.

L-cystein as a stabilizer gives the biggest particles with a mean particle size of 2.8 nm, whereas thioglycerine seems to stabilize the smallest particles with an apparent mean diameter of only 0.9 nm. The fact that the smallest particle size of the distribution is 0 nm stems from the insufficient diffusion correction of the *s* distribution. Therefore, the real distributions can be considered to be narrower. The difference between the two particle size distributions obtained with thioglycerine acid as stabilizer in Figs. 3, 4 is due to the different reaction conditions. Assuming that the contribution of the stabilizer to the total particle diameter is 0.5 nm in all three cases the core consists of just one Cd atom and one S atom for thioglycerine, 72 atoms of Cd and S, respectively, in the thiolactic acid case and 128 atoms of Cd and S, respectively, for L-cysteine. Despite the fact that the assumption may not hold, this calculation

shows that the measurements allow access to a size range where the critical nucleus is expected. The apparent mean diameters of the L-cysteine and thiolactic acid stabilized particles differ by 0.4 nm. It is clearly shown that even small differences in the particle size distributions can be distinguished with the synthetic boundary crystallization method. The method is therefore sensitive to the effects of small changes in the chemical structure of the stabilizing molecules (Fig. 5).

Conclusions and outlook

Synthetic boundary crystallization ultracentrifugation is suitable for studying the influence of template molecules, stabilizers or any other kind of additives on the particle sizes of inorganic colloids. The combination of UV-vis and DLS data gives access to particle sizes and densities. As this is possible starting from critical crystal nuclei, crystallization mechanisms become potentially accessible yielding information about biomineralization and biomimetic processes and colloidal stabilization in general. Thus it can be expected that not only basic knowledge about the technically and scientifically important crystallization processes can be derived using this method, but furthermore, that it can serve as a fast test to compare the effectiveness of different additives.

For future studies the direct combination of DLS and sedimentation velocity measurements in the AUC is to be attempted so as to determine particle size and density in the one experiment.

Acknowledgements The authors would like to acknowledge the Max Planck Society and the Deutsche Forschungsgemeinschaft (SFB 448) for financial support and Anna Peytcheva for the DLS measurements.

References

1. Mann S, Archibald DD, Didymus JM, Douglas T, Heywood BR, Meldrum FC, Reeves NJ (1993) Science 261: 1286
2. Mann S (ed) (1995) Biomimetic materials chemistry, 1st edn. VCH New York
3. LaMer VK, Dinegar RH (1950) J Am Chem Soc 72: 4847
4. Cölfen H, Pauck T (1997) Colloid Polym Sci 275: 175
5. Fojtik A, Weller H, Koch U, Henglein A (1984) Ber Bunsenges Phys Chem 88: 969
6. Svedberg T, Rinde H (1924) J Am Chem Soc 46: 2677
7. Rinde H (1928) PhD thesis. Upsala, Sweden
8. Cölfen H, Pauck T, Antonietti M (1997) Progr Colloid Polym Sci 107: 136
9. Börger L, Cölfen H, Antonietti M (1999) Colloids Surf A accepted for publication.
10. Hersh R, Schachman HK (1955) J Am Chem Soc 77: 5228
11. Kemper DL, Everse J (1973) In: Hirs CHW, Timasheff SN (eds) Active enzyme centrifugation in methods in enzymology, vol 27., Enzyme Structure Part D, 67 Academic Press, New York
12. Lechner MD, Mächtle W (1999) Prog Colloid Polym Sci This issue
13. Baldwin RL, Williams JW (1950) J Am Chem Soc 72: 4325
14. Williams JW, Baldwin RL, Saunders W, Squire PG (1952) J Am Chem Soc 74: 1542
15. Gralen N, Lagermalm G (1951) J Phys Chem 56: 514
16. van Holde KE, Weischet WO (1978) Biopolymers 17: 1387
17. Fujita H (1975) Foundations of ultra-centrifugal analysis. Wiley, New York
18. Weller H (1993) Angew Chem Int Ed Engl 32: 41
19. Weller H, Schmidt HM, Koch U, Fojtik A, Baral S, Henglein A, Kunath W, Weiss K, Dieman E (1986) Chem Phys Lett 124: 557
20. Schnablegger H, Glatter O (1991) Appl Opt 30: 4889

21. Nosaka Y, Ohta N, Fukuyama T, Fujii N (1993) J Colloid Interface Sci 155: 23

22. Vossmeyer T, Katsikas L, Giersig M, Popovic IG, Diesner K, Chemseddine A, Eychmüller A, Weller H (1994) J Phys Chem 98: 7665

23. Weast RC, Astle MJ, Beyer WH (eds) (1984) CRC Handbook of chemistry and physics, 65th edn. CRC Press, Boca Raton

24. Elias HG (1990) Makromoleküle, 5th edn. Hüthig and Wepf, Basel, pp 114–115

Progr Colloid Polym Sci (1999) 113:29–36
© Springer-Verlag 1999

D. Kisters
H. Cölfen
W. Borchard

A comparison of evaluation methods for sedimentation velocity data of polymer solutions

D. Kisters · W. Borchard (✉)
Institut für Physikalische
und Theoretische Chemie
Angewandte Physikalische Chemie
Gerhard-Mercator-Universität
Lotharstrasse 1
D-47057 Duisburg, Germany

H. Cölfen
Max-Planck-Institut für Kolloid- und
Grenzflächenforschung
Kolloidchemie, Am Mühlenberg
D-14424 Potsdam, Germany

Abstract In this paper, three different methods for the calculation of the sedimentation coefficients s_2 of the polymer from sedimentation velocity runs of bovine serum albumin solutions are compared: firstly, the calculation of s_2 by means of the time-dependent movement of the boundary solvent/solution; secondly, the calculation of s_2 by means of the time-dependent movement of a hypothetical infinitely sharp boundary due to Goldberg; thirdly, the determination of s_2 with respect to the time-dependent movement of the local centre of mass of the polymer component. The results indicate that the sedimentation coefficients due to Goldberg must be about twice as large as the values due to the centre-of-mass method. The calculated values, derived from the experimental and simulated data, are in very good agreement with the theoretical relations derived in this work. The results show clearly that the method of the peak maximum of the Schlieren curve and the method of Goldberg yield nearly the same values for s_2 and the value of infinite dilution s_2^0, as it is expected for native polymers having diffusion coefficients of rather low values. A comparison between s_2 and s_2^0 due to the method of Goldberg and the corresponding values due to the centre-of-mass method showed clearly the important influence of the choice of the reference system. The resulting differences, attributable to a transition from the laboratory system (Goldberg and peak maximum) to the centre-of-mass system are explained in this paper.

Key words Sedimentation coefficient · Goldberg · Centre of mass and moving boundary methods

Introduction

The analysis of sedimentation data of polymer solutions, which can be derived by means of velocity runs in an Analytical Ultracentrifuge (AUC), is a well-known tool to calculate molar masses, sedimentation and diffusion coefficients, etc. of polymer molecules [1–3].

The characterization of thermoreversible gels by use of analytical ultracentrifugation is quite difficult in comparison to that of polymer solutions [4, 5]. In the past only a few researchers tried to calculate sedimentation coefficients of gels using the AUC.

Johnson and coworkers [6–9] calculated sedimentation coefficients of thermoreversible gels by measuring the time-dependent movement of the boundary gel/solvent at high centrifugal forces. A general problem at this time was the low intensity of the light sources used. Due to this, it was not possible to detect the gradient of concentration in the whole gel phase during the measurements and there was no alternative to the method mentioned above.

A few years ago, the AUC Beckman model E at the department of Applied Physical Chemistry in Duisburg was equipped with a new Schlieren optical system [10].

The advantage of this set-up was the relatively high light intensity which was available for the measurements. Due to the improved optical system, it was possible to detect the gradient of concentration of the polymer component in the complete gel phase during the sedimentation runs. Hinsken [11] showed for κ-carrageenan gels that sedimentation of the polymer takes place before the meniscus gel/solvent begins to move towards the bottom of the cell; therefore, the method of Johnson cannot be used to describe the sedimentation of the polymer component in the gel.

Borchard and Hinsken [12] introduced a new evaluation method to calculate the sedimentation coefficient of gels. They calculated the sedimentation coefficients of κ-carrageenan gels by detection of the time-dependent movement of the centre of mass of the polymer component in the gel during the run. In the case of κ-carrageenan and gelatin gels the sedimentation coefficients, which were calculated by the centre-of-mass method, showed a nearly quadratic dependence on the height of the gel column.

The intention of our work is to find a universal evaluation method for sedimentation velocity experiments, which can be applied not only to sedimentation data of polymer solutions but also to thermoreversible gels. In this case it would be possible to describe a polymer/solvent system which forms a gel at a defined polymer concentration over the whole concentration range with one evaluation method.

In the present paper we have calculated the sedimentation coefficients s_2 of bovine serum albumin (BSA) solutions of different concentrations by means of three different evaluation methods. The first two are typical procedures to evaluate data of polymer solutions whereas the centre-of-mass method has not been applied to polymer solution data until now. The aim of this procedure was to get an idea in which manner changes in the evaluation method have an influence on the resulting sedimentation coefficients.

It will be shown that changes in the reference system have an important influence on the sedimentation coefficients of BSA.

Theory

For polymer solutions the sedimentation coefficient of component 2, s_2, is defined as the ratio of the velocity of the particles of component 2 and the centrifugal acceleration.

$$s_2 = \frac{dr_2/dt}{\omega^2 r_2} \quad , \tag{1}$$

where r_2 is the distance of component 2 from the axis of rotation and ω is the angular velocity.

The sedimentation coefficient s_2 describes the velocity of component 2 in a unit field and is usually measured in Svedberg units. Although the velocity of the particles of component 2 is only indirectly detectable, the average velocity of all particles can be measured in the following manner.

In a sedimentation velocity experiment the centrifugal acceleration is so high that the diffusion of the particles can be neglected because sedimentation is much more rapid than diffusion; therefore, the solute will be depleted in the region of the meniscus and a boundary will be formed between the solvent and the solution phase. In the solution phase the so-called plateau region will be formed, where the polymer concentration does not depend on the distance from the axis of rotation r. The boundary moves towards the bottom of the cell with increasing time and can be detected directly using UV absorption or interference optics. The relations mentioned above are presented schematically in Fig. 1.

If Schlieren optics are used, the boundary region appears as a peak in the concentration gradient/distance curve and the position of the boundary can be detected as the maximum of the peak.

Due to the sector shape of the cell the solution in the plateau region will be diluted with increasing time because of radial movement; therefore, the concentration in the plateau zone can be described as a function of time [1, 2].

$$c_{P,2} = c_{0,2} \exp(-2\omega^2 s_2 t) \quad , \tag{2}$$

where $c_{P,2}$ is the plateau concentration of component 2 and $c_{0,2}$ is the initial concentration of component 2.

From Eq. (2) the so-called square dilution rule can be derived:

$$\left(\frac{c_{P,2}}{c_{0,2}} \right) = \left(\frac{r_m}{r_{bnd}} \right)^2 \quad , \tag{3}$$

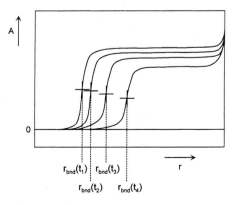

Fig. 1 Schematical representation of the movement of a boundary versus the radial distance during a sedimentation velocity run observed with UV absorption optics at four different times ($t_1 < t_2 < t_3 < t_4$)

where r_m is the position of the meniscus vapour/solution and r_{bnd} is the position of the boundary solvent/solution.

By means of Eq. (3) it is possible to eliminate the dilution effect for a sector-shaped cell during a run.

The use of Eq. (1) yields an apparent sedimentation coefficient $(s_2)_{app}$ which depends on temperature, concentration and pressure.

The denominator of Eq. (1), $\omega^2\,dt$, must be written as an integral of time $\int_0^t \omega^2\,dt$ due to the fact that the centrifuge needs some time to accelerate the rotor to the desired constant rotational speed [13].

$$(s_2)_{app} = \frac{\ln(r^*/r_0^*)}{\int_0^t \omega^2\,dt} \quad , \tag{4}$$

where r^* is the position of a specific point according to the different evaluation methods and r_0^* is the position of the specific point at time $t=0$.

The pressure increases with the distance from the axis of rotation, r, and this dependence of $(s_2)_{app}$ can be eliminated using the following equation [1]:

$$(s_2)_{app} = s_2 \left\{ 1 + B_1 \left[\left(\frac{r^*}{r_0^*} \right)^2 - 1 \right] \right\} \quad , \tag{5}$$

where B_1 is a parameter for the pressure effect.

A plot of $(s_2)_{app}$ against $(r^*/r_0^*)^2 - 1$ yields as the intercept the sedimentation coefficient of component 2, s_2, which depends only on temperature and concentration.

According to Fujita [1] the concentration dependence of s_2 is eliminated by means of Eq. (6):

$$\frac{1}{s_2} = \frac{1}{s_2^0}(1 + k_S c_2) \tag{6}$$

or the equivalent linearized form $s_2 = s_2^0(1 - k_S c_2)$, which usually yields good results for spherical particles.

s_2^0 is the sedimentation coefficient of component 2 at infinite dilution and can be derived by plotting $1/s_2$ against the concentration c_2. The factor k_S is independent of concentration and yields further information about the particles.

To compare the sedimentation coefficients determined in different solvents and at different temperatures the sedimentation coefficient can be standardized to water as solvent at a standard temperature of, for example, 20 °C, by means of Eq. (7). This standardization takes differences in density and viscosity into account. The result is the standardized, concentration-independent sedimentation coefficient $s_{20,w}^0$ [1–3].

$$s_{20,w}^0 = s_2^0 \frac{(1 - \tilde{V}_2 \rho_{20,w})\,(\eta_{\vartheta,s})}{(1 - \tilde{V}_2 \rho_{\vartheta,s})\,(\eta_{20,w})} \tag{7}$$

where $\tilde{V}_2$ is the partial specific volume of the solute, $\rho_{20,w}$ is the density of water at 20 °C, $\rho_{\vartheta,s}$ is the density of the solvent at the measuring temperature ϑ, $\eta_{\vartheta,s}$ is the viscosity of the solvent at the temperature ϑ and $\eta_{20,w}$ is the viscosity of water at 20 °C.

For the case where the boundary solvent/solution remains symmetrical with respect to the position of the maximum of the Schlieren curve during the run, the detection of the time-dependent movement of the boundary or the peak maximum is a sufficiently accurate tool to calculate the sedimentation coefficients of component 2 with Eq. (1). In this case it follows that $r^* = r_{bnd}$ and $r_0^* = r_m$.

If the boundary solvent/solution is broadened and becomes asymmetrical during the run, the detection of the maximum of the gradient curve cannot be used for the determination of the average velocity of the particles.

In this case the diffusion of the polymer molecules is no longer negligible and the sedimentation coefficient, s_2, of the solute can be determined by calculation of the second moment of the gradient curve, introduced by Goldberg in 1953 [14].

The square root of the second moment of the gradient curves yields a position in the cell which can be described as a hypothetical infinitely sharp boundary.

This boundary at $\bar{r}_2$ moves with the same velocity as the molecules ahead and due to this the diffusion is mainly eliminated.

The second moment can be calculated by the following equation [3, 14]:

$$\bar{r}_2{}^2 = \frac{\int_{r_m}^{r_P} r^2 (dc_2/dr)\,dr}{\int_{r_m}^{r_P} (dc_2/dr)\,dr} \quad , \tag{8}$$

where r_P is a constant position in the plateau region and r_m is the position of the meniscus vapour/solution.

Using an optical system which yields directly the mass concentration of the polymer, c_2, (UV or interference optics), Eq. (8) can be written as [14]

$$\bar{r}_2{}^2 = r_P^2 - \frac{2}{c_{P,2}} \int_{r_m}^{r_P} c_2 r\,dr \quad . \tag{9}$$

By means of Eq. (8) it is possible to calculate the position of the infinitely sharp boundary $\bar{r}_2$. Substitution of r^* and r_0^* with the resulting time-dependent values of $\bar{r}_2$ in the Eqs. (4)–(6) then yields s_2^0 according to the method of Goldberg. It is very important to mention that the reference system using the method of the peak maximum as well as the method due to Goldberg is the laboratory system.

Another procedure which can be applied to sedimentation data of polymer solutions is the detection of the time-dependent movement of the local centre of mass of the polymer component as it was done in the past for thermoreversible gels [11]. In this case the reference system is the centre-of-mass system in contrast to the other evaluation methods mentioned previously.

The movement of the local centre of mass is a clear measure of the mean movement of the molecules in the solution with time; therefore, the definition of a sedimentation coefficient on this basis is possible.

The position of the local centre of mass, $r_{2,S}$, between the position of the meniscus vapour/solution, r_m, and a constant position in the plateau region, r_P, of component 2 in a sector-shaped cell can be described as the ratio of the integrals of the masses $dm_{2,i}$ of component 2, weighted with the distance from the axis of rotation and the mass of component 2 in the cell between r_m and r_P [16].

$$r_{2,S} = \frac{\int_{r_m}^{r_P} r_i \, dm_{2,i}}{\int_{r_m}^{r_P} dm_{2,i}} = \frac{\int_{r_m}^{r_P} r_i c_{2,i} \, dV_i}{\int_{r_m}^{r_P} c_{2,i} \, dV_i} \quad , \tag{10}$$

where

$$dm_{2,i} = c_{2,i} \, dV_i \quad , \tag{11}$$

r_i is the distance from the axis of rotation, r_P is the constant position in the plateau region during a series of measuring times and r_m is the position of the meniscus vapour/solution.

The volume dV_i is a function of the height of the cell, h, r_i and the half-sector angle, ϕ. The cell height does not depend on r_i and can be eliminated. The sector shape of the cell can be taken into consideration by means of a factor $H = (\sin \phi / \phi)$. Due to this, Eq. (10) can be rearranged to give [15]

$$r_{2,S} = \frac{\int_{r_m}^{r_P} c_{2,i} r_i^2 \, dr_i}{\int_{r_m}^{r_P} c_{2,i} r_i \, dr_i} \frac{\sin \phi}{\phi} \quad , \tag{12}$$

By means of Eq. (12) it is possible to calculate the position of the local centre of mass $r_{2,S}$, if the mass concentration $c_{2,i}$ as a function of r_i is measured.

The sedimentation coefficient s_2^0 due to the centre-of-mass method is derived by means of Eqs. (4)–(6), and substituting r^* and r_0^* with the corresponding values of $r_{2,S}$.

Experimental

For all experiments, we applied a Beckman Optima XL-I AUC equipped with UV/VIS absorption optics and an on-line Rayleigh interferometer. The experiments were performed at 50 000 rpm and 25 °C in self-made titanium double-sector centrepieces. The scanning wavelengths were 230 and 280 nm. BSA (fraction V, initial fraction by heat shock, purity greater than 98%) was obtained from Sigma and was dissolved in an aqueous 0.5 N NaCl solution.

Results

According to Beer's law the measured values of the absorption are directly proportional to the BSA con-

centration. Therefore all calculations in this paper were directly carried out with the measured values of the absorption in absorption units (AU).

General remark

During the calculations of the sedimentation coefficients according to the three different evaluation methods mentioned in the Theory section it became apparent, that two different procedures are possible. The first one is to calculate s_2 directly by means of Eqs. (4) and (5) using the measured values for r^* and r_0^*, according to the corresponding method. The second one is to use interpolated values of r^*. For that, r^* was plotted against the run-time integral and a linear regression was carried out. The result was an equation which describes the time dependence of r^*. By means of this equation the interpolated values, r_{int}^*, were calculated and the sedimentation coefficients were also derived by use of these interpolated values.

Figure 2 shows, for the case of the movement of the peak maximum, that the use of the interpolated values yields a dependence of $(s_2)_{app}$ on the corresponding values of $(r^*/r_0^*)^2 - 1$ with a relatively low scattering at low values of this function.

An explanation for this can be derived by the use of a simple error propagation of Eqs. (4) and (5). The maximum error in $(s_2)_{app}$ caused by the error Δr^* is as follows:

For Eq. (4)

$$\Delta(s_2)_{app} \sim (1/r^*)\Delta r^* \quad ; \tag{13}$$

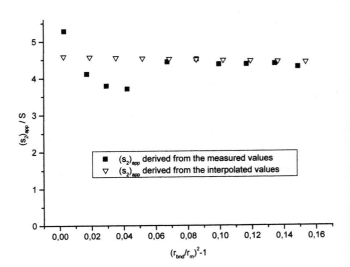

Fig. 2 Dependence of $(s_2)_{app}$ on $(r_{bnd}/r_m)^2 - 1$ using the measured and interpolated values of r^* (moving of the peak maximum; $c_{BSA} = 1.5$ g/l); see text

For Eq. (5)

$$\Delta(s_2)_{app} \sim (r^*)\Delta r^* \quad . \tag{14}$$

It results from this that it is better to carry out the interpolation of the measured values of r^* and to calculate $(s_2)_{app}$ by means of the interpolated values, because then the influence of Δr^* towards $\Delta(s_2)_{app}$ can be minimized. Thus, all calculations were done with the interpolated values.

Considerations of the dilution effect

Usually the dilution effect can be eliminated by means of Eq. (3). For the samples with small BSA concentrations it was very difficult to detect the position of the peak maximum due to the noise of the measurements; therefore, the dilution effect was eliminated by means of the following procedure. First the time-dependent concentration in the plateau region was determined by calculating the mean value in this region. Then these value for $c_{P,2}$ were plotted against the run-time integral giving a nearly linear line (Fig. 3); therefore, a linear regression was carried out. The intercept of this function yields the initial BSA concentration directly.

After that $c_{0,2}$ was divided by the values for $c_{P,2}$. The result was a factor F for each time a measurement was performed. To eliminate the dilution effect, the concentration c_2 as a function of the distance r from the axis of rotation was multiplied by the F factors over the whole range of r, resulting in the conversion of $c_{P,2}$ into $c_{0,2}$ for each time. Table 1 shows the dependence of the resulting F factors on the run-time integral and $c_{P,2}$ for a concentration of 2 g/l.

Table 1 Time-dependent dilution in the plateau region. $c_{0,2} = 1.263$ absorption units (AU)

$\int \omega^2 dt/10^9\,s$	$c_{P,2}/AU$	F
1.437	1.2408	1.01800
2.984	1.2278	1.02878
4.635	1.2110	1.04302
6.28	1.1900	1.06145
7.921	1.1667	1.08262
9.56	1.1444	1.10378
11.21	1.1153	1.13262

A comparison between the application of the square dilution rule (Eq. 3) for concentrations of 2–0.5 g/l and the procedure mentioned above showed that the deviations for different initial concentrations $c_{0,2}$ calculated by both methods are in the range of about 0.5%, which is not explicitly shown in the table. For these concentrations it was possible to detect the position of the peak maximum; therefore, $c_{0,2}$ was derived for all samples by means of the time-dependent decrease in $c_{P,2}$ and not by applying the square dilution rule.

The sedimentation coefficients due to the three methods

All calculations mentioned in this section were performed for the measured curves as well as for the curves which were corrected for the dilution effect. The evaluation of the experimental data using the method of the movement of the peak maximum was quite difficult because of the presence of a bad signal-to-noise ratio (S/N).

In all cases the measured absorption/distance curves were differentiated and the resulting derivations were treated with a smoothing procedure to improve the S/N ratio. The calculation of the sedimentation coefficient was only done for five BSA concentrations (2/1.75/1.5/1/0.8 g/l). The lower concentration could not be evaluated by means of the peak-maximum method.

In the cases of the Goldberg method and the centre-of-mass method the experimental curves were integrated directly by means of Eq. (9) (Goldberg) and Eq. (12) (centre of mass). For both methods all concentrations were used. The sedimentation coefficients, s_2, according to a distinct method were calculated by means of Eqs. (4) and (5) for each concentration using the calculated values of r^* and r_0^* corresponding to the method used.

The extrapolation of s_2 to the hypothetical BSA concentration, $c_2 = 0$ g/l, was done by means of Eq. (6). The results of this extrapolation due to all three methods are shown in Fig. 4. $s_{20,w}^0$ was calculated for all three methods by means of Eq. (7).

During the calculations it became apparent, that the correction of the experimental data, due to the dilution effect, has no influence on the resulting sedimentation coefficients s_2 and s_2^0 for all three methods.

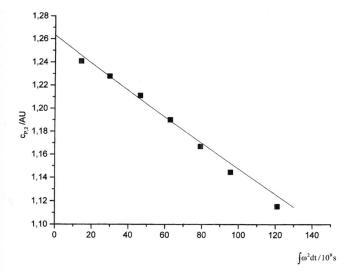

Fig. 3 Time-dependent dilution in the plateau region. $c_{0,2} = 1.263$ absorption units (AU)

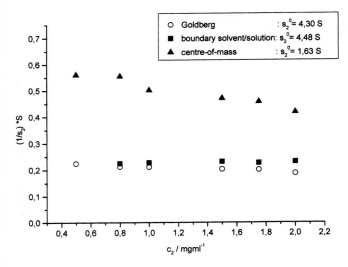

Fig. 4 Interpolation of the values of $(1/s_2)$ by means of Eq. (6) due to the different evaluation methods; see text

Discussion

The calculations have shown clearly for all three methods that the dilution effect does not have to be taken into consideration for these BSA solutions of a nearly monodisperse polymer because the resulting values for s_2 and s_2^0 do not change. It was shown that it is possible to calculate the apparent sedimentation coefficients $(s_2)_{app}$ by means of two different procedures to eliminate the pressure-dependence.

Table 2 comprises the results of the calculations due to the different evaluation methods.

The differences between the sedimentation coefficients due to the peak-maximum method and the Goldberg method are very small as it was expected for high-molar-mass native polymers.

It is quite interesting that the differences between the Goldberg and the centre-of-mass method are rather large. It follows from the experimental results that the sedimentation coefficient, calculated by means of the Goldberg method, is proportional to the values derived from the centre-of-mass procedure.

$$s_2(\text{Goldberg}) = A_{\exp}\, s_2\,(\text{centre of mass}) \ , \tag{15}$$

where $A_{\exp}$ is a factor due to the change in the reference system.

From the calculations $A_{\exp}$ can be determined to be $A_{\exp} \approx 2.5$.

This means that changes in the reference system will produce these large changes in s_2. In the case of the method of Goldberg and that of the peak maximum the reference system which has to be taken into consideration is the laboratory system, and in the case of the other method it is the centre-of-mass system.

To explain this relation we first assume a rectangular measuring cell with a constant concentration $c_{P,2}$ between positions r_{t1} and r_P at time t_1. After a time Δt the boundary at r_{t1} has moved to r_{t2}. Due to the shape of the cell, the concentration in the plateau region does not change during this time interval.

The position of the local centre of mass, $r_{2,S}$, at times t_1 and $(t_1 + \Delta t)$ can be derived using the following equations:

$$r_{2,S}(t_1) = r_P - \frac{r_P - r_{t1}}{2} \ , \tag{16}$$

$$r_{2,S}(t_1 + \Delta t) = r_P - \frac{r_P - r_{t2}}{2} \ . \tag{17}$$

The movement of the centre of mass, $\Delta r_{2,S}$, during the time interval Δt can be calculated by means of Eqs. (16) and (17):

$$\begin{aligned}\Delta r_{2,S} &= r_{2,S}(t_1 + \Delta t) - r_{2,S}(t_1) \\ &= \tfrac{1}{2}(r_{t2} - r_{t1}) = \tfrac{1}{2}\Delta r \ . \end{aligned} \tag{18}$$

Due to the procedure of Goldberg, r_{t1} and r_{t2} can be assigned to the positions of the infinitely sharp boundary with respect to time. Therefore, Eq. (18) shows for a rectangular measuring cell that if the boundary moves a distance Δr towards the bottom of the cell in the time interval Δt, the centre of mass will move only a distance $(\Delta r/2)$. In this case the sedimentation coefficient, calculated by means of the method of Goldberg, would be twice as much as the value derived from the centre-of-mass method because the slope for the time-dependent movement of r is directly proportional to s_2 (see Eq. 1).

The same assumptions can be applied to a sector-shaped measuring cell. In this case the local centre of mass in such a cell with a constant plateau concentration $c_{P,2}$ is defined by Eq. (19) [15]:

$$r_{2,S} = \frac{2}{3}\frac{r_P^3 - r^3}{r_P^2 - r^2}\frac{\sin\phi}{\phi} \ , \tag{19}$$

where r is the position of r_{t1} or r_{t2} (see Fig. 5) and ϕ is the half-sector angle in radian.

Table 2 Comparison of the sedimentation coefficients s_2, s_2^0 and $s_{20,w}^0$ due to the different evaluation methods

$c_2/(\text{g/l})$	s_2/S (peak maximum)	s_2/S (Goldberg)	s_2/S (centre of mass)	$A_{\exp}$ (see text)
2	4.33	5.36	2.39	2.24
1.75	4.43	4.99	2.19	2.28
1.5	4.33	4.95	2.13	2.32
1	4.40	4.75	1.99	2.38
0.8	4.45	4.71	1.80	2.62
0.5		4.46	1.78	2.50
0	4.48	4.30	1.63	2.63
$s_{20,w}^0/S$	4.40	4.22	1.60	2.64

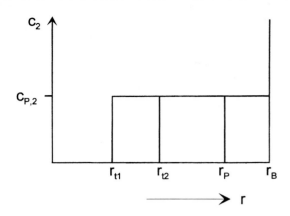

Fig. 5 Schematic concentration/distance relation for an infinitely sharp boundary in a rectangular/sector-shaped cell; see text

Plots of the values of $r_{2,S}$ with respect to r (with constant r_P) in the range in which r and r_P are usually measured (6.1–7.0 cm) are shown in Fig. 6. In this range the plots can be fitted to a very good approximation by means of linear equations.

The use of these equations yields the dependence between $r_{2,S}$ and r for the case of a sector-shaped cell and a constant plateau concentration:

$$r_{2,S} = Cr + B \quad , \tag{20}$$

where $C = f(r_P)$ and $B = f(r_P)$.

The differentiation of Eq. (20) with respect to time (with constant r_P) yields

$$\left(\frac{\mathrm{d}r_{2,S}}{\mathrm{d}t}\right)_{r_P} = C\left(\frac{\mathrm{d}r}{\mathrm{d}t}\right)_{r_P} \quad . \tag{21}$$

As already mentioned, r can be assigned to the position of the Goldberg boundary; therefore, Eqs. (20) and (21)

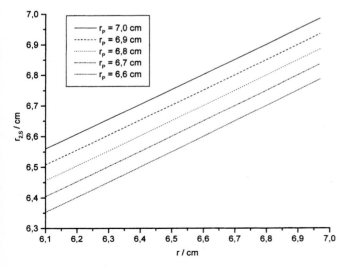

Fig. 6 Dependence of the position of the local centre of mass $r_{2,S}$, versus the position of the Goldberg boundary due to Eq. (19)

can be used to determine the relation between the sedimentation coefficients due to the Goldberg and centre-of-mass method

$$s_2(\text{centre of mass}) = \frac{(\mathrm{d}r_{2,S}/\mathrm{d}t)_{r_P}}{\omega^2 r_{2,S}} = \frac{C(\mathrm{d}\bar{r}_2/\mathrm{d}t)_{r_P}}{\omega^2(C\bar{r}_2 + B)} \quad , \tag{22}$$

$$s_2(\text{Goldberg}) = \frac{(C\bar{r}_2 + B)}{C\bar{r}_2} s_2 \text{ (centre of mass)} \tag{23}$$
$$= A_{\text{theo}} s_2(\text{centre of mass}) \quad ,$$

where

$$A_{\text{theo}} = f(r_P, \bar{r}_2) = \frac{(C\bar{r}_2 + B)}{C\bar{r}_2} \quad . \tag{24}$$

By means of Eq. (23) the relation between the sedimentation coefficients based on the two methods can be derived. Due to the fact that the pressure dependence of the sedimentation coefficients for both methods was eliminated by extrapolation to the position of the meniscus of the vapour/solution, r_m, (Eq. 5), this position must be used to calculate A_{theo}:

$$A_{\text{theo}} = f(r_P, r_m) = \frac{(Cr_m + B)}{Cr_m} \quad . \tag{25}$$

Equation (25) was used to calculate A_{theo} for the values of r_P which were used to derive the sedimentation coefficient due to the different methods, with the experimentally determined positions of r_m:

$$r_P = 6.8\,\text{cm} \quad A_{\text{theo}} = 2.13$$
$$r_P = 6.9\,\text{cm} \quad A_{\text{theo}} = 2.16$$
$$r_P = 7.0\,\text{cm} \quad A_{\text{theo}} = 2.18 \quad .$$

The results show that the dependence of A_{theo} on changes in r_P is rather small ($\sim 2\%$) and therefore this dependence can be neglected.

The same result is obtained if Eq. (19) is developed into a geometrical series. These theoretical calculations showed clearly that the differences between s_2 or s_2^0 due to the different evaluation methods are caused only by the choice of the reference systems; therefore, it is possible to convert the values derived by means of a distinct method into each other.

The difference between the calculated values for A_{exp} and the theoretically determined value of $A_{\text{theo}} = 2.16$ are in the range of about 15%.

To prove the theoretically determined relation, we also calculated the sedimentation coefficient, based on the centre-of-mass method, by means of Eq. (19) for four concentrations with the previously determined values of $\bar{r}_2$ and $c_{P,2}$.

The resulting values for s_2 and A_{exp}^* due to the different methods are given in Table 3.

Table 3 shows that the resulting values for A_{exp}^* correspond very well to the theoretically derived values, if the local centre of mass is calculated by means of

Table 3 Comparison of s_2 and s_2^0 due to the Goldberg and the centre-of-mass methods, calculated by means of Eq. (19)

c_2/(g/l)	s_2 /S (Goldberg)	$A_{theo}*s_2$/S (centre of mass)	A_{exp}^*
2	5.36	5.06	2.16
1.5	4.95	4.71	2.15
1	4.75	4.53	2.14
0.5	4.46	4.24	2.15

Table 4 Comparison of the factors A_{sim} between the sedimentation coefficients due to the Goldberg and the centre-of-mass method, for the simulated sedimentation velocity data

Parameters	Claverie simulation a	Claverie simulation b
s_2/S	5	5
$D/(10^{-7}$ cm^2/s)	7	1
$c_{0,2}$/AU	1.0	1.0
Speed/rpm	50 000	50 000
r_B/cm	7.2	7.2
r_m/cm	6.0	6.0
$T/°C$	20	20
A_{sim}	2.30	2.25

The sedimentation coefficients of these simulated data due to the Goldberg method and the centre-of-mass method were calculated by means of the previously mentioned equations: a factor A_{sim} between these values for s_2 was calculated.

Table 4 shows that A_{sim} corresponds well to the theoretically determined value for A_{theo}. The differences are in the range of about 4%.

Conclusions

The results of this research show that it is possible to apply the centre-of-mass method to sedimentation velocity data of polymer solutions. The differences between the theoretically derived relations and the experimental results are in the range of about 15% and must be assigned to two effects:

1. Diffusion during a run and by this the broadening of the boundary solvent/solution
2. Noise in the measurements.

Due to this, the application to polymer solutions is limited at the moment.

The relations, derived from the simulated data, are in good agreement with the theoretical relation. In the future we want to determine the influences of these two effects on the sedimentation coefficients using more sets of simulated sedimentation velocity data. For this case the deviations of the sedimentation coefficients according to the centre-of-mass method between the experimentally determined and the theoretically derived values can be used to obtain information about the poly-dispersivity of the solute.

Acknowledgements We thank the Deutsche Forschungsgemeinschaft for financial support. Furthermore we thank J.S. Philo for publishing the program Svedberg.

Eq. (19). The differences between A_{exp}^* and A_{theo}^* are in the range of about 1%.

Furthermore we performed Claverie simulations by means of the Svedberg (version 5.01) program to simulate sedimentation velocity data of a monodisperse polymer for two different sets of parameters, which are given in Table 4. Details of this program are given in the literature [16, 17]. The idea was to confirm the theoretically derived relation between the sedimentation coefficients due to the different evaluation methods.

For these sets of parameters the theoretical factor A_{theo} can be calculated by means of Eq. (25). For $r_m = 6.0$ cm and $r_P = 6.9$ cm one obtains $A_{theo} = 2.20$.

References

1. Fujita H (1975) Foundations of ultracentrifugal analysis. Wiley, New York
2. Williams JW (1972) Ultracentrifugation of macromolecules. Academic Press, New York
3. Schachman HK (1959) Ultracentrifugation in biochemistry. Academic Press, New York
4. Cölfen H (1995) Colloid Polym Sci 273: 1101–1137
5. Cölfen H (1999) Biotechnol Genet Eng Rev 16: 87–140
6. Johnson P, Metcalfe JC (1967) Br Polym J 3: 423–447
7. Johnson P (1964) Proc R Soc and Ser A 278: 527
8. Johnson P, King RW (1968) J Photogr Sci 16: 82
9. Johnson P (1971) J Photogr Sci 19: 49
10. Cölfen H, Borchard W (1994) Anal Biochem 219: 321–334
11. Hinsken H (1998) Dissertation. Duisburg
12. Borchard W, Hinsken H (1997) Prog Colloid Polym Sci 107: 172–179
13. Lechner DM, Mächtle W (1995) Prog Colloid Poly Sci 99: 120–124
14. Goldberg RJ (1953) J Phys Chem 57: 194–202
15. Hütte (1971) Des Ingenieurs Taschenbuch Band I Mechanik, vol 29. Ernst & Sohn, Berlin
16. Philo JS (1994) In: Schuster TM, Laue TM (eds) Modern Analytical Ultracentrifugation. Birkhauser, Boston, pp 156–170
17. Philo JS (1997) Biophys J 72: 435–444

Progr Colloid Polym Sci (1999) 113:37–43
© Springer-Verlag 1999

M.D. Lechner
W. Mächtle

Determination of the particle size distribution of 5–100-nm nanoparticles with the analytical ultracentrifuge: consideration and correction of diffusion effects

M.D. Lechner (✉)
Physical Chemistry, University
Barbarastrasse 7
D-49069 Osnabrück, Germany
e-mail: lechner@rz.uni-osnabrueck.de
Tel.: +49-541-9692819
Fax: +49-541-9691205

W. Mächtle
Kunststofflaboratorium, Polymer Physics
BASF AG, D-67056 Ludwigshafen
Germany

Abstract Small nanoparticles in the diameter range 5–100 nm are used in large quantities for very different applications. A precise determination of the important physical quantities which are responsible for the application properties such as diameter, particle size distribution, density, and (specific) surface area is therefore essential. Analytical ultracentrifugation is one of the best methods for the characterization of such nanoparticles as they are fractionated in the analytical ultracentrifuge cell during sedimentation. This paper presents the theoretical foundations for the determination of the physical quantities of nanoparticles with special consideration of the diffusion effect with respect to the particle size distribution measurements in the analytical ultracentrifuge. Besides several extrapolation procedures a new method for the correction of the sedimentation coefficient and its distribution is presented. Several examples demonstrate that the suggested diffusion-broadening correction procedures work. The procedure which gives the best values depends on the particular nanoparticle.

Key words Analytical ultracentrifugation · Particle size distribution · Nanoparticles · Diffusion correction · Theoretical consideration

Introduction

Nanoparticles are colloidal materials with diameters from 5 to 100 nm, mostly dispersed in water. They are produced in large quantities for very different applications such as lacquers, dyes, cosmetics, food, magnetic storage materials (tapes, disks, hard disks), catalysts, and print materials (toners, inks). The development of new nanoparticles and the modification of available nanoparticles with new application properties needs considerable scientific effort. Important physical quantities of nanoparticles are the diameter, d, the mass, m, the molar mass, M, the volume, V, the density, $\rho = m/V = M/V_m$ (V_m = molar volume), the specific volume, $v = 1/\rho$, the surface area, O, the specific surface area, $o = O/m$, the differential mass and number distribution, $w(d)$ and $x(d)$, and the integral mass and number distribution, $W(d)$ and $X(d)$, of the physical quantities mentioned above, especially the particle size distribution (PSD). Nanoparticles usually have a spherical structure; therefore, the relationships between the physical quantities are easy: $V = \pi d^3/6$, $O = \pi d^2$, $m = \pi d^3 \rho/6$, and $M = N_A \pi d^3 \rho/6$. For catalysts the (specific) surface of the material is an important quantity and kinetics studies of the production of nanoparticles need number distributions of the physical quantities instead of mass distributions. The most reliable experimental methods for the characterization of nanoparticles are analytical ultracentrifugation (AUC) and dynamic light scattering. AUC has the advantage that the particles are fractionated during sedimentation, therefore, different sizes of the particles may be detected with high precision, i.e., the complete particle size distribution at high resolution. During this sedimentation the diffusion of the particles plays no important role if the particle diameter is larger than 50 nm; however, if the particles are smaller one has to consider the diffusion-broadening of the sedimenting nanoparticles' boundary if one wishes to get a correct AUC-PSD. The measurement of the PSD via AUC without diffusion-broadening correction is a well-established method [1].

This paper demonstrates the techniques for the determination of the physical quantities of nanoparticles using AUC with a Schlieren (or interference) optics detector and introduces a new diffusion-broadening correction method for very small nanoparticles.

Theoretical considerations

The basis to estimate PSDs with an AUC is first to measure the distribution of the sedimentation coefficient, S, of the nanoparticle inside an AUC cell. Then, this S distribution is transformed into the particle diameter distribution (equivalent to the PSD) using the well-known Stokes' law [1]:

$$d = [18\eta_1 S/(\rho_2 - \rho_1)]^{1/2}; \quad S = (1/\omega^2)\mathrm{d}\ln r/\mathrm{d}t , \quad (1)$$

where η_1 is the viscosity of the solvent, S is the sedimentation coefficient, ρ_2 is the density of the particle, ρ_1 is the density of the solvent (dispersing agent), $\omega = 2\pi N$ is the angular frequency of the rotor, r is the distance from the axis of rotation, and t is the running time. The differential distribution function of the sedimentation coefficient $g(S,t,c_0)$ and the integral distribution function $G(S,t,c_0)$

$$g(S,t,c_0) = \partial G(S,t,c_0)/\partial S \quad \text{and} \quad G(S,t,c_0) = c(r,t)/c_0 , \quad (2)$$

with concentration $c(r,t)$ at distance r and time t and initial concentration, c_0, may then be calculated according to Bridgman [2]

$$g(S,t,c_0) = (1/c_0)[\partial c(r,t)/\partial r](r/r_\mathrm{m})^2 r \int_0^{t'} \omega^2\,\mathrm{d}t \quad (3)$$

or to Stafford [3]

$$g(S,t,c_0) = (1/c_0)[\partial c(r,t)/\partial t](r/r_\mathrm{m})^2 t \int_0^{t'} \omega^2\,\mathrm{d}t/\ln(r_\mathrm{m}/r) \quad (4)$$

using the measured (often with Schlieren optics) concentration distribution, $c(r,t)$, of the nanoparticles inside the ultracentrifuge cell as functions of running time, t, and cell radius, r. The height of the Schlieren peak curve at radius position r is directly proportional to $(\partial c/\partial r)_t$. Use of the AUC calibration constant and integration of the Schlieren peak curve yields the complete concentration distribution $c(r, t)/c_0$. Equations (3) and (4) include the dilution rule for sector-shaped cells, $c/c_0 = (r_\mathrm{m}/r)^2$.

For the experimental determination of the PSD of very small particles of $d < 50$ nm in an AUC it has to be taken into account that sedimentation and diffusion of the particles superimpose. That means, the broadening of the sedimenting (free) particle boundary is not only due to a broad PSD, but also to a diffusion movement of the particles perpendicular to this boundary. This additional diffusion-broadening becomes stronger the smaller the particles are. To calculate the PSD either both effects need to be separated or the PSD needs to be calculated from combined sedimentation–diffusion equations. This may be done in three different ways to take into account the diffusion:

1. Extrapolation procedures. All kinds of running time extrapolation procedures [3–7] take into consideration the fact that the sedimentation of a particle is proportional to the running time (see Eq. 1), whereas the diffusion is proportional to the square root of the time, $t^{1/2}$. That means extrapolation to t at infinity yields pure sedimentation and extrapolation to t at zero yields pure diffusion. The correct form of the differential distribution function of S without diffusion-broadening, $g(S,c_0)$, or its integral function $G(S,c_0)$ could be derived by investigating the asymptotic behaviour of the boundary-spreading equation [4] for values of times reaching infinity. All extrapolation procedures give true values of S together with its integral or differential distributions. From this the diameter of the particle with its distribution and all other physical quantities may be calculated.

2. Application of Lamm's equation. The problem has been solved only with respect to monodisperse particles [4, 8]. Therefore this method is up to now not applicable to very polydisperse particles.

3. Correction procedure (new). This means correction of the apparent sedimentation coefficient distribution with respect to an average diffusion coefficient. As far as we know, this procedure, which we propose in this paper, has not been applied up to now.

Finally for all three ways it is known that the diameter, d, and its differential and integral distribution functions, $w(d)$ and $W(d)$, may be calculated according to Eq. (1) and the measured S distributions

$$W(d) = G(S) \quad \text{and} \quad w(d)\mathrm{d}d = g(S)\mathrm{d}S . \quad (5)$$

In the following we describe only two ways, the extrapolation procedures and the new correction procedure, in more detail. We do not apply Lamm's equation because it fails for polydisperse particles.

Extrapolation procedures to infinite running time

The theory of Gosting [5] extrapolates (via Schlieren optics) the measured apparent differential distribution of S, $g(S,t,c_0)$, at specified values of S to $1/t \to 0$. His results read

$$g(S,t,c_0) = g(S,c_0)$$
$$+ \frac{D}{\omega^4 r_m^2} \frac{1 - \exp(-2S\omega^2 t)}{2S\omega^2 t} \frac{1}{t} \frac{d^2 g(S,t,c_0)}{dS^2} + \cdots ,$$
(6)

where D is the (average) diffusion coefficient of the nanoparticles. The derivation was made under the condition $2S\omega^2 t \ll 1$. For small values of $2S\omega^2 t$ ($2S\omega^2 t < 0.1$) the factor $[1 - \exp(-2S\omega^2 t)]/(2S\omega^2 t)$ reaches unity. Equation (6) may then be simplified to

$$g(S,t,c_0) = g(S,c_0) + \frac{D}{\omega^4 r_m^2} \frac{1}{t} \frac{d^2 g(S,t,c_0)}{dS^2} + \cdots .$$
(7)

A similar equation to Eq. (7) was derived by Fujita [4] with respect to the integral distribution $G(S,c_0)$ instead of the differential distribution

$$G(S,t,c_0) = G(S,c_0) + \frac{D}{\omega^4 r_m^2} \frac{1}{t} \frac{d^2 G(S,t,c_0)}{dS^2} + \cdots .$$
(8)

Gralen and Lagermalm [6] suggest plotting the Schlieren or interference optically measured apparent sedimentation coefficient, S_{app}, versus $1/t$ at specified values of the integral distribution $G(S,t,c_0)$. Extrapolation of S_{app} to $1/t \to 0$ yields S and $G(S,c_0)$. The theoretical background to this procedure was given by Fujita [4].

van Holde and Weischet [7] demonstrated theoretically and practically with several examples that it would be better to plot S_{app} versus $1/t^{1/2}$ at specified values of $G(S,t,c_0)$. Their result reads [7]

$$S_{app} = S - \frac{D^{1/2}}{r_m} \frac{2}{\omega^2} \text{erf}^{-1}[1 - 2G(S,t,c_0)]t^{-1/2} ,$$
(9)

where $\text{erf}^{-1}(x)$ is the inverse error function with $\text{erf}(x) = (2/\pi^{1/2}) \int_0^x \exp(-s^2) ds$ and $\text{erf}^{-1}[\text{erf}(x)] = x$. The examples of van Holde and Weischet show that even at considerably short running times the plot $S_{app} = f(1/t^{1/2})$ gives straight lines which extrapolate well.

The two theories of Gralen–Lagermalm–Fujita and van Holde–Weischet are similar as they extrapolate the apparent sedimentation coefficient S_{app}, at specified values of the integral distribution of S, $G(S,t,c_0)$ to $1/t \to 0$ and to $1/t^{1/2} \to 0$, respectively.

Equations (7)–(9) hold under the following preconditions:

1. The factor $D/(\omega^4 r_m^2)$ must be sufficiently small and the PSD (and so the diffusion coefficient distribution) is not very broad.
2. The running time must be large enough for extrapolation but is not allowed to increase indefinitely.
3. Several experimental Schlieren curves, $(\partial c/\partial r)_t = f(r)$, or interference optically measured $c(r,t) = f(r)$

at several running times are needed for extrapolation.

Under these conditions there may be an intermediate range of running times where plots of $g(S,t,c_0)$ or $G(S,t,c_0)$ or S_{app} versus $1/t$ or $1/t^{1/2}$ should be linear and the diffusion-broadening corrected $g(S,c_0)$ or $G(S,c_0)$ could be obtained by extrapolating $g(S,t,c_0)$ or $G(S,t,c_0)$ to $1/t = 0$ at specified values of S (Gosting, Fujita) [9] or by extrapolating S_{app} versus $1/t$ (Gralen–Lagermalm–Fujita) or $1/t^{1/2}$ (Holde–Weischet) at specified values of $G(S,t,c_0)$.

Correction of the apparent sedimentation coefficient distribution with respect to the diffusion coefficient (new method)

In the case where the PSD (and that also means the S distribution) is not very broad and the sedimentation of the particles dominates the diffusion, we propose a new diffusion-broadening correction method. The basic idea is first to measure the apparent S distribution. From this we calculate the average sedimentation coefficient, $\bar{S}$. In the case of monomodal distributions the peak maximum of the distribution curve should be prefered for $\bar{S}$, whereas for multimodal distributions (e.g., the distribution has several peaks) it would be better to use the median of the S distribution for $\bar{S}$. Then we transform $\bar{S}$ with Eq. (1) into an average diameter, $\bar{d}$. We then use the well-known Stokes–Einstein relation

$$\bar{D} = kT/(3\pi\eta_1 \bar{d}) ,$$
(10)

where k is the Boltzmann constant, T is the temperature, and η_1 is the viscosity of the solvent, and calculate the average diffusion coefficient, $\bar{D}$, of the nanoparticles under investigation. Using $\bar{D}$ we then calculate the diffusion-broadening of the sedimentation boundary of hypothetical monodisperse nanoparticles with the diameter $\bar{d}$ using the well-known diffusion equations [10]

$$c(r,t)/c_0 = (1/2)\left\{1 - \text{erf}\left[(r - r_m)/(4Dt)^{1/2}\right]\right\}$$
(11)

or

$$(\partial c/\partial r)_t/c_0 = [1/(4\pi Dt)] \exp\left[-(r - r_m)^2/(4Dt)\right] .$$
(12)

We subtract this diffusion-broadening from the measured apparent $c(r,t)$ or $(\partial c/\partial r)_t$ curves, or the $G(S,t,c_0)$ or $g(S,t,c_0)$ distributions to get $G(S,c_0)$ or $g(S,c_0)$.

All extrapolation procedures and the new diffusion correction procedure are less accurate for broad and multimodal distributions. The reason is that both methods rely on an average diffusion coefficient. This difficulty could be partly overcome by choosing the rotor speed of the AUC to be as high as possible: this procedure would minimize the diffusion effect in comparison to the sedimentation effect.

Normally the sedimentation and diffusion of spherical nanoparticles are concentration-independent under usual measuring conditions, $c_0 < 20$ g/l, and therefore $g(S) = g(S, c_0)$.

Results and discussion

The measurements were performed with a model E AUC with Schlieren optics (Beckman Instruments, Palo Alto). The evaluations of the Schlieren photographs were done with a comparator, a digital graphics table, and a computer.

The characterization of nanoparticles requires the whole distribution curve. For nanoparticles it would be better to use mass-distribution-based p quantiles (preferably deciles) instead of number average, mass average, and z average d values. The reason is that they are more stable against outliers. In Figs. (1)–(3) the following p quantiles and range parameter are used:

$d_{10} = 0.1$ quantile = 1st decile
$d_{50} = 0.5$ quantile = 5th decile = median
$d_{90} = 0.9$ quantile = 9th decile
$(d_{90} - d_{10})/d_{50} =$ normalized interdecile range

The new diffusion-broadening correction procedure proposed is demonstrated and tested with a strong monodisperse, well-defined, sphere-shaped nanoparticle sample, the turnip yellow mosaic virus (TYMV; see electron microscopy [ELMI] photograph, Fig. 1a with an ELMI diameter of $d_{ELMI} \approx 29$ nm) in a buffer solution. Electron microscopy, light scattering and X-ray scattering yields a diameter of 25–29 nm [11]. Sedimentation measurements yield $S \approx 110$ Sv and (using Stokes' law, Eq. 1) $d \approx 20$ nm [10]. Calculation of the unhydrated TYMV sphere yields 22.2 nm [12]. The reason for the differences in diameter is that TYMV is hydrated; therefore, the density of the solute ρ_2 becomes smaller and the diameter larger. As is seen in Fig. 1b the original Schlieren curve was corrected according to the dilution rule for sector-shaped cells. The mean diffusion coefficient, D_{max}, was determined in the following way: the maximum value of $(\partial c / \partial r)_t$ allows S_{max} and d_{max} to be determined according to Eq. (1) and D_{max} according to Eq. (10). The measurements on the TYMV particles show a sedimentation coefficient of $S_{max} = 117.3$ Sv. Using Stokes' law and the known unhydrated particle density, $\rho_2 = 1.51$ g/cm^3, a diameter of $d_{max} = 20.3$ nm is calculated. The diffusion curve for $D_{max} = 2.13 \times 10^{-7}$ cm^2/g was calculated according to Eq. (12) and then subtracted at specified values of $(\partial c / \partial r)_t$ from the dilution-corrected Schlieren curve. The solid line in Fig. 1b represents the dilution- and diffusion-corrected Schlieren curve. The calculated diffusion curve in Fig. 1b is nearly identical

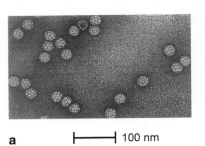

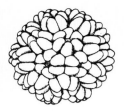

a

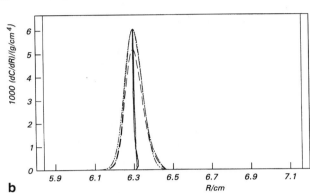

b

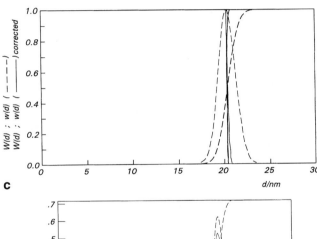

c

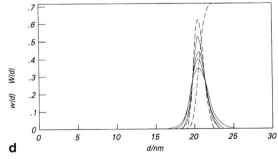

d

Fig. 1a–d Turnip yellow mosaic virus (*TYMV*) in buffer; $c_0 = 0.7$ g/l; $\rho_1 = 0.997$ g/cm^3; $\rho_2 = 1.36$ g/cm^3; $\eta_1 = 0.01003$ g/(cm s); $N = 12,000$ min^{-1}. **a** Electron microscopy photograph and model structure. **b** Schlieren optics curve at $t = 70$ min: original curve (–––); correction for dilution (– - –); diffusion curve (– - - –); correction for dilution and diffusion (———). **c** Particle size distribution, correction method: without diffusion correction (– – –); with diffusion correction (———). **d** Particle size distribution, extrapolation method: distribution curves at times $t = 50, 70, 90, 110$ min (———) extrapolated distribution for $1/t \rightarrow 0$ (– – –) (Eq. 7)

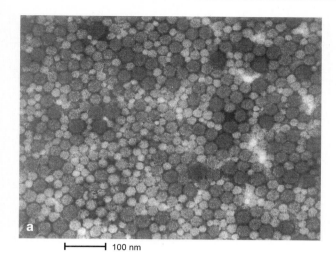

a

|← 100 nm →|

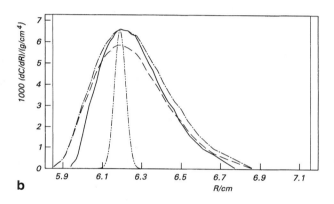

b

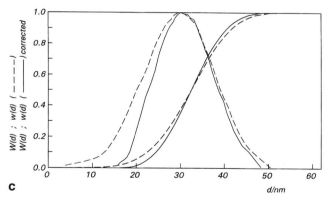

c

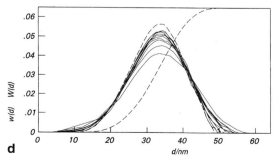

d

Fig. 2a–d Polystyrene latex in water; $c_0 = 3.015$ g/l; $\rho_1 = 0.997$ g/cm^3; $\rho_2 = 1.055$ g/cm^3; $\eta_1 = 0.00891$ g/(cm s); $N = 25{,}000$ min^{-1}. **a** Electron microscopy photograph. **b** Schlieren optics curve at $t = 45$ min: original curve (– – –); correction for dilution (– - –); diffusion curve (– - - –); correction for dilution and diffusion (——). **c** Particle size distribution, correction method: without diffusion correction (– – –); with diffusion correction (——). **d** Particle size distribution, extrapolation method: distribution curves at times $t = 15, 20, 25, 30, 35, 40, 45, 50, 55, 60$ min (——); extrapolated distribution for $1/t \rightarrow 0$ (– – –) (Eq. 7)

with the dilution-corrected Schlieren curve and so the subtraction results naturally in a nearly vertical line with respect to the dilution- and diffusion-corrected curve, as expected for monodisperse nanoparticles. Therefore, the corrected normalized interdecile range is 0.0 as it should be. This is strong proof that our proposed diffusion-broadening correction works.

The corrected sedimentation coefficient distribution and the diameter distribution (equivalent to the PSD) function may then be calculated according to Eqs. (1)–(5). The calculated PSD functions are shown in Fig. 1c; the dashed lines are the uncorrected curves and the solid lines the dilution- and diffusion-corrected ones. The solid lines in Fig. 1d represent the PSD curves at running times $t = 50, 70, 90,$ and 110 min and the dashed curves the extrapolated differential and integral PSD at a running time $1/t = 0$. The extrapolation was done according to the extrapolation procedure of Gosting [5, 9]. The new correction procedure is compared with the extrapolation procedure of Gosting in Table 1. This comparison shows that our correction method is better, i.e., closer to the true PSD.

The next nanoparticle example shows no monodisperse particles, instead it shows the normal case, a broadly distributed, small polystyrene latex in water with a diameter of approximately 30 nm, as demonstrated by the ELMI photograph of this polystyrene latex (Fig. 2a). Figure 2b demonstrates that the influence of diffusion on the broadening of the Schlieren curve of this 30-nm latex is much smaller than the influence of the diameter distribution of the particle. This is the normal case as the rotor speed and the density difference of the solute and the solvent should be adjusted in such a way that the sedimentation of the particle overwhelms the diffusion. Figure 2c and Table 1 show that the proposed new correction procedure and the extrapolation procedure of Gosting [5], Fig. 2d, yield nearly the same results and that the calculated interdecile range, $(d_{90} - d_{10})/d_{50}$, exhibits only small differences with respect to the new diffusion correction method and to Gosting's extrapolation method. The reason for the differences may be that both methods are approximations: Gosting's extrapo-

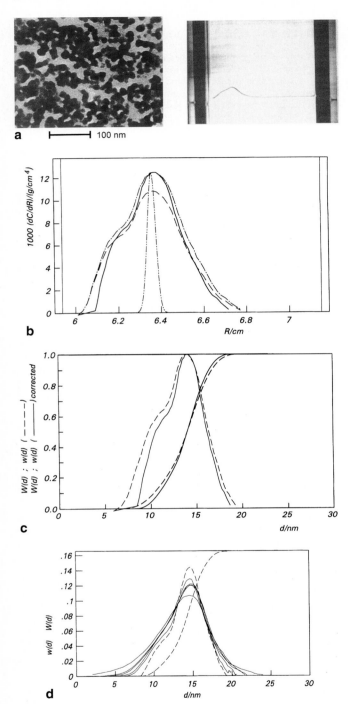

b Schlieren optics curve axis label: $1000 \; |dC/dR|/[g/cm^4]$, R/cm

c $W(d); w(d)$ (– – –); $W(d); w(d)$ (——) corrected, d/nm

d $w(d) \; W(d)$, d/nm

Fig. 3a–d Colloidal silica in water; $c_0 = 6.61$ g/l; $\rho_1 = 0.997$ g/cm³; $\rho_2 = 2.367$ g/cm³; $\eta_1 = 0.00891$ g/(cm s); $N = 26{,}000$ min⁻¹. **a** Electron microscopy photograph and Schlieren photograph at $t = 14$ min. **b** Schlieren optics curve at $t = 14$ min: original curve (– – –); correction for dilution (– - –); diffusion curve (– - - –); correction for dilution and diffusion (——). **c** Particle size distribution, correction method: without diffusion correction (– – –); with diffusion correction (——). **d** Particle size distribution, extrapolation method: distribution curves at times $t = 7, 9, 10, 12, 14$ min (——); extrapolated distribution for $1/t \to 0$ (– – –) (Eq. 7)

Table 1 Comparison of the calculated particle size distribution with the correction method and the extrapolation method

Average values	Uncorrected	Corrected (new method)	Extrapolated ($1/t \to 0$) (Gosting)
Turnip yellow mosaic virus			
d_{10}/nm	19.2	20.3	19.8₅
d_{50}/nm	20.4	20.3₅	20.6
d_{90}/nm	21.6	20.5	21.4₅
$(d_{90} - d_{10})/d_{50}$	0.11	0.00	0.07
Polystyrene latex			
d_{10}/nm	22.0	24.3	23.7
d_{50}/nm	32.2	32.3	33.1
d_{90}/nm	41.8	40.6	41.5
$(d_{90} - d_{10})/d_{50}$	0.61	0.50	0.53
Colloidal silica			
d_{10}/nm	10.3₅	10.8₅	11.1₅
d_{50}/nm	14.0	14.0	14.4
d_{90}/nm	16.7₅	16.4₅	16.8₅
$(d_{90} - d_{10})/d_{50}$	0.45	0.39	0.39

lation method and our diffusion-broadening correction method use an average value of the diffusion coefficient, D, and Gosting's extrapolation method is additionally only applicable within a restricted running time window [4].

This 30-nm latex example demonstrates that the diffusion-corrected, i.e., "true", particle size distribution of very small nanoparticles is considerably narrower than the uncorrected 'wrong' distribution.

Figure 3 and Table 1 show that the diffusion-broadening correction procedure is also applicable for multimodal nanoparticle size distribution functions. The particle investigated is colloidal silica (see ELMI photograph, Fig. 3a with an ELMI average diameter of $d_{ELMI} \approx 15$ nm and a particle density of $\rho_2 = 2.367$ g/cm³). On the right-hand side of Fig. 3a an original Schlieren photograph at $t = 14$ min is shown. Original and the corrected Schlieren curves of this SiO₂ sample are shown in Fig. 3b. The calculated PSDs using our diffusion-broadening correction method are shown in Fig. 3c and those using Gosting's [5] extrapolation method are shown in Fig. 3d. Both methods yield nearly the same corrected PSDs. The evaluations in Fig. 3 demonstrate that the AUC in conjunction with sophisticated evaluation techniques is able to detect multimodal PSDs in this very small nanoparticle size range.

Conclusions

This paper demonstrates that the AUC-PSD method only gives precise PSDs in the range $d < 50$ nm if the diffusion-broadening is corrected. Among the extrapolation methods discussed Gosting's method gives the best values when a Schlieren optics detector is used. Our correction procedure is equivalent to the extrapolation method but is easy to handle as only one experimental curve at one running time is needed. In the case of very small PSDs our diffusion-broadening correction procedure gives better results than the extrapolation procedures (see TYMV example, Fig. 1).

Acknowledgements Particular thanks are due to BASF AG, Ludwigshafen, and to the Fonds der Chemischen Industrie for financial support.

References

1. Mächtle W (1984) Makromol Chem 185:1025–1039
2. Bridgman WB (1942) J Am Chem Soc 64:2349
3. Stafford WF (1992) In: Harding SE, Rowe AJ, Horton JC (eds) Analytical ultracentrifugation in biochemistry and polymer science. Royal Society of Chemistry, Cambridge, pp 359–393
4. Fujita H (1975) Foundations of ultracentrifugal analysis. Wiley, New York
5. Gosting LJ (1952) J Am Chem Soc 74:1548–1552
6. Gralen N, Lagermalm G (1952) J Phys Chem 56:514–523
7. van Holde KE, Weischet WO (1978) Biopolymers 17:1387–1403
8. Schuck P, MacPhee CE, Howlett GJ (1998) Biophys J 74:466–474
9. (a) Lechner MD, Mächtle W (1995) Prog Colloid Polym Sci 99: 120–124; (b) Lechner MD, Mächtle W (1995) Prog Colloid Polym Sci 99:125–131
10. Tanford C (1961) Physical chemistry of macromolecules. Wiley, New York
11. Katouzian-Safadi M, Berthet-Colominas C, Witz J, Kruse J (1983) Eur J Biochem 137:47–55
12. Durchschlag H, Zipper P (1997) Prog Colloid Polym Sci 107:43–57

Progr Colloid Polym Sci (1999) 113 : 44–49
© Springer-Verlag 1999

INNOVATIONS IN DATA ANALYSIS

H. Cölfen
K. Schilling

Pseudo synthetic boundary experiments: a new approach to the determination of diffusion coefficients from sedimentation velocity experiments

H. Cölfen (✉) · K. Schilling
Max Planck Institute of Colloids
and Interfaces
Colloid Chemistry
Am Mühlenberg
D-14476 Golm, Germany
e-mail: helmut.coelfen@mpikg-golm.
mpg.de
Tel.: +49-331-5679513
Fax: +49-331-5679502

Abstract In this paper, a method is presented which efficiently separates diffusion-boundary-broadening from polydispersity in sedimentation velocity experiments. This can be achieved by the already well-known extrapolation of the apparent sedimentation coefficient distributions acquired at different times to infinite time to remove the effect of diffusion. The so-derived diffusion-corrected sedimentation coefficient distribution (SCD) is then subtracted from each experimental scan in the sedimentation coefficient domain to yield the boundary-broadening exclusively by diffusion in the radial domain. By this means, experimental scans are transferred into a pseudo synthetic boundary experiment – the classical experiment for the determination of diffusion coefficients – and can be evaluated by the various, already well established methods for synthetic boundary experiments. In principle, even diffusion coefficient distributions are accessible. The advantages of the method presented are
1. That any imperfections in the layering process of synthetic boundary experiments resulting in zero-time corrections are avoided.
2. That even extremely large particles such as micelles or vesicles which would sediment in any synthetic boundary experiment can be investigated in terms of diffusion coefficient distributions.
3. One single experiment yields the sedimentation coefficient distribution and the corresponding pseudo synthetic boundary experiment (diffusion coefficient distribution).
4. The combination of sedimentation and diffusion coefficient distributions from one experiment can, in principle, yield the density or molar mass distribution, both of which are important quantities especially for highly complex mixtures.

Key words Sedimentation velocity · Polydispersity · Diffusion · Sedimentation coefficient distribution · Boundary-broadening

Introduction

Due to fractionation capabilities, sedimentation velocity experiments in the analytical ultracentrifuge are the method of choice for the investigation especially of polydisperse samples; however, the determination of sedimentation coefficient distributions is complicated by diffusion-broadening which leads to a narrowing of the apparent sedimentation coefficient distribution with time. As sedimentation proceeds with time but diffusion only with the square root of time, extrapolation of the apparent sedimentation coefficient distributions to infinite time yields the sedimentation coefficient distribution corrected for diffusion effects. Several classical procedures are available for this purpose, namely those of Baldwin and Williams [1, 2], Gralen and Lagermalm [3], the van Holde/Weischet method [4] or Fujita's linear extrapolation to $1/t = 0$ [5]. Nevertheless, very often one is also interested in the diffusion coefficient as a complimentary physicochemical quantity or even better its distribution. There are some more advantageous methods than analytical ultracentrifugation (AUC) for the determination of diffusion coefficients (dynamic light scattering) [6] or diffusion coefficient distributions (flow-field flow fractionation) [7]. However, although these methods are faster and more convenient than AUC, it is highly desirable to gain as much information as possible from one experiment as it is well known that the combination of quantities from different methods might bear a high error potential. One example is the determination of the molar mass via the Svedberg equation [8] especially for polydisperse systems.

On the other hand, the systems of interest in polymer and colloid science are becoming increasingly complex, thus requiring a fractionating method which can yield as much information on each species as possible. The multidetector ultracentrifuge presented as science fiction in 1991 [9] would be one step in this direction – the other principal step, however, would be to extract as much information as possible from one experiment. Here it is especially advantageous that, in principle, sedimentation velocity experiments yield information about the sedimentation coefficient as a quantity dependent on molar mass/size and density as well as on the diffusion coefficient as a quantity only dependent on molar mass/size. Combining these quantities from one experiment could yield the density and the molar mass of the sample. Thus, it would be possible to derive the following quantities from only a single sedimentation velocity experiment:

1. Sedimentation coefficient distribution
2. Diffusion coefficient distribution
3. Molar mass distribution
4. Density distribution

Therefore, spending effort on the determination of the diffusion coefficient distribution from sedimentation velocity experiments appears promising because this is the key towards the density distribution and the molar mass distribution from AUC velocity experiments via the combination of sedimentation and diffusion coefficient distributions. The first steps towards this goal have already been made, but only for average values and not distributions [5, 8, 10, 11]. Furthermore, for polydisperse systems the diffusion coefficient from sedimentation velocity experiments is usually found to be too high due to the fact that polydispersity is not separated from diffusion and, thus, the boundary-spreading due to polydispersity is added to that of diffusion when calculating the diffusion coefficient. More recent approaches apply fitting of sedimentation velocity profiles to approximate solutions of the Lamm equation [12],

but despite their merits for mono- or paucidisperse systems, these approaches also do not yet see the light at the end of the tunnel when dealing with polydispersity [13–17]. Other new approaches work with correction of the sedimentation coefficient distribution for diffusion with an estimate for the diffusion coefficient [18]. However, no method has yet been presented which is in principle able to extract the diffusion coefficient distribution from a sedimentation velocity experiment simultaneously to the sedimentation coefficient distribution. This paper is a first step towards this goal by transforming conventional sedimentation velocity profiles into pseudo synthetic boundary concentration profiles.

Theory

The classical van Holde/Weischet method [4] is one possibility for deriving a diffusion-corrected sedimentation coefficient distribution (SCD). This is done by selecting a fixed number of data points in equal concentration intervals from the normalized experimental scans corrected for radial dilution and transformed into the SCD.

If these data sets are acquired for experimental scans taken at different runtimes, they can be plotted as the apparent sedimentation coefficients, s_w^*, versus the inverse root of the runtime yielding the typical van Holde/Weischet plot (Fig. 1).

Subsequently, slices of data with the same concentration value are selected and extrapolated to infinite time. The diffusion-corrected SCD is derived by linear extrapolation to the zero inverse root of the runtime. Equation (1) gives the mathematical description of the extrapolation:

$$s_w^* = s - \frac{2\sqrt{D}}{r_m \omega^2} \cdot \Phi^{-1}(1 - 2w) \cdot \frac{1}{\sqrt{t}} \ , \tag{1}$$

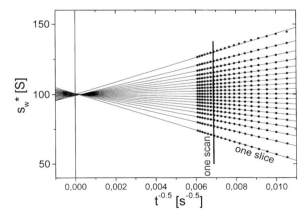

Fig. 1 Typical vanHolde/Weischet plot of a sedimentation velocity experiment with a monodisperse system

where D is the diffusion coefficient, ω the angular velocity, r_m the radius of the meniscus, w the relative concentration and t the runtime. Φ^{-1} is the inverse Gaussian error function.

In the classical van Holde/Weischet approach, D is calculated from the slopes of the regression lines and plotted as a function of the corresponding relative concentration, w. The result should be a diffusion coefficient distribution, but the singularity at $w = 0.5$ where the inverse Gaussian error function is not defined causes the center of the boundary to be not evaluable (Fig. 2a). About one-third of the data must be disregarded to obtain a good mean value that is unfortunately based on the outer regions of the boundary that are not perfectly described by the Gaussian error function [4]. We therefore suggest, alternatively, a global fit to the slopes of the van Holde/Weischet plot with D as a fit parameter, thus avoiding the problem of the singularity at $w = 0.5$, regarding all data, but obtaining only a mean value for D. In this paper we will refer to this approach as the modified van Holde/Weischet method. The difference between the classical and the modified van Holde/Weischet method is shown in Fig. 2.

The greater reliability of D by the modified van Holde/Weischet method becomes obvious. Whereas the mean value obtained by the classical approach depends greatly on the number of datapoints disregarded around $w = 0.5$, the fit does not suffer from the singularity at this value. Since both calculations are based on the same model, this is the only difference and the fit results are not more precise than those obtained with the classical method.

Nevertheless, this evaluation procedure can be significantly improved if the extrapolated SCD at infinite runtime from the van Holde/Weischet analysis is subtracted from the SCD for each experimental scan. The remaining width of the distribution now only reflects the boundary-broadening due to diffusion. Integrating these distributions and transforming them back into the cell radius domain yields a data set characteristic of a synthetic boundary experiment (Fig. 3).

These data can now be treated like those derived from a conventional synthetic boundary experiment; therefore, we call this method the "pseudo synthetic boundary" experiment. The data in Fig. 3 contain, in principle, information about the diffusion coefficient distribution; however, methods to derive diffusion coefficient distributions from synthetic boundary experiments have not yet been established and are the subject of current investigations.

It must be stressed that other methods for correcting SCDs for the effect of diffusion can be applied as well. Which extrapolation method to infinite time is best suited for polydisperse systems is beyond the scope of the present paper.

Materials and methods

Simulated data were created using both Borries Demeler's Ultrascan II, version 3.0 for UNIX software and Svedberg by John

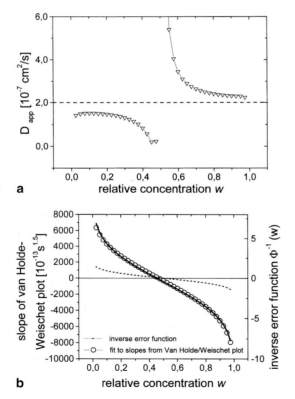

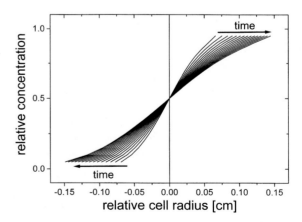

Fig. 2 a) Classical van Holde/Weischet evaluation of the diffusion coefficient for a simulated datafile ($D = 2.0 \times 10^{-7}$ cm^2/s) yielding $D = 2.20 \times 10^{-7}$ cm^2/s and b) modified van Holde/Weischet method with fit of the slopes of the van Holde/Weischet plot with D as a floating parameter yielding $D = 1.87 \times 10^{-7}$ cm^2/s

Fig. 3 Pseudo synthetic boundary scans for a monodisperse species generated by subtraction of the diffusion corrected SCD from the SCDs corresponding to each experimental scan

Philo. Data files were simulated for a monodisperse system with $s = 100$ S, $D = 2 \times 10^{-7}$ cm^2/s. A polydisperse system was created to be similar to the real systems investigated: the mean values were $s_{\mathrm{w}} = 82.6$ S, $D = 0.98 \times 10^{-7}$ cm^2/s. A polydispersity of 10% was simulated by overlaying 11 datasets in a Gaussian concentration ratio.

For real data, we investigated a 64-nm polystyrene latex with a polydispersity of approximately 10% in aqueous dispersion as a model system for hard-sphere particles.

Measurements were carried out in a Beckman Optima XL-A/XL-I analytical ultracentrifuge at 5000 rpm and 25 °C using the on-line Rayleigh interferometer. The accessible detection range was limited by the detection systems. The concentration of the latex ranged between 2 and 9 g/l.

A correction for k_{s} was applied by extrapolation of $1/s_{\mathrm{w,app}}$ versus concentration. For the latex as a hard sphere, however, the correction is marginal as expected and was finally omitted. All data were corrected for radial dilution. Further corrections were not applied.

The diffusion coefficient was derived using the following methods:

1. Evaluation of the boundary-spreading via Gaussian fits to raw data and the SCD and calculation of the diffusion coefficient from the squared standard deviation of the Gaussian fits [19].
2. Evaluation of D as described in the original paper by van Holde and Weischet [4].
3. Evaluation of D via the error function fit for the van Holde/Weischet analysis (see Theory).
4. Pseudo synthetic boundary experiment and evaluation via the method of Chervenka [20].
5. Pseudo synthetic boundary experiment and evaluation via the method described by Ralston [21].

Results and discussion

Simulated data

The new approach for the separation of boundary-spreading from polydispersity and diffusion (pseudo synthetic boundary experiment) was first tested for a set of simulated datafiles for a monodisperse and a polydisperse species (see Materials and methods). The results are presented in Table 1.

These results clearly show that D is overestimated for polydisperse systems by the classical evaluation of boundary-spreading via Gaussian fits. The evaluation of D by the classical and the improved van Holde/Weischet method, on the other hand, yields too low D values. Although all datapoints were used in the modified van Holde/Weischet evaluation yielding a safer estimate of D, its value is still too low.

Pseudo synthetic boundary experiments, however, clearly yield the best result, especially for polydisperse systems, regardless of whether they are analyzed by the Chervenka or the Ralston method. Furthermore, it is interesting to note that the plots of the squared boundary-spreading versus time almost pass through the origin indicating that zero-time corrections are not necessary for pseudo synthetic boundary experiments. This is important as in conventional synthetic boundary experiments, layering imperfections make these corrections necessary while D multiplied with the zero-time correction should be less than 10^{-4} cm^2 to get valid results [22]. Nevertheless, the plots even for simulated monodisperse samples do not show an exact linear dependence but a parabolic one (Fig. 4). Theory would predict the plots to be linear, the slope being the 0.275 fold of D [20] or the 4π fold, respectively.

The reason for this is not yet clear. Although fitting the monodisperse data to a parabola instead of a regression line yields the exact diffusion coefficient of 2×10^{-7} cm^2/s regardless of whether it is evaluated by the Chervenka or the Ralston method, the same does not work for the polydisperse system. Here, the regression lines yield the best result (Table 1). As most practical systems exhibit polydispersity, we will therefore keep the linear regression until the reason for the bending of the squared boundary-spreading versus runtime can be clarified theoretically.

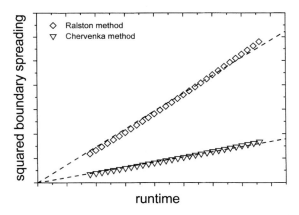

Fig. 4 Pseudo synthetic boundary experiment for the simulated monodisperse single species dataset evaluated with the Chervenka method [20] and the method described by Ralston [21]. The parabolic curve becomes evident

Table 1 Results from various methods for the evaluation of the diffusion coefficient for simulated mono- and polydisperse systems

	Theoretical result	Gaussian fits	van Holde/Weischet classical	van Holde/Weischet modified	Pseudo synthetic boundary Chervenka	Pseudo synthetic boundary Ralston
Monodisperse	2.00	2.27	1.91	1.89	2.20	2.22
Polydisperse	0.98	2.99	0.77	0.73	0.96	0.89

48

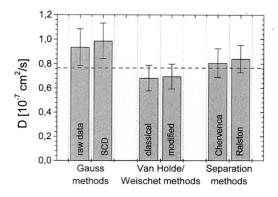

Fig. 5 Evaluation of D at 25 °C from a sedimentation velocity experiment with a polystyrene latex ($d = 64 \pm 6$ nm, $D = 0.77 \times 10^{-7}$ cm^2/s)

Polystyrene latex as a monodisperse hard sphere

As a next test for the various methods, the diffusion coefficient was evaluated for a 64 ± 6-nm polystyrene latex (data derived by dynamic light scattering) as a relatively well-defined real system. The results are displayed in Fig. 5.

Here, it becomes obvious that the classical analysis of boundary-broadening by Gaussian fits again yields systematically too high diffusion coefficients, resulting from the fact that the boundary-broadening due to diffusion is not separated from that due to polydispersity. On the other hand, evaluation of the same data set by the classical van Holde/Weischet method yields systematically too low values for an unknown reason. The pseudo synthetic boundary experiment, however, yields a satisfactory result if the average value is considered. Nevertheless, the error is quite high and is clearly evident in Fig. 5.

Conclusions and outlook

It was demonstrated that normal sedimentation velocity profiles can be converted into a pseudo synthetic boundary experiment by subtraction of the diffusion-corrected SCD from the SCD corresponding to each experimental scan. This exclusively yields the contribution of diffusion to boundary-broadening. The pseudo synthetic boundary experiment can then be evaluated by the classical methods available. Big advantages of this method compared to classical synthetic boundary experiments are that any imperfections in the layering process resulting in zero-time corrections are avoided and that furthermore even very large samples sedimenting at low speeds can be investigated; however, a drawback of the method is the high data quality required.

Although the method suggested relies on the correct determination of the diffusion-corrected SCD, which can be a problem for polydisperse and nonideal systems by means of the van Holde/Weischet analysis, this is not a principal weakness of the pseudo synthetic boundary experiment. Other, more suitable procedures for the determination of the diffusion-corrected SCD of polydisperse or nonideal systems may be found which are beyond the scope of the present paper.

The method suggested is well suited to yield correct diffusion coefficients even in the case of polydispersity. Also, it accounts much better for concentration dependencies of the sedimentation coefficient. However, one of the biggest advantages of the pseudo synthetic boundary experiment is seen in the fact that the diffusion coefficient distribution is, in principle, accessible in analogy to a real synthetic boundary experiment. This is a first step towards the determination of this practically important distribution which in combination with the SCD can then yield density and molar mass distributions from only a single sedimentation velocity experiment. This would be a most valuable tool for the investigation of very complex colloidal and polymeric systems.

Acknowledgements Dr. Borries Demeler is acknowledged for leaving his program Ultrascan in the newest version at our disposition. Dr. John Philo is acknowledged for providing the program Svedberg free of charge. Furthermore, we thank the Max Planck Society for financial support of this project.

References

1. Baldwin RL, Williams JW (1950) J Am Chem Soc 72:4325
2. Williams JW, Baldwin RL, Saunders W, Squire PG (1952) J Am Chem Soc 74:1542
3. Gralen N, Lagermalm G (1951) J Phys Chem 56:514
4. van Holde KE, Weischet WO (1978) Biopolymers 17:1387
5. Fujita H (1975) Foundations of ultracentrifugal analysis. Wiley, New York
6. Berne BJ, Pecora R (1976) Dynamic light scattering. Wiley, New York
7. Cölfen H, Antonietti M (1999) In: Schmidt M (ed) Advances in Polymer Science Vol. 150; Special Volume: New Developments in Polymer Analytics. Springer, Berlin Heidelberg, New York
8. Svedberg T, Pedersen KO (1940) The ultracentrifuge. Clarendon, Oxford
9. Mächtle W (1991) Prog Colloid Polym Sci 86:111
10. Fujita H (1959) J Phys Chem 63:1092
11. Goldberg RJ (1953) J Phys Chem 57:194
12. Lamm O (1929) Ark Math Astron Fys 21B 2:1
13. Philo J (1994) In: Schuster TM, Laue TM (eds) Modern analytical ultracentrifugation. Birkhäuser, Boston, p 156
14. Behlke J, Ristau O (1997) Biophys J 72:428
15. Philo J (1997) Biophys J 72:435

16. Demeler B, Saber H (1998) Biophys J 74:444
17. Schuck P (1998) Biophys J 75:1503
18. Lechner MD, Mächtle W (1999) Progr Colloid Polym Sci 113:37–43
19. Baldwin RL (1957) Biochem J 65:503
20. Chervenka CH (1969) A manual of methods for the analytical ultracentrifuge. Spinco Division, Beckman Instruments, Palo Alto
21. Ralston G (1993) Introduction to analytical ultracentrifugation, Beckman Instruments, Fullerton
22. Creeth JM, Pain RH (1967) Progr Biophys Mol Biol 17:217

Progr Colloid Polym Sci (1999) 113:50–56
© Springer-Verlag 1999

K. Schilling
H. Cölfen

Application of the solvent density variation method to sedimentation velocity experiments on biological systems

K. Schilling · H. Cölfen (✉)
Max Planck Institute of Colloids and
Interfaces, Colloid Chemistry, Am
Mühlenberg, D-14476 Golm, Germany
Tel.: +49-331-5679513
Fax: +49-331-5679502
e-mail: helmut.coelfen@mpikg-golm.
mpg.de

Abstract The solvent density variation method has proved useful for the simultaneous determination of the partial specific volume of biopolymers and the molar mass/hydrodynamic size, applying small sample quantities in the microgram range. For this purpose, the equilibrium method suggested by Schachman turned out to be best. If more-complex multicomponent or impure systems need to be characterized, fractionating sedimentation velocity experiments are better suited. Problems may occur if the sedimenting boundaries show extensive diffusion-broadening. Therefore, an extra term has been introduced into the underlying equation taking diffusion-broadening into account. This improved solvent density variation method for sedimentation velocity experiments is able to analyse not only monodisperse samples but furthermore multimodal or polydisperse samples both with or without significant diffusion-broadening. Even if the boundary shows only one apparent step for a bimodal mixture due to diffusion-broadening, the improved method is well able to resolve the density as well as the hydrodynamic radii of both species. This method has been tested on a number of simulated data files as well as on an impure apoferritin preparation. The partial specific volume for apoferritin was in good agreement with literature data as well as the protein hydration if the shape is provided. This result suggests that the solvent density variation method is a fast and convenient method for the determination of the partial specific volume, the frictional coefficient and thus protein hydration exclusively from sedimentation velocity experiments.

Key words Sedimentation velocity · Density variation method · Partial specific volume · Complex mixtures · Protein hydration

Introduction

In order to convert a sedimentation coefficient distribution (SCD) into a particle size distribution (PSD) or molar mass distribution, the density of the solute is required. The conversion of multimodal distributions or SCDs of polydisperse species with varying density presents significant problems, especially if the density of the solute is unknown. A strategy for handling this problem was suggested in the 1950s for sedimentation velocity [1–5] and later by Edelstein and Schachman [6] for sedimentation equilibrium. The first approaches applied several sedimentation velocity experiments in solvents of differing densities so that the zero-buoyancy point could be determined. The newer approach [6] consists of the combination of two measurements in chemically similar solvents of different densities (e.g. H_2O and D_2O) to obtain a set of two equations with two unknowns for each radial position in the ultracentrifuge cell; thus, both the density and the particle size can be obtained. Schachman gives several advantages for applying the equilibrium method [6]:

1. All problems connected with transport processes are eliminated, such as possible differences in frictional ratios in the different solvents, the influence of temperature and of solvent viscosity.
2. Smaller sample amounts are required.
3. The problem of determining sedimentation coefficients of potentially very small particles with sufficient precision is avoided.

These arguments are certainly valid for many pure biological single-component systems. Temperature influences are now only of minor importance with the modern generation of XL-I ultracentrifuges.

Sedimentation velocity, however, offers the benefit of a dispersive measurement which is especially convenient for mixtures or when impurities are present. The transport properties derived from the same experiment can be used for further evaluations as shown in the last section of this paper.

The sedimentation velocity approach has been refined and has proved to be a powerful characterization technique in industrial product control and other applications [7, 8].

Combination of corresponding sedimentation coefficients in the two solvents s_1 and s_2 (s_1 and s_2 correspond to the same concentration) will lead to the following expressions for the density, ρ_p, and the diameter, d_p, of this species [7, 8]:

$$\rho_P = \frac{s_1\eta_1\rho_2 - s_2\eta_2\rho_1}{s_1\eta_1 - s_2\eta_2} \tag{1}$$

$$d_P = \sqrt{\frac{18(s_2\eta_2 - s_1\eta_1)}{\rho_1 - \rho_2}}, \tag{2}$$

where ρ_1, ρ_2 and η_1, η_2 are the densities and viscosities of the solvents. These equations are valid only for hard-sphere particles. Swollen particles require modifications [9]. Extended evaluations have been derived for polymer coils, proteins and micelles (exhibiting excess properties) and will be described in a forthcoming paper. Here, we will focus on the influence of diffusion on the evaluation of solvent density variation experiments.

An interesting application of the solvent density variation method is the determination of the partial specific volume, $\bar{v}$, the sedimentation coefficient and the molar mass for proteins and other biological systems which are often available only in small quantities but which cannot be reasonably purified or which are present as multispecies systems (heterologous interacting systems, etc.).

The conventional determination of $\bar{v}$, however, requires large sample amounts if carried out by macroscopic density measurements such as in a density oscillation tube [10]. Furthermore, this method can only give an average $\bar{v}$ as no sample fractionation takes place.

An increment method for theoretical calculation of $\bar{v}$ from the amino acid composition of proteins has been developed by Perkins [11], but this approach is not applicable for proteins of unknown composition or for systems forming heterologous complexes with other compounds or undergoing structural changes under the conditions of interest. For such problems, especially when investigating complex systems, the density variation method for sedimentation velocity is the method of choice; however, diffusion-broadening of the sedimenting boundary is a significant error source making the application of the density variation method impossible if this effect is considerable. This paper describes an extension of the basic equation (Eq. 1) correcting for diffusion-broadening, thus making the method suitable for more systems of practical interest. The method is demonstrated for simulated data and for real data on the protein apoferritin.

Theory

Van Holde and Weischet [12] have suggested the following equation as a first approximation to describe the displacement of the sedimentation coefficient, s, from the apparent sedimentation coefficient, s_w^*, where w is a value between 0 and 1 describing the normalized concentration for the s value in question

$$s_w^* = s - \frac{2\sqrt{D}}{r_m\omega^2} \, \Phi^{-1}(1 - 2w) \, \frac{1}{\sqrt{t}} \, . \tag{3}$$

r_m is the meniscus position, ω the angular velocity of the rotor, t the runtime for the scan of interest and $\Phi^{-1}(x)$ the inverse Gaussian error function. The latter represents the shape of the sedimentation boundary; spreading increases towards the plateaus.

We have applied this approach to reduce the influence of diffusion on the evaluation of the results of solvent density variation experiments: this is demonstrated with simulated data.

The correction can be performed by transforming both SCDs using Eq. (3) before applying Eq. (1). We have found it more convenient to introduce the correction into Eq. (1), thus obtaining the following extended expression for the particle density:

$$\rho_P = \frac{s_1\eta_1\rho_2 - s_2\eta_2\rho_1 + (\eta_1\rho_2 - \eta_2\rho_1)\,\sigma}{s_1\eta_1 - s_2\eta_2 + (\eta_1 - \eta_2)\,\sigma}, \tag{4}$$

with σ defined as

$$\sigma = \frac{2\sqrt{D}}{r_m\omega^2} \, \Phi^{-1}(1 - 2w)\frac{1}{\sqrt{t}} \, . \tag{5}$$

Here D must be estimated, whereas r_m can be taken from the experimental scans. It should be noted that σ does not need to be equal for the two measurements compared: provided that the scans were taken simultaneously, D and r_m might still be different. It will be shown, however, that a rough estimate for both parameters provides satisfactory results and so σ does not need to be separated into σ_1 and σ_2 contributions. For the sake of a simpler expression, this form has been given. If data was acquired by independent runs, however, the separation of σ is necessary.

Materials and methods

Simulated data were produced using Ultrascan II, version 3.0 for UNIX, by Borries Demeler. Sedimentation and diffusion coefficients were chosen to resemble the natural protein ovalbumin and are compared to the values found from experiment in the following section. These data were simulated with and without a contribution from diffusion. In addition, a second species with s and D values close to those in the monodisperse distribution was added in a 50:50 concentration ratio.

Real data were acquired using horse spleen apoferritin purchased from Sigma without further purification. Dialysis against buffer, normally a common procedure, was avoided in order to keep any impurities. Measurements were carried out at 25 °C in 1 M NaCl solution. For water an angular velocity of 30 000 rpm was applied; for D_2O 40 000 rpm. The Rayleigh interferometer was applied simultaneously to the UV/VIS absorption optics scanning at 280 nm.

Results and discussion

Simulated density variation experiments

Bimodal system without boundary-broadening by diffusion

The capability of the improved solvent density variation method to resolve two species very similar in size and density is demonstrated in Fig. 1. For these data, no diffusion-broadening was added. The s values were selected such that the smaller species sediments more rapidly than the larger species to show that this unusual behaviour presents no difficulties to our program. As

this case is more complicated than the straightforward monodisperse system without diffusion, we do not explicitly treat the latter system here.

As in all subsequent plots the left figure shows the integral sedimentation coefficient distributions where the right, solid line always represents the measurement in the solvent of lower density. In the right plot the density distribution is plotted in circles (right axis) and the integral (dashed line, left axis) and differential (solid line) distributions as a function of the particle size. The dotted line indicates density values extrapolated in order to obtain a complete particle size axis. This extrapolation is rather arbitrary since the integral distributions have nearly reached the upper and lower plateaus, respectively, and so these particle densities and sizes are barely populated. For the particle density distribution, only the large symbols are relevant, as only these are densities populated.

For this simulation, the same parameters were used as in a later example given in Table 1, with the exception that no diffusion was added. The evaluation yields the correct results to the third decimal digit although the peaks for both species in the PSD were not baseline-resolved.

Monomodal system with boundary-broadening by diffusion

Evaluated without diffusion correction. For particles with sizes below 10 nm, diffusion-broadening becomes significant even for high rotor speeds. This affects the shape of the sedimentation boundaries such that the resulting particle density distribution is broadened. The uncorrected evaluation is shown in Fig. 2.

The density value in the centre of the distribution is in poor agreement with the expected value (1.381 g/ml found, 1.343 g/ml given). Furthermore, the density increases throughout the distribution from 1.25 to 1.5 g/ml, which is clearly unacceptable for a simulated data file.

Fig. 1 Evaluation results for a bimodal mixture (simulated data). This data contains no diffusion-broadening

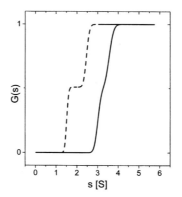
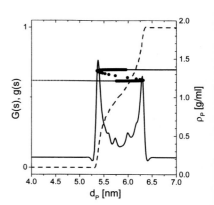

Fig. 2 Evaluation results for a monodisperse system (simulated data). Diffusion-broadening results in an incorrect density distribution

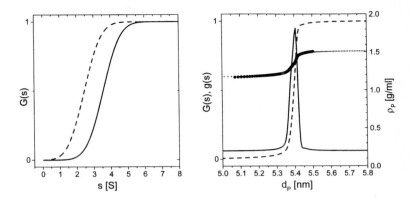

Evaluated with diffusion correction. The result of the evaluation with consideration of diffusion effects is shown in Fig. 3. Here, the particle density is found to be 1.349 g/ml, in good agreement with the given value.

To completely eliminate the diffusion contribution from the evaluation, a correction similar to Eq. (4) would have to be applied to the determination of particle sizes which creates the x-axis of Fig. 3. Since these are data for a monodisperse system, however, such a correction would cause the integral distribution to collapse into a single point. For this reason, diffusion correction was omitted for particle sizes, and the Gaussian distribution reflects not polydispersity, but diffusion-broadening. The centre of the distribution, however, is correct. The same is valid for all following examples.

The procedure described is quite robust with respect to the assumed diffusion coefficient. An initial guess of the correct order of magnitude will provide a good mean value for the particle density. The particle density distribution will be more or less broadened, but the effect is relatively small.

The resulting particle density distributions corresponding to that shown in Fig. 3 for several estimates of D are given in Fig. 4.

It should be noted again that Eq. (4) does not take into account the fact that the diffusion coefficients are

not identical for both solvents. Since usually H_2O ($\eta^{25\,°C} = 0.89$ cP) and D_2O ($\eta^{25\,°C} = 1.09$ cP) serve as the two solvents for the solvent density variation method an error of 18 % is transferred to the diffusion coefficient. Figure 4 shows that a wrong estimate for D of this magnitude can be neglected. Nevertheless, the true D (and thus σ) should be used since this is no problem, and different r_m, ω and t values may make this separation necessary as well.

A bigger problem may exist in the application of the same D for correction of multiple species. Figure 4 suggests that a constant diffusion coefficient can be a reasonable assumption since the van Holde extrapolation itself and the rough guess for the diffusion coefficient introduce by far larger errors that do not seriously affect the density results.

Bimodal system with boundary-broadening by diffusion

The method described in the previous section was also applied to a simulated sedimentation velocity experiment with the same mixture of two species, this time taking the diffusion coefficients into account. Whereas the evaluation without diffusion correction yields incorrect density values as already observed for the monomodal case, the evaluation with diffusion correction

Fig. 3 Evaluation of the same data as in Fig. 2. This time, diffusion effects are accounted for and the density is nearly constant. The particle size axis is not corrected for diffusion (see text)

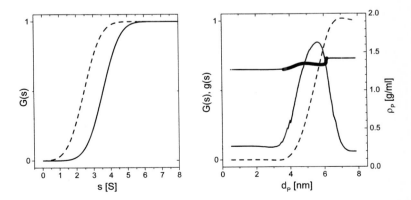

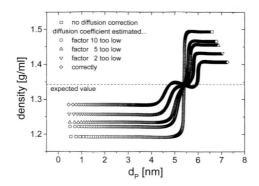

Fig. 4 Particle density distributions reflecting diffusion effects for several estimates of *D*. Too high estimates would yield analogous curves

works satisfactorily. The parameters are given in Table 1; the plot is shown in Fig. 5.

In the raw data in the left plot, diffusion-broadening causes the two steps in the SCD to be imperceptible (cf. Fig. 1). This is an extreme situation where most of the conventional analysis techniques would suggest a one-component system with diffusion-broadening. Our algorithm, however, clearly separates the two species and yields the input diameters and densities with high precision.

The integral shape in the right plot appears somewhat distorted, especially when compared to the same data in Fig. 1 where there was no contribution from diffusion. The reason for this artifact is due to the fact that the sedimentation boundaries of the two species overlap due to boundary-spreading. Although the contributions from the two species are separated, rearrangement of the contributions with increasing particle size can

produce some artifacts. In our evaluation, we therefore included intermediate plots (not shown) before rearranging the particle size axis. In this manner, the mean particle size of each species can be determined with much higher reliability.

Apoferritin

To test the evaluation method for real systems, apoferritin was selected as an impure biopolymer. The raw data reveal a low-molecular-weight impurity that especially appears in absorption optics. Although the impurity does not present a problem, interference data were more reliable because of a better signal-to-noise ratio. The SCDs derived from both detection systems are displayed on the left-hand side in Fig. 6. The right plot shows the effect of significant diffusion-broadening for the differential distributions obtained from interference optics, resulting in a sharpening of the SCDs with time.

The results show that the impurity does not present any difficulties to our evaluation method. In this particular case, this is also due to the fact that the impurity sediments more slowly and that the distributions are well separated. The resulting plot is given in Fig. 7.

The value of $\bar{v}$ found from the absorption data is 0.727 ml/g; that from the interference data is 0.733 ml/g. Other valuable information besides $\bar{v}$ can also be derived from the solvent density variation sedimentation velocity experiment. Since hydration water affects the frictional coefficient in the Svedberg equation, the frictional ratio will differ from unity although the particle is spherical. On the other hand, the Perrin function [13]

Table 1 Given parameters and evaluation results for the bimodal, diffusion-broadened system in Fig. 5. The index 20,w refers to water at 20 °C

	Given			s_1 (S)	s_2 (S)	Found		
	ρ (g/ml)	d (nm)	$D_{20,w}$ (10^{-7} cm^2/s)			ρ (g/ml)	d (nm)	$D_{20,w}$ (10^{-7} cm^2/s)
Species 1	1.343	5.39	6.69	3.55	2.45	1.340	5.57	6.59
Species 2	1.214	6.29	5.75	3.00	1.50	1.203	6.30	5.83

Fig. 5 Evaluation results for a bimodal system exhibiting diffusion-broadening. Diffusion effects hide the step in the sedimentation coefficient distribution on the *left*, but the species are well separated by the evaluation (on the *right*)

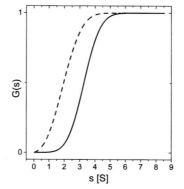
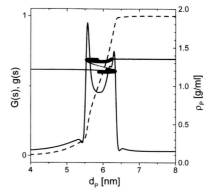

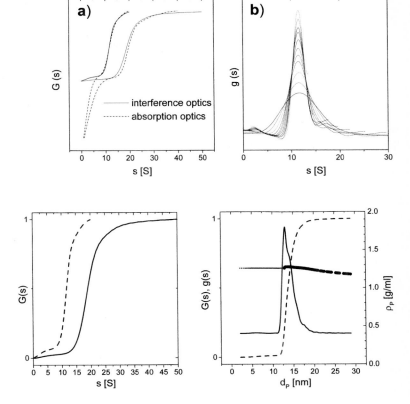

Fig. 6 a Raw data for apoferritin in H_2O and in D_2O revealing some low molecular impurities and **b** diffusion-broadening after calculation of $g(s)$ from the D_2O data

Fig. 7 Density evaluation for apoferritin taking diffusion into account

that takes hydration into account will have a value of unity. Thus, calculating the frictional ratio as

$$\frac{f}{f_0} = \left(\frac{M}{N_A} \frac{(1 - \bar{v}\rho_0)}{6\pi\eta_0 s_{20,w}^0}\right) \sqrt[3]{\frac{4\pi N_A}{3\bar{v}M}} \tag{6}$$

and inserting the result into the Perrin function

$$P = \frac{f}{f_0} \frac{1}{\sqrt[3]{1 + \delta/\bar{v}\rho_0}} \tag{7}$$

will yield the protein hydration, δ. For the molecular mass, we used a published value of 440 000 g/mol [14]. The results are summarized in Table 2.

The hydration as well as $\bar{v}$ is in good agreement with independently derived results [15]. The theoretical prediction from the amino acid sequence is much higher as it implies a monomolecular coverage of the protein, which is not correct as apoferritin is a hollow sphere assembled of 24 dimeric subunits. As the water entrapped in the cavity does not contribute to the frictional behaviour of apoferritin as found by Byron [15], the experimentally determined hydration can only be that on the outer surface of apoferritin rather than the theoretically predicted hydration for monomolecular coverage. The same was stated by Byron [15] for the difference between experimental and modelled parameters.

Conclusions and outlook

It could be shown that modification of the basic equation for the solvent density variation method taking diffusion into account extends the possibilities of sedimentation velocity experiments. Now, density distributions of biopolymers can be simultaneously aquired from PSDs and molar mass distributions using fast sedimentation velocity experiments even with complex mixtures with considerable diffusion-broadening of the sedimenting boundary. As only microgram sample

Table 2 Hydration of apoferritin by combination of $\bar{v}$ with the sedimentation coefficient and an independently derived molar mass in comparison to independent results and theoretical calculations

	$S_{20,w}$ (S)	$\bar{v}$ (ml/g)	Hydration (%)
Measured	19.0	0.733	21
Bead model + small-angle neutron scattering + analytical ultracentrifugation [15]	18.3	0.728	20
Theoretical from amino acid composition [11]	–	0.736	41

amounts are required, the sedimentation velocity method is recommended for all biopolymers which

- Cannot be purified or are complex mixtures.
- Degrade rapidly so no sedimentation equilibrium run can be performed with sufficient precision.
- Need to be checked simultaneously for sample homogeneity.

For all pure and monodisperse systems, the equilibrium method suggested by Edelstein and Schachman [6] remains the method of choice due to its higher precision for these systems.

Nevertheless, sedimentation velocity experiments have the further advantage that they provide the sedimentation coefficient as a quantity dependent on friction. With the simultaneous determination of $\bar{v}$, hydrodynamic quantities such as the (otherwise difficult to access) hydration become available. Together with new evaluation methods for a more precise diffusion coefficient from sedimentation velocity experiments [16] this paper is a step towards a fast characterization of complex mixtures by multiple physicochemical quantities.

Acknowledgements We thank Borries Demeler, University of Texas, for leaving the newest versions of his ULTRASCAN software at our disposition free of charge. The Max Planck Society is acknowledged for financial support of this project.

References

1. Sharp DG, Beard JW (1950) J Biol Chem 185: 247
2. Cheng PY, Schachman HK (1955) J Polym Sci 16: 19
3. Katz S, Schachman HK (1955) Biochim Biophys Acta 18: 28
4. Martin WG, Cook WH, Winkler CA (1956) Can J Chem 34: 809
5. Martin WG, Cook WH (1959) Can J Chem 37: 1662
6. Edelstein SJ, Schachman HK (1973) Methods Enzymol 27: 82
7. Mächtle W (1984) Macromol Chem 183: 1025
8. Müller HG, Herrmann F (1995) Progr Colloid Polym Sci 99: 114
9. Schilling K (1999) PhD thesis. Potsdam
10. Kratky O, Leopold H, Stabinger H (1969) Z Angew Phys 27: 273
11. Perkins SJ (1986) Eur J Biochem 157: 169
12. van Holde KE, Weischet WO (1978) Biopolymers 17: 1387
13. Perrin F (1936) J Phys Radium 7: 1
14. Pauck T, Cölfen H (1998) Anal Chem 70: 3886
15. Byron O (1997) Biophys J 72: 408
16. Cölfen H, Schilling K (1999) Progr Colloid Polym Sci 113: 44–49

Progr Colloid Polym Sci (1999) 113:57–61
© Springer-Verlag 1999

A.D. Molina-García

Hydrostatic pressure in ultracentrifugation (revisited)

A.D. Molina-García
Inst. del Frío, CSIC,
Ciudad Universitaria,
E-28040 Madrid, Spain
e-mail: ifrm111@if.csic.es
Tel.: +34-91-5445607
Fax: +34-91-5493627

Abstract Standard analytical ultracentrifugation is able to generate a significant hydrostatic pressure at the bottom of a spinning cell. In effect, a pressure gradient is created along its axis which, at maximum angular speeds, can reach nearly 50 MPa (around 500 atm). This pressure is high enough to alter the properties of some specially sensitive biological systems. Although this phenomenon can lead to important mistakes if the possible effects of hydrostatic pressure are not duly considered, it is possible to study the effect of the moderately high pressures generated in ultracentrifugation on biological systems of interest, such as macromolecular complexes or protein assemblies.

Key words Hydrostatic pressure · Ultracentrifugation · Protein denaturation

Introduction

When a standard analytical ultracentrifugation cell is spun at high speed significant hydrostatic pressure is developed with a gradient that is a function of the radial position in the cell and which is a maximum at the bottom of the cell. This effect was observed and used by some researchers more than 20 years ago [1–6]. Many other studies report the observation of pressure effects when biological systems are submitted to centrifugal forces [7–11].

In the author's opinion, the potential of ultracentrifugation for the study of macromolecules under hydrostatic pressure has not been fully exploited. Newer and more precise measuring instruments with higher resolution data will be able to extend the applicability of this technique. Additionally, the possible effects of pressure should not be neglected, or mistakes are likely to be made with the precise interpretation of analytical ultracentrifugation data. Here simple mathematical modelling with Mathlab (MathlabWorks) will help to introduce, or rather, revisit the subject.

The generation of pressure

The increase in hydrostatic pressure arises from two combined factors. One is the development of a centrifugal acceleration gradient that increases with increasing radius relative to the different points in the ultracentrifugation cell from the meniscus to the bottom of the cell. The second factor is the increase in the height of the solution column exerting its pressure above a given radial point of the cell. So, neglecting all other possible modulating factors such as the variation of the density of the solution in different regions (due both to the compressibility of the solution and to the migration of the solute to the bottom of the cell) as the sedimentation process progresses, the dependence of hydrostatic pressure on cell radius is described by Eq. (1).

$$P = \omega^2 \rho (r_o h + h^2/2) \tag{1}$$

where P is the pressure (above atmospheric pressure), ω the angular velocity, ρ the density of the fluid column, r_o the radial distance from the axis of rotation to the

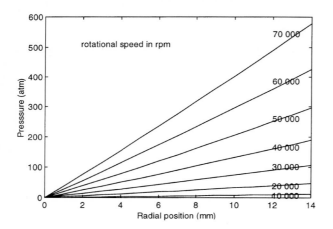

Fig. 1 Calculated variation of hydrostatic pressure with radius and rotational speed. Experimental parameters: $\rho = 1$ g/ml, $r_o = 7$ cm, cell length $= 14$ mm

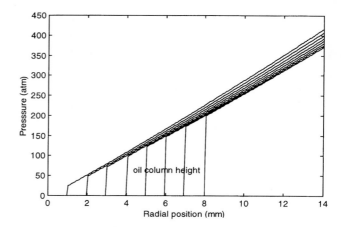

Fig. 2 Calculated variation of the pressure obtained with the height of an overlaid column of inert oil of density $= 0.85$ g/ml. Experimental parameters: $\omega = 60,000$ rpm, $\rho = 1$ g/ml, $r_o = 7$ cm, cell length $= 14$ mm

meniscus, and h the radial position in the cell: the liquid column height.[1]

The shape of the relationship between hydrostatic pressure and radius is shown in Fig. 1. The data approximately correspond to a standard ultracentrifugation cell: a radial distance at the meniscus of 7 cm and a cell length of 14 mm. The maximum speed practically achievable in an ultracentrifuge lies around 60,000 rpm. A solution of density 1 g/ml gives a gradient from 1 atm to nearly 450 atm at the bottom of the cell.[2]

The changes in the pressure gradient with rotational speed can also be seen in Fig. 1. An obvious way to check if hydrostatic pressure is having any role in the behaviour of the sample is to change the rotational speed, whereby the effects of pressure will disappear in an experiment at lower speed.

Other methods to increase the pressure obtained or to modify its distribution can be applied or at least conceived. For example, the pressure gradient affecting the solution will begin at higher levels if an inert oil column is placed on top of the solution. The variation of the pressure obtained with the height of a column of inert oil (Fig. 2) can also be used to check the effect of pressure, and to distinguish between the latter and the effects of sedimentation time.

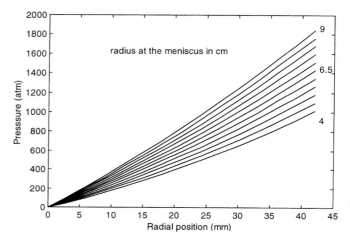

Fig. 3 Calculated variation of pressure with the radial position at the meniscus, r_o. Experimental parameters: $\omega = 60,000$ rpm, $\rho = 1$ g/ml, cell length $= 42$ mm (a cell 3 times longer than a standard one)

In Fig. 3 it is shown that if the radial position of the cell can be altered (using different rotors or a single one with several possible radial positions), the test can be achieved at a fixed speed. This figure also shows how pressure can be increased by the development of special cells with longer columns.

Also density-increasing agents can be employed, although care has to be taken not to alter the properties of the sample by these cosolutes. The variation of the pressure with the density of the solution is shown in Fig. 4. If D_2O or nonsedimenting additives are employed, these density and pressure distributions will be stable. Sedimenting solutes, such as sucrose, will soon migrate and create a density gradient with a more complex pressure distribution.

---

[1] Eq. (1) is equivalent to Eq. (1) of Ref. [4] or to Eq. (3) of Ref. [6]. The parameter h was introduced to give a more intuitive view of the differential pressure effect as related to the position inside the cell

[2] In preparative ultracentrifugation, i.e. zonal ultracentrifugation, where the radius is more than twice that considered here and where the liquid column height is between 4 to 8 times that of an analytical cell, one could expect a pressure 10–20-fold reported here. Consequently the effects of pressure would be much more noticeable; however, these effects would have to be studied after the centrifugation experiment had been stopped as no optical system is available for studies during the run

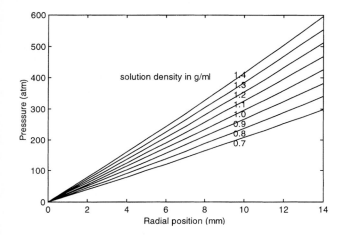

Fig. 4 Calculated variation of pressure with solution density. Experimental parameters: $\omega = 60{,}000$ rpm, $r_o = 7$ cm, cell length $= 14$ mm

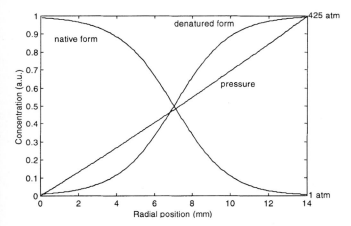

Fig. 5 Simulation of the distribution along the cell length of the concentration of native and denatured forms of an imaginary protein submitted to ultracentrifugation. The protein is assumed to suffer a two-state denaturation process whose equilibrium is ruled by pressure following Eq. (2). The equilibrium parameters are $K_o = 0.01$ and $\Delta V = 553$ ml/mol. The experimental parameters of the ultracentrifugation are $\omega = 60{,}000$ rpm, $\rho = 1$ g/ml, $r_o = 7$ cm and cell length $= 14$ mm

Effect of hydrostatic pressure in proteins

The highest pressures currently found in standard ultracentrifugation experiments lie at about 400 atm. Are they enough to cause a significant effect in biological systems? The answer is both yes and no. Most protein systems are believed not to be affected by pressure up to at least 1000–2000 atm [12, 13].[3] In some cases, however, the system is sensitive to lower pressures, and though the

midpoint of a transition promoted by pressure may be higher than 400 atm the effect can be observed in an ultracentrifuge below 400 atm [6, 14, 15]. Such is the case for those systems whose pressure effects have already been studied through the use of an ultracentrifuge [1–5].

The effects of pressure on proteins can be of several kinds: denaturation, through unfolding, changes in the aggregation state of assemblies, disulphide bridge rearrangements, ligand association alterations, etc. [6, 12, 13]. I will only refer here to the effects of pressure on the folding–unfolding and aggregation–dissociation equilibria. Both processes are driven by the same forces: always with the displacement of the equilibrium towards the form with the smaller volume (which includes that of the solvent affected by the change). In many cases this means that the form favoured by pressure is the unfolded and dissociated state, because there is a tendency to fill the inter- and intramolecular voids to improve packing, to relax forced bonds and to allow a higher degree of favourable interactions with the solvent. Nevertheless, there are examples of the opposite behaviour [16]. The equation ruling this equilibrium is of the type

$$\ln K = \ln K_o + P\,\Delta V/RT \qquad (2)$$

where K and K_o are the equilibrium constants (defined in this example as unfolding constants) at a given pressure, P, and at atmospheric pressure, respectively, and where ΔV is the molar volume increase [6] in the folding/aggregation process.[4]

Through a very simple simulation (Fig. 5) we can show how an imaginary protein with ΔV of about 550 ml/mol and K_o of 0.01 would behave in an ultracentrifugation cell. These parameters were sought so that the two-state change from native to denatured form takes place within the range of pressures achieved in the cell. This increase in volume is larger than those reported for many proteins (i.e., ΔV for haemoglobin 78–91 ml/mol [18], for phosphoglycerate kinase 52–66 ml/mol [14], for mutant glutathione S-transferase 18–32 ml/mol [15]). It must be noted, however, that the parameter in Eq. (2) is the molar volume increase and thus is independent of the protein size: a large macroaggregate would certainly be sensitive to such moderate pressures [6].

Observation in the ultracentrifuge

Two kinds of phenomena will be observed in an ultracentrifugation experiment on a sample that is affected by moderate hydrostatic pressure. One would be the species distribution in the cell before net matter transport has taken place (Fig. 6). Of course to observe

---

[3] I will refer only to proteins, though pressure effects can also be considered in other cases, such as nucleic acids and membranes

[4] This equation is easily obtained by calculating $(\mathrm{d}\Delta G/\mathrm{d}P)_T$, and is similar to Eq. (3) of Ref. [12] or to Eq. (3) of Ref. [17]

coefficient, s, is smaller and the solute will not proceed towards the bottom of the cell as quickly as it enters this region (still with a higher s). This will produce a distortion in the boundary.

However, this is not very realistic, because an equilibrium governed by pressure described by Eq. (2) would more likely be established. This is modelled in Fig. 8a, where the formation of a slower species is found to cause a bend in the plot of s versus radial distance (not shown) caused by the reduction in the speed of the solute migration.

In some examples the situation may be the opposite and the isometric monomers would travel faster than the large dissociating aggregate (e.g., anisometric). As can be seen in the model Fig. 8b, the effect would be the bending of the sedimentation coefficient plot in the opposite direction. It should be noted that possible hydrodynamic and convection effects that could be produced by the alterations in sedimentation have also been neglected.

Conclusions

By following the sedimentation behaviour (without the need of a difference in optical properties) it is possible to determine the sedimentation coefficient of both reactants and products by extrapolating to the different ends of the cell. The disturbances caused by the concentration of an inhomogeneous distribution could be avoided by using only the first scans, and working at different pressures using columns of inert oil. The equilibrium parameters can be calculated by repeating the experiment at different speeds.

Even when many proteins are not well suited to study by this technique (or to introduce complications in a standard ultracentrifugation experiment where pressure effects are not accounted for), it must be remembered that the stability of proteins and aggregates is a complex phenomenon governed by many factors, such as pH, ionic strength, temperature, denaturants, etc., and that one can adjust these variables to try and place a given protein in a situation whereby the study of hydrodynamic pressure effects in an ultracentrifuge cell can be undertaken.

Acknowledgements I thank J.M. Malpica, A. Rowe, O. Byron and J. Taylor for their useful comments.

References

1. Josephs R, Harrington WF (1967) Proc Natl Acad Sci USA 58: 1587
2. Van Diggelen OP, Oostrom H, Bosch L (1973) Eur J Biochem 39: 511
3. Poto EM, Wood HG (1977) Biochemistry 16: 1949
4. Marcum JM, Borisy GG (1978) J Biol Chem 253: 2852
5. Morel JE, Garrigos M (1982) Biochemistry 21: 2679
6. Harrington WF, Kegeles G (1973) Methods Enzymol 27: 306
7. Wattiaux R, Wattiaux-De Coninck S, Ronveaux-Dupal MF (1971) Arch Int Physiol Biochim 79: 214
8. Rosin MP, Zimmerman AM (1977) Mutat Res 44: 207
9. Wattiaux-De Coninck S, Dubois F, Wattiaux R (1977) Biochim Biophys Acta 471: 421
10. Champeil P, Buschlen S, Guillain F (1981) Biochemistry 20: 1520
11. Halle D, Yedgar S (1988) Biophys J 54: 393
12. Weber G, Drickamer HG (1983) Q Rev Biophys 16: 89
13. Silva JL, Weber G (1993) Annu Rev Phys Chem 44: 89
14. Cioni P, Strambini GB (1994) J Mol Biol 242: 291
15. Atkins WM, Dietze EC, Ibarra C (1997) Protein Sci 6: 873
16. Jung C, Hui Bon Hoa G, Davydov D, Gill E, Heremans K (1995) Eur J Biochem 233: 600
17. Prevelige PE Jr, King J, Silva, JL (1994) Biophys J 66: 1631
18. Pin S, Royer CA, Gratton E, Alpert B, Weber G (1990) Biochemistry 29: 9194

Progr Colloid Polym Sci (1999) 113:62–68
© Springer-Verlag 1999

D.R. Hall
S.E. Harding
D.J. Winzor

The correct analysis of low-speed sedimentation equilibrium distributions recorded by the Rayleigh interference optical system in a Beckman XL-I ultracentrifuge

D.R. Hall · D.J. Winzor (✉)
Centre for Protein Structure
Function and Engineering
Department of Biochemistry
University of Queensland
Brisbane, Queensland 4072, Australia
e-mail: winzor@biosci.uq.edu.au
Tel.: +61-7-33652132
Fax: +61-7-33654699

S.E. Harding
NCMH unit, University of Nottingham
Sutton Bonington
Leicestershire LE12 5RD, UK

Abstract The molecular mass of ovalbumin, a well-characterized protein, has been determined from low-speed sedimentation equilibrium distributions recorded by the absorption and Rayleigh optical systems of a Beckman XL-I ultracentrifuge in order to assess the reliability of various procedures for analyzing the Rayleigh interferometric records. Despite assertions to the contrary, the present results demonstrate the importance of establishing a concentration distribution in terms of absolute fringe displacement, $J(r)$ versus r, before quantitative analysis of the distribution is atttempted in terms of the basic sedimentation equilibrium expression for a single solute: this consideration is particularly important in experiments where the concentration at the meniscus is sizeable in relation to the concentration difference across the equilibrium distribution. In that regard the incorporation of a synthetic boundary experiment into the protocol seems to provide the most reliable means of ascertaining the absolute concentration distribution from Rayleigh interferometric records of low-speed sedimentation equilibrium experiments, and is an essential prerequisite for the quantitative characterization of interacting systems.

Key words Low-speed sedimentation equilibrium · Molecular mass determination · Rayleigh interference distributions · Synthetic boundary · Ovalbumin

Introduction

Irrespective of whether sedimentation equilibrium is used for molecular mass measurement or for the characterization of macromolecular interactions, the quantitation is based on analysis of the dependence of solute concentration upon radial distance. In that regard the corresponding distribution in terms of absorbance, $A_\lambda(r)$ versus r, may be substituted for the corresponding variation in solute concentration, $c(r)$, because of the direct proportionality between the two ordinate parameters. However, the Rayleigh interference optical system provides a measure of the difference between the solute concentration at radial distance r and that at the air-liquid meniscus, r_a. In an interferometric record of a sedimentation equilibrium distribution the number of Rayleigh fringes observed, $j(r)$, ranges between zero (at r_a) and $j(r_b)$ at the other extremity of the solution column subjected to centrifugation. From the viewpoint of defining the Rayleigh counterpart of the absolute concentration distribution, the required radial dependence is that of $J(r) = j(r) + J(r_a)$, where $J(r_a)$ is the solute concentration (expressed in terms of Rayleigh fringes) at the air-liquid meniscus. Only in the event that a sufficiently high rotor speed is used to ensure a value of essentially zero for $J(r_a)$ may $j(r)$ be equated with $J(r)$ [1].

Although various methods have been devised for the measurement of $J(r_a)$ in low-speed [2, 3] sedimentation equilibrium experiments, there is now a widespread belief that the sophisticated nonlinear regression programs installed in the current-generation analytical ultracentrifuges have rendered redundant the necessity to determine experimentally the magnitude of $J(r_a)$. Instead, it is merely regarded as an additional parameter that emerges from analysis of the radial dependence of the concentration difference data, $j(r)$ versus r, in terms of the basic sedimentation equilibrium expression,

$$[j(r) + J(r_a)] = J(r_a) \exp[M_A \phi_A (r^2 - r_a^2)] \tag{1a}$$

$$\phi_A = (1 - \bar{v}_A \rho_s)\omega^2/(2RT) , \tag{1b}$$

where M_A and $\bar{v}_A$ are the molecular mass and partial specific volume, respectively, of solute A, ρ_s is the solvent density and is ω the angular velocity in an experiment conducted at temperature T.

The present investigation examines the validity of this supposition in the simplest application of low-speed sedimentation equilibrium experiments, namely determination of the molecular mass of a homogeneous solute (ovalbumin). By demonstrating the fallacy of the approach in instances where the magnitude of $J(r_a)$ is sizeable in relation to $j(r_b) = J(r_b) - J(r_a)$, we reinforce the desirability of establishing the absolute concentration distribution (in Rayleigh-fringe terms) before embarking upon its quantitative analysis.

Experimental

Prior to one series of sedimentation equilibrium experiments the crystalline preparation of ovalbumin (Sigma grade V) was dissolved in acetate–chloride buffer (0.005 M sodium acetate–0.005 M acetic acid–0.200 M sodium chloride), pH 4.6, $I = 0.205$, conditions under which ovalbumin is essentially isoelectric [4, 5]. Aliquots (0.5 ml) of protein solution were then subjected to zonal chromatography on a column (2 × 20 cm) of Sephadex G-200 preequilibrated with the same buffer. This exclusion chromatography step not only served to remove any contaminating material with markedly different size characteristics but also to provide an ovalbumin solution in dialysis equilibrium with the buffer to be used in sedimentation equilibrium studies. Phosphate–chloride buffer (0.0128 M Na2HPO4–0.0115 M KH2PO4–0.0500 M NaCl), pH 7.0, $I = 0.10$, was used in a second series of sedimentation equilibrium experiments on ovalbumin.

Sedimentation equilibrium experiments were conducted in filled-epon double-sector cells in which fluorocarbon FC-43 had been included to generate a well-defined outer extremity of the protein solution subjected to sedimentation equilibrium. The solutions of ovalbumin in acetate–chloride buffer (150 μl, 0.6 or 0.3 g/l) were centrifuged for 24 h in a Beckman XL-I ultracentrifuge operated at 9000 rpm and 20 °C: those in phosphate–chloride buffer (1.24 g/l) were centrifuged at 17 000 rpm. Solute distributions were recorded at intervals of 4 h by both the Rayleigh and absorbance optical systems. For the latter measurements a wavelength of 280 nm was used to monitor the protein distribution, which was then corrected for baseline irregularities by means of the corresponding optical record obtained previously for the same cell with water in both sectors. On completion of the preliminary run the cell remained tightened throughout the water removal, and refilling stages. An interferometric pattern was also recorded for the preliminary run to

provide a means of correcting for radial fluctuation in Rayleigh response across the cell. For equilibrium runs on ovalbumin in acetate–chloride buffer an absorbance distribution was recorded immediately after attainment of rotor speed in the actual sedimentation equilibrium experiment to facilitate location of the hinge point (r_h) as the radial distance at which the protein concentration (absorbance) remained invariant throughout the sedimentation equilibrium experiment. A synthetic boundary experiment was used to obtain the initial protein concentration in terms of Rayleigh fringes (J_o).

Protein distributions were analyzed in terms of the basic sedimentation equilibrium equation for a homogeneous solute with molecular mass M_A and a partial specific volume $\bar{v}_A$, namely [6, 7]

$$z_A(r) = z_A(r_a) \exp[M_A \phi_A (r^2 - r_a^2)] , \tag{2}$$

which defines the thermodynamic activity $z_A(r)$ of solute at radial distance r in terms of that, $z_A(r_a)$, at the air–liquid meniscus r_a. The fact that the distribution of macromolecular solute is governed by its thermodynamic activity under conditions of constant chemical potential of solvent [8, 9] dictates the use of solutions that are in dialysis equilibrium with solvent – a requirement met in the present study.

In the application of Eq. (2) to the experimental distributions for ovalbumin in acetate–chloride buffer (pH 4.6) the simplification has been made that the thermodynamic activities may be replaced by the corresponding concentrations (or parameters directly related thereto) on the grounds that assumed thermodynamic ideality is a reasonable approximation for such dilute solutions of an uncharged protein [10]. To avoid the distortion of experimental error introduced by use of its linear logarithmic transform, nonlinear regression analysis (least-squares minimization) of the data in terms of Eq. (2) by means of SCIENTIST (Micromath Scientific Software, Salt Lake City) was used to obtain the parameter related to $z_A(r_a)$, $A_{280}(r_a)$ or $J(r_a)$, and the product $M_A \phi_A$ as the two curve-fitting parameters. Conversion of the latter to a molecular mass was based on a calculated buffer density [11] of 1.0110 g/ml and a partial specific volume of 0.736 ml/g for ovalbumin – the value obtained on the basis of a 3% carbohydrate content ($\bar{v}$ of 0.60 ml/g) and a calculated partial specific volume [12] of 0.740 ml/g for the protein component from its amino acid sequence [13]. Use of the reported partial specific volume of 0.748 ml/g for ovalbumin [14] leads to 5% higher estimates of its molecular mass.

In the experiment at neutral pH the anionic nature of ovalbumin enhances the magnitude of the second virial coefficient sufficiently to necessitate consideration of the molecular mass obtained by the above procedure as an apparent parameter, M_A^{app}, that is related to the true value, M_A, by

$$M_A^{app} = M_A/(1 + 2B_{AA}c_A/M_A) , \tag{3}$$

where c_A, taken as the mean of $c_A(r_a)$ and $c_A(r_b)$, was calculated from the absolute fringe displacement and the relationship (for proteins) $c_A = J/3.33$ [15]. The magnitude of the second virial coefficient, B_{AA}, was calculated on the basis of the expression

$$2B_{AA} = 32\pi N R_A^3/3 + [Z_A^2/(2I)](1 + 2\kappa R_A)/(1 + \kappa R_A)^2 \tag{4}$$

for a spherical solute with radius R_A and net charge Z_A [8]. Respective values of 2.92 nm [16] and −16 [17] for these two parameters lead to an estimate of 1060 l/mol for $2B_{AA}$.

Results and discussion

Because the absorption optical records of the distributions in these sedimentation equilibrium experiments were well within the concentration range covered by the

Lambert–Beer limiting law, their analysis is considered first to provide a reliable estimate of the molecular mass as a standard with which to compare the values obtained by various treatments of the Rayleigh interference records of the same equilibrium distributions.

Analysis of absorbance distributions

The absorbance optical record obtained at 280 nm for ovalbumin in acetate–chloride buffer (Fig. 1) provides definitive estimates of the radial extremities (r_a, r_b) of the sedimentation equilibrium distribution, despite the aberrant absorbance measurements in the vicinity of r_b. This aberration reflects the use of an incremental step in radius that is smaller than the radial width of the cell segment being scanned; however, the consequences of the aberration are readily removed from the analysis by the indicated truncation of the data set. In view of the inherent uncertainty of absorbance measurements in the vicinity of the air–liquid meniscus, absolute reliance upon a measured estimate of $A_{280}(r)$ was avoided by nonlinear regression analysis of the data in terms of Eq. (2) to obtain a best-fit value of this parameter as well as of $M_A\phi_A$ for ovalbumin. The consequent estimate of 0.379 (± 0.003) for $A(r_a)$, where the uncertainty is expressed as 2SD, is a seemingly reasonable description of the meniscus absorbance. The corresponding best-fit estimate of 0.2115 (± 0.0036) Da for $M_A\phi_A$ translates into a molecular mass, M_A, of 45 300 (± 800) Da, a value confirmed by that of 44 700 (± 1700) from the sedimentation equilibrium run on 0.3 g/l ovalbumin. These estimates of the molecular mass are in reasonable agreement with the value of 44 000 Da that is obtained on the basis of the amino acid sequence [13] and an octasaccharidic chain comprising six mannose and two

N-acetylaminoglucosamine residues attached to Asn-292.

Direct analysis of Rayleigh solute distributions

Having established that the absorbance records of sedimentation equilibrium distributions for ovalbumin conform with the known molecular characteristics of the protein, we now examine a range of procedures for extracting the molecular mass from the same distributions recorded as Rayleigh fringe patterns, $j(r)$ versus r, which range from an ordinate reading of zero at the air–liquid meniscus (r_a) to $j(r_b)$ at the other extremity of the solution column (Fig. 2a). The first procedure to be examined concerns the acceptability of nonlinear regression analysis in terms of Eq. (1) to obtain $J(r_a)$ and $M_A\phi_A$ as two curve-fitting parameters – an analysis that would clearly obviate the necessity to establish experimentally the magnitude of the former parameter. Such analysis of the sedimentation experiment with the higher loading concentration of protein in acetate–chloride buffer certainly provides a reasonable description of the

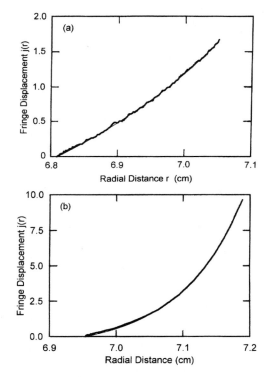

Fig. 2a, b Analysis of sedimentation equilibrium distributions by direct curve-fitting of Rayleigh interference data for ovalbumin to Eq. (1a) in order to obtain $J(r_a)$ as well as the buoyant molecular mass, $M_A\phi_A$. **a** Ovalbumin (0.6 g/l) in acetate–chloride buffer (pH 4.6) subjected to centrifugation at 9000 rpm and 20 °C. **b** Ovalbumin (1.24 g/l) in phosphate buffer (pH 7.0) subjected to centrifugation at 17 000 rpm and 20 °C

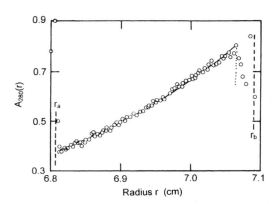

Fig. 1 Evaluation of the buoyant molecular mass, $M_A\phi_A$, from the spectrophotometric record of a sedimentation equilibrium distribution for ovalbumin (acetate–chloride buffer, pH 4.6; 9000 rpm; 20 °C) by analysis in terms of Eq. (2) with absorbance, $A_{280}(r)$, substituted for thermodynamic activity, $z_A(r)$

$[r, j(r)]$ data (Fig. 2a) in that it leads to well-defined estimates of 1.45 (± 0.06) and 0.2306 (± 0.0066) Da for $J(r_a)$ and $M_A\phi_A$, respectively; however, the consequent estimate of 49 400 (± 1400) Da for M_A is high – a manifestation of the failure of the analysis to extract the correct value of $J(r_a)$ from analysis of the radial dependence of $j(r)$ in terms of Eq. (1). Corresponding treatment of the sedimentation equilibrium data for ovalbumin in phosphate–chloride buffer (Fig. 2b) also yielded an estimate of $J(r_a)$, 0.92 (± 0.02) fringes, that was proven subsequently to be incorrect: in this case the estimate of M_A was 44 800 (± 300) Da.

As noted by Minton [18], treatments such as those described above for the analyses of the absorbance as well as of the Rayleigh distributions do not take into account the interdependence of $J(r_a)$ and $M_A\phi_A$, which are regarded as independent parameters. In the analysis of absorbance distributions the analogous treatment is commonly referred to as "floating the baseline", and although we prefer to avoid this practice, it may not lead to serious error. However, the situation for Rayleigh optical records can be more serious because, as already noted, the optical record is of a concentration relative to that at the meniscus: the magnitude of the offset to be estimated can therefore be quite large.

Use of the interdependence of $J(r_a)$ and $M_A\phi_A$ was noted by Teller et al. [19] and formed the basis of a procedure for extracting $J(r_a)$ given by Creeth and Harding [20, 21] which has been incorporated into the MSTARI algorithm for molecular weight analysis from Rayleigh interferometric records [22, 23]. The basis of the method is the expression

$$j(r)/(r^2 - r_a^2) = J(r_a)M_A\phi_A$$
$$+ [2M_A\phi_A/(r^2 - r_a^2)]\int_{r_a}^{r} j(r)r\,dr \quad . \quad (5)$$

The dependence of $j(r)/(r^2 - r_a^2)$ upon $[\int j(r)r\,dr]/(r^2 - r_a^2)$ has an ordinate intercept of $J(r_a)M_A\phi_A$ and a slope of $2M_A\phi_A$: their ratio thus has the potential to provide an estimate of $J(r_a)$. An equivalent procedure was described some 12 years later by Minton [18], who wrote Eq. (5) in the form

$$J(r_a) = \left[\int_{r_a}^{r} j(r)r\,dr\right] \Big/ \Big\{ [\{\exp[M_A\phi_A(r^2 - r_a^2)] - 1\}$$
$$/(2M_A\phi_A)] - (r^2 - r_a^2)/2 \Big\} \quad ,$$
$$(6)$$

which clearly states the interdependence of $J(r_a)$ and the buoyant molecular mass.

The problem in obtaining $J(r_a)$ has been that the limiting slope and the intercept at the meniscus are required – a region where the data are at their noisiest. In practice the experimenter has to interactively select a series of data ranges for performing the fit, and then "select" the most appropriate fit: consequently, $J(r_a)$ is obtained with an accuracy of only about 0.17 fringe data [21]. For experiments where the fringe increment between meniscus and base is relatively large, $j(r_b)$ greater than four fringes, this does not lead to significant problems, but for low fringe increments errors in molecular mass determination of 10% or more can accrue. This is born out by the experiments on ovalbumin. For the experiment in phosphate–chloride buffer at 1.24 g/l and 17 000 rpm (Fig. 3a), with a large fringe increment (meniscus to base) of nearly ten fringes, a reasonably accurate value of the meniscus concentration [$J(r_a) = 1.0$, cf. 1.04] is returned: the resultant M_A^{app} of 42 500 (Fig. 3b) translates into a molecular mass of 44 200 Da; however, for the low-fringe-increment experiment (0.6 g/l, 9000 rpm) with a fringe increment (meniscus to base) of only 1.7 fringes, the MSTARI analysis (Fig. 4a) returns a value of 1.4 (cf. 1.6) for $J(r_a)$, which gives rise to an estimate of 48 600 (± 1000) Da from subsequent analysis of the molecular mass (Fig. 4b).

For the quantitative analysis of interacting systems an accurate delineation of the absolute concentration distribution is essential. We therefore conclude this investigation by emphasizing the ease and rapidity with

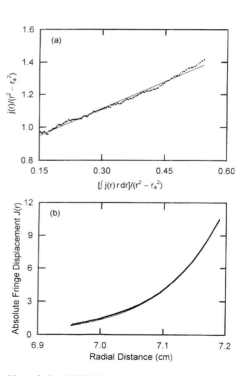

Fig. 3a, b Use of the MSTAR approach to evaluate $J(r_a)$ and M_A from a Rayleigh record of the sedimentation equilibrium distribution (17 000 rpm, 20 °C) for ovalbumin in phosphate–chloride buffer, pH 7.0. **a** Evaluation of $J(r_a)$ by means of Eq. (5). **b** Consequent radial dependence of the absolute fringe displacement, $J(r) = J(r_a) + j(r)$

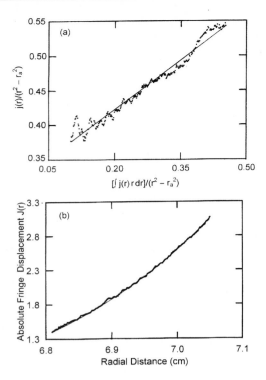

Fig. 4a, b Corresponding application of the MSTAR approach for **a** the determination of $J(r_a)$ and hence **b** the delineation of the sedimentation equilibrium distribution (9000 rpm, 20 °C) in terms of absolute fringe displacement $J(r)$ from the Rayleigh pattern for ovalbumin in acetate–chloride buffer, pH 4.6

which an unequivocal concentration distribution may be obtained. In other words we revert to the original contention [24, 25] that analysis of low-speed sedimentation equilibrium experiments is best performed by conducting a separate synthetic boundary experiment on the solution used for the sedimentation equilibrium run in order to obtain an accurate value of the meniscus concentration $J(r_a)$ and hence a reliable description of the solute distribution, $J(r)$ versus r, when the Rayleigh optical system is used to record the distribution.

Analysis based on the conduct of a synthetic boundary run

Selection of the upper limit of integration in Eqs. (5) or (6) as r_b, the base of the solution column, eliminates the need for concomitant evaluation of $M_A \phi_A$ and $J(r_a)$, because the integral now describes the amount of solute accounted for by the Rayleigh pattern. In as much as the total amount of solute in the cell is given by the corresponding integral at the onset of centrifugation, when $J(r)$ is the loading concentration, J_o, throughout the column length, it follows that the measurement of J_o in a separate synthetic boundary experiment allows $J(r_a)$ to be obtained from the relationship

$$J(r_a) = \left[J_o - 2 \int_{r_a}^{r_b} j(r) \mathrm{d}r^2 \right] \bigg/ \left(r_b^2 - r_a^2 \right) . \qquad (7)$$

This is equivalent to the corresponding expression

$$J(r_a) = J_o - \left[j(r_b) r_b^2 - \int_0^{j(r_b)} r^2 \, \mathrm{d}j \right] \bigg/ \left(r_b^2 - r_a^2 \right) \qquad (8)$$

developed by Richards and Schachman [24] on the same basis of mass conservation of solute. Unequivocal delineation of the absolute concentration distribution, $J(r)$ versus r, is then obtained as $J(r) = J(r_a) + j(r)$.

As noted by Richards and Schachman [24], a more direct procedure for determining the absolute concentration distribution entails identification of the hinge point – the radial position (r_h) at which the solute concentration remains invariant throughout the approach to sedimentation equilibrium, and hence the radial position at which $J(r) = J_o$. In the absence of a convenient procedure for identifying the hinge point by the white-light-fringe method [24], we located r_h by overlaying the absorption distribution obtained at the onset of centrifugation upon that obtained at sedimentation equilibrium [26]. The absolute concentration distribution in Rayleigh-fringe terms is then obtained as $J(r) = J_o + [j(r) - j(r_h)]$.

In keeping with the findings of Richards and Schachman [24], the application of either procedure to the present Rayleigh record of the sedimentation equilibrium distribution for 0.6 g/l ovalbumin in acetate–chloride buffer leads to a common radial dependence of $J(r)$ (Fig. 5a). Furthermore, the corresponding nonlinear regression analysis in terms of Eq. (2) leads to a best-fit estimate of 0.2130 (± 0.0005) Da for $M_A \phi_A$, which signifies a molecular mass estimate of 45 600 (± 1000) Da that essentially duplicates the value obtained by analysis of the distribution recorded by the absorption optical system. For the high-fringe-increment run (1.24 g/l, 17 000 rpm) in phosphate–chloride buffer, the correct $J(r_a)$ is now 1.04 (Fig. 5b), and analysis in terms of Eq. (2) yields a slightly lower molecular mass of 43 600 Da.

Once $J(r_a)$ has been correctly identified by the synthetic boundary procedure, it can be directly entered into programs such as MSTARI [23] for general molecular weight analysis or PSI [27] for the accurate quantitative analysis of interacting systems based on the $\psi_i(r)$ function [26]: these programs will now evaluate $J(r_a)$ based on input of the correct value for J_o.

Clearly, the establishment of an absolute concentration distribution, $J(r)$ versus r, is mandatory for reliable molecular mass determination from Rayleigh records of low-speed sedimentation equilibrium distributions, $j(r)$ versus r, when $J(r_a)$ is sizeable in relation to $J(r_b) - J(r_a)$. In as much as molecular mass determination for a single noninteracting solute represents the least stringent

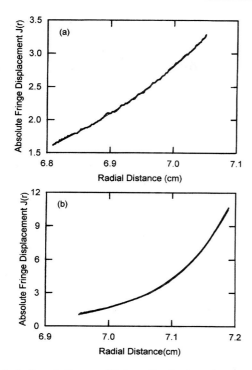

Fig. 5a, b Sedimentation equilibrium distributions in terms of unequivocal absolute fringe displacement $J(r)$ for ovalbumin solutions in **a** acetate–chloride buffer, pH 4.6, and **b** phosphate–chloride buffer, pH 7.0: respective rotor speeds were 9000 and 17 000 rpm in these experiments conducted at 20 °C

application of sedimentation equilibrium, the same caveat applies to analyses of sedimentation distributions reflecting either solute self-association or interaction between dissimilar reactants. However, for extremely heterogeneous systems such as polysaccharides or synthetic polymers it may not be possible to relate J_o to $J(r_a)$ via the conservation of mass of rule because of inability to measure the fringe increment $j(r_b) = J(r_a) -$

$J(r_b)$ as the result of steeply rising fringes at the cell base: in these cases the unconstrained Creeth–Harding algorithm [21], despite its lower accuracy, has to suffice.

Concluding remarks

This investigation of molecular mass determination by low-speed sedimentation equilibrium has demonstrated that conversion of the Rayleigh interferometric optical record of the concentration distribution into one in terms of absolute fringe displacement is a prerequisite for its quantitative analysis. There is thus little substance to a commonly held viewpoint that the meniscus concentration, $J(r_a)$, should be determinable as an additional curve-fitting parameter to emanate from nonlinear regression analysis of the Rayleigh distribution, $j(r)$ versus r, in terms of Eq. (1). In that regard a disconcerting finding is the fact that the statistical uncertainty of the incorrect estimates of M_A thereby derived is no · greater than that associated with the correct estimate; hence the fact that there is no obvious indication that the direct curve fitting has yielded an erroneous estimate of the molecular mass. It is hoped that the results of this investigation may convince experimenters of the desirability of reinstating the synthetic boundary run into the routine protocol for the conduct of low-speed sedimentation equilibrium studies in which the Rayleigh optical system is to be used to record the concentration distribution.

Acknowledgements Financial support (to D.J.W.) from the Australian Research Council is gratefully acknowledged, as is the receipt (by D.R.H.) of a University of Queensland Postgraduate Award. The Nottingham aspects of the work (S.E.H.) are funded by the United Kingdom Biotechnology and Biomolecular Sciences Research Council and the Engineering and Physical Sciences Research Council.

References

1. Yphantis DA (1964) Biochemistry 3:297–317
2. Van Holde KE, Baldwin RL (1958) J Phys Chem 62:734–743
3. Creeth JM, Pain RH (1967) Prog Biophys Mol Biol 17:219–287
4. Perlmann G (1952) J Gen Physiol 35:711–726
5. Creeth JM, Winzor DJ (1962) Biochem J 83:566–574
6. Haschemeyer RH, Bowers WF (1970) Biochemistry 9:435–445
7. Milthorpe BK, Jeffrey PF, Nichol LW (1975) Biophys Chem 3:169–176
8. Wills PR, Winzor DJ (1992) In: Harding SE, Rowe AJ, Horton JC (eds) Analytical ultracentrifugation in biochemistry and polymer science. Royal Society of Chemistry, Cambridge, pp 311–330
9. Wills PR, Comper WD, Winzor DJ (1993) Arch Biochem Biophys 300:206–212
10. Ogston AG, Winzor DJ (1974) J Phys Chem 79:2496–2500
11. Laue TM, Shah BD, Ridgeway TM, Pelletier SM (1992) In: Harding SE, Rowe AJ, Horton JC (eds) Analytical ultracentrifugation in biochemistry and polymer science. Royal Society of Chemistry, Cambridge, pp 90–125
12. Cohn EJ, Edsall JT (1943) Proteins, amino acids and peptides as dipolar ions. Reinhold, New York, pp 370–381
13. Nisbet AD, Saundry RH, Moir AJ, Fothergill LA, Fothergill JE (1981) Eur J Biochem 115:335–345
14. Dayhoff MO, Perlmann GE, MacInnes DA (1952) J Am Chem Soc 74:2515–2517
15. Voelker P (1995) Prog Colloid Polym Sci 99:162–166
16. Shearwin KE, Winzor DJ (1990) Eur J Biochem 190:523–529
17. Nichol LW, Siezen RJ, Winzor DJ (1978) Biophys Chem 9:47–55

18. Minton AP (1994) In: Schuster TM, Laue TM (eds) Modern analytical ultracentrifugation. Birkhäuser, Boston, pp 81–93

19. Teller DC, Horbett TA, Richards EG, Schachman HK (1969) Ann NY Acad Sci 164:66–101

20. Creeth JM (1980) Biochem Soc Trans 8:520–521

21. Creeth JM, Harding SE (1982) J Biochem Biophys Methods 7:25–34

22. Harding SE, Horton JC, Morgan PJ (1992) In: Harding SE, Rowe AJ, Horton JC (eds) Analytical ultracentrifugation in biochemistry and polymer science. Royal Society of Chemistry, Cambridge, pp 275–294

23. Cölfen H, Harding SE (1997) Eur Biophys J 25:333–346

24. Richards EG, Schachman HK (1959) J Phys Chem 63:1578–1591

25. Richards EG, Teller DC, Schachman HK (1968) Biochemistry 7:1054–1076

26. Winzor DJ, Jacobsen MP, Wills PR (1998) Biochemistry 37:2226–2233

27. Cölfen H, Winzor DJ (1997) Prog Colloid Polym Sci 107:36–42

Progr Colloid Polym Sci (1999) 113:69–75
© Springer-Verlag 1999

INNOVATIONS IN DATA ANALYSIS

D.J. Winzor
M.P. Jacobsen
P.R. Wills

Allowance for thermodynamic nonideality in the analysis of sedimentation equilibrium distributions reflecting complex formation between dissimilar reactants

D.J. Winzor (✉)
Centre for Protein Structure, Function and
Engineering
Department of Biochemistry
University of Queensland, Brisbane
Queensland 4072, Australia
e-mail: winzor@biosci.uq.edu.au
Tel.: +61-7-33652132
Fax: +61-7-33654699

M.P. Jacobsen
Macromolecular Interaction Facility
Department of Biochemistry
and Biophysics
University of North Carolina School of
Medicine, Chapel Hill
NC 27599-7045, USA

P.R. Wills
Department of Physics
University of Auckland
Auckland, New Zealand

Abstract General quantitative expressions are developed to take rigorous statistical-mechanical account of the effects of thermodynamic nonideality in sedimentation equilibrium distributions reflecting interaction between dissimilar macromolecular reactants. These quantitative expressions form the basis of an analysis which yields global estimates of the equilibrium constant(s) and the corresponding reference thermodynamic activities of free reactants in several experiments. Simulated data for 1:1 complex formation between dissimilar reactants are used to illustrate the procedure, which is currently unable to take advantage of the third virial coefficient terms to nonideality because of the lack of expressions for the excluded-volume contribution of three-body clusters of dissimilar species. In the absence of experimental studies where chemical interaction is sufficiently weak to warrant the use of concentrations commensurate with significant thermodynamic nonideality effects, published data on lysozyme self-association have been analysed to demonstrate the experimental application of the global analysis.

Key words Thermodynamic non-ideality · Sedimentation equilibrium · Protein self-association · Heterogeneous association

Introduction

Interest in the characterization of macromolecular interactions has been boosted greatly by the introduction of the new-generation analytical ultracentrifuges – the Beckman XL-A and XL-I instruments. To accommodate this interest procedures have been established for the analysis of sedimentation equilibrium distributions reflecting macromolecular self-association [1] as well as complex formation between dissimilar macromolecular reactants [2–5]; however, none of these procedures has taken adequate account of the consequences of thermodynamic nonideality. This deficiency has now been rectified for the analysis of sedimentation equilibrium distributions reflecting solute self-association [6, 7], but there still remains the problem of allowing for effects of thermodynamic nonideality in instances where complex formation is between dissimilar reactants. In a continuation of studies designed to characterize macromolecular interactions by direct analysis of sedimentation equilibrium distributions [5–8], this investigation presents a means of rigorous allowance for effects of nonideality in complex formation between dissimilar reactants on the statistical-mechanical basis of excluded volume [9, 10].

Theoretical considerations

Although the solute distribution in a sedimentation equilibrium experiment is recorded in terms of concentration, the thermodynamic origins of the data dictate their description in terms of a basic sedimentation equilibrium expression written in terms of thermodynamic activity [11, 12]; however, the nature of the activity remained obscure until this decade, when it was established that the pertinent quantity is z_i, the thermodynamic activity defined under conditions of constant temperature and chemical potential of solvent [6, 13, 14]. Advantage may therefore be taken of theory developed in the context of osmotic pressure, for which the same operational constraints apply to the thermodynamic activity of solute that governs its magnitude.

We commence this statistical-mechanical analysis of the thermodynamics of a macromolecular solution with the traditional expression for the dependence of osmotic pressure (Π) of a noninteracting mixture of two macromolecules (A and B) upon the molar concentrations (C_i) of the two solute components,

$$\Pi/(RT) = C_A + C_B + B_{AA}C_A^2 + B_{BB}C_B^2$$
$$+ B_{AB}C_A C_B + B_{AAA}C_A^3 + B_{BBB}C_B^3$$
$$+ B_{AAB}C_A^2 C_B + B_{ABB}C_A C_B^2 + \cdots . \qquad (1)$$

Less familiar is the corresponding expression in terms of thermodynamic activities rather than concentration [10],

$$\Pi/(RT) = z_A + z_B + b_{AA}z_A^2 + b_{BB}z_B^2$$
$$+ b_{AB}z_A z_B + b_{AAA}z_A^3 + b_{BBB}z_B^3$$
$$+ b_{AAB}z_A^2 z_B + b_{ABB}z_A z_B^2 + \cdots . \qquad (2)$$

The two sets of virial coefficients in these expressions for osmotic pressure exhibit the following interrelationships [10]:

$$b_{ii} = -B_{ii}; \quad i = A, \ B \qquad (3)$$

$$b_{ij} = -B_{ij}; \quad j \neq i \qquad (4)$$

$$2b_{iii} = 4B_{ii}^2 - B_{iii}; \quad i = A, \ B \qquad (5)$$

$$2b_{iij} = 4B_{ij}B_{ii} + B_{ij}^2 - B_{iij}; \quad j \neq i \qquad (6)$$

These coefficients allow the specification of molar concentrations, C_i, in terms of thermodynamic activities via the expression [10]

$$C_i = z_i[\partial(\Pi/RT)/\partial z_i]_{T,z_j}; \quad i = A, \ j = B \text{ or } i = B, \ j = A$$
$$= z_i + 2b_{ii}z_i^2 + b_{ij}z_i z_j + 3b_{iii}z_i^3$$
$$+ 2b_{iij}z_i^2 z_j + b_{ijj}z_i z_j^2 + \cdots . \qquad (7)$$

Apart from the introduction of numerical constants as the result of the differentiation, the expression for total solute concentration is clearly of the same form as Eq. (2), and we may therefore proceed with the expression for Π/RT in order to demonstrate the manner in which chemical interaction is accommodated.

From a thermodynamic viewpoint complex formation does not affect the number of components, whereupon it follows that Eq. (2) continues to describe the dependence of osmotic pressure upon the molar thermodynamic activities of the two solute components.

However, upon adoption of that viewpoint the formation of complexes must be regarded as a special manifestation of thermodynamic nonideality [6, 10], and the equilibrium constants expressing the activities of complex species in terms of those for reactants must be incorporated into the magnitudes of the thermodynamic coefficients, which now become constitutive parameters. Because we think of chemically reacting systems in terms of interacting species rather than interacting components, the task at hand is to relate the description of osmotic pressure, and also the total concentration of a component, in terms of species activities to the thermodynamic description in terms of components. The former becomes

$$\Pi/(RT) = z_A + z_B + \bar{b}_{AA}z_A^2 + \bar{b}_{BB}z_B^2 + \bar{b}_{AB}z_A z_B + \bar{b}_{AAA}z_A^3$$
$$+ \bar{b}_{BBB}z_B^3 + \bar{b}_{AAB}z_A^2 z_B + \bar{b}_{ABB}z_A z_B^2 + \cdots , \qquad (8)$$

where the overbar notation is used to denote that the virial coefficients are now constitutive quantities reflecting both physical and chemical interactions. The counterpart of Eq. (7) is

$$\bar{C}_i = z_i + 2\bar{b}_{ii}z_i^2 + \bar{b}_{ij}z_i z_j + 3\bar{b}_{iii}z_i^3$$
$$+ 2\bar{b}_{iij}z_i^2 z_j + \bar{b}_{ijj}z_i z_j^2 + \cdots . \qquad (9)$$

For illustrative purposes we consider a mixture comprising an acceptor A that interacts with macromolecular ligand B to form a series of complexes C ($\equiv$AB), D ($\equiv$AB$_2$), etc. The thermodynamic activities of these complexes may be written in terms of stoichiometric equilibrium constants (K_{AB}, K_{ABB}, etc.) and the thermodynamic activities of the two reactants raised to the appropriate powers as

$$z_C = K_{AB}z_A z_B \qquad (10)$$

$$z_D = K_{ABB}z_A z_B^2 . \qquad (11)$$

Expression of the osmotic pressure in terms of the coefficients of the species rather than components leads to the relationship

$$\Pi/(RT) = z_A + z_B + z_C + z_D$$
$$+ b_{AA}z_A^2 + b_{BB}z_B^2 + b_{AB}z_A z_B$$
$$+ b_{AAB}z_A^2 z_B + b_{ABB}z_A z_B^2 + b_{AAA}z_A^3 + b_{BBB}z_B^3$$
$$+ b_{AC}z_A z_C + b_{BC}z_B z_C + \cdots , \qquad (12)$$

where the virial coefficients b_{AA}, b_{AB}, etc., simply describe physical (or excluded-volume) interactions between the designated species. Incorporation of Eqs. (10) and (11) into Eq. (12) then gives

$$\Pi/(RT) = z_A + z_B + b_{AA}z_A^2 + b_{BB}z_B^2 + (b_{AB} + K_{AB})z_A z_B$$
$$+ b_{AAA}z_A^3 + b_{BBB}z_B^3 + (b_{AAB} + b_{AC}K_{AB})z_A^2 z_B$$
$$+ (b_{ABB} + b_{BC}K_{AB} + K_{ABB})z_A z_B^2 + \cdots . \qquad (13)$$

Comparison of Eq. (13) with Eq. (8), the corresponding description in terms of the thermodynamic coefficients, reveals the identity of the coefficients for terms involving solely z_A or z_B ($\bar{b}_{AA} = b_{AA}$, $\bar{b}_{BB} = b_{BB}$, $\bar{b}_{AAA} = b_{AAA}$, $\bar{b}_{BBB} = b_{BBB}$); however, coefficients involving products of z_A and z_B are related by the expressions

$$\bar{b}_{AB} = b_{AB} + K_{AB}$$
$$= K_{AB} - B_{AB} \quad (14)$$

$$\bar{b}_{AAB} = 2b_{AAB} + b_{AC}K_{AB}$$
$$= 2B_{AB}B_{AA} + (1/2)(B_{AB}^2 - B_{AAB}) - B_{AC}K_{AB} \quad (15)$$

$$\bar{b}_{ABB} = b_{ABB} + b_{BC}K_{AB} + K_{ABB}$$
$$= K_{ABB} + 2B_{AB}B_{BB} + (1/2)(B_{AB}^2 - B_{ABB}) - B_{AC}K_{AB} , \quad (16)$$

where the second expression in each case reflects the incorporation of Eqs. (3)–(6).

Experimental estimation of the thermodynamic coefficients $\bar{b}_{AB}$, $\bar{b}_{ABB}$, etc., thus opens up a possible means of determining the various equilibrium constants provided that magnitudes can be assigned to the species virial coefficients b_{AB}, b_{ABB}, etc. In that regard expressions are available [9] for estimating the magnitudes of all three second virial coefficients (B_{AA}, B_{BB}, B_{AB}) but for only two of the third virial coefficients (B_{AAA} and B_{BBB}). Consequently, although Eqs. (14)–(16) signify a potential for the evaluation of equilibrium constants from the magnitudes of second and third virial coefficients, their application to include the use of b_{AAB} and b_{ABB} terms describing three-body clusters must await a means of assigning values to B_{AAB} and B_{ABB}. We therefore restrict consideration to the characterization of a 1:1 interaction from results obtained over the concentration range where nonideality is described satisfactorily by Eq. (13) truncated at second virial coefficient terms.

Analysis of sedimentation equilibrium distributions

As previously noted, the basic sedimentation equilibrium expression for a given solute describes its distribution in terms of thermodynamic activity defined under conditions of constant temperature and chemical potential of the solvent. The dependence of that thermodynamic activity, z_i, upon radial distance, r, is therefore given by the relationship [5, 6, 11, 12]

$$z_i(r) = z_i(r_F)\psi_i(r) , \quad (17)$$

where r_F is an arbitrary reference radial position and

$$\psi_i(r) = \exp[M_i\phi_i(r^2 - r_F^2)] \quad (18)$$

$$\phi_i = (1 - \bar{v}_i\rho_s)\omega^2/(2RT) . \quad (19)$$

M_i and $\bar{v}_i$ denote the solute molecular weight and partial specific volume respectively, ρ_s is the solvent density

[6, 13, 14] and ω is the angular velocity in an experiment conducted at temperature T. For any given experiment the ψ function for acceptor (A) can be expressed in terms of that for ligand (B) by means of the relationship [5]

$$\psi_A(r) = [\psi_B(r)]^p; \quad p = (M_A\phi_A)/(M_B\phi_B) . \quad (20)$$

For a series of sedimentation equilibrium experiments in which the total molar concentration of each of the two components ($\bar{C}_A$, $\bar{C}_B$) has been determined as a function of radial distance, the combination of Eqs. (9) and (20) gives

$$\bar{C}_A(r) = z_A(r_F)[\psi_B(r)]^p + 2\bar{b}_{AA}[z_A(r_F)]^2[\psi_B(r)]^{2p} + \bar{b}_{AB}z_A(r_F)z_B(r_F)[\psi_B(r)]^{p+1} + \cdots \quad (21)$$

$$\bar{C}_B(r) = z_B(r_F)\psi_B(r) + 2\bar{b}_{BB}[z_B(r_F)]^2[\psi_B(r)]^2 + \bar{b}_{AB}z_A(r_F)z_B(r_F)[\psi_B(r)]^{p+1} + \cdots . \quad (22)$$

In as much as $z_A(r_F)$ and $z_B(r_F)$ are both constants within a given experiment, $\bar{C}_A(r)$ and $\bar{C}_B(r)$ are given by specific multinomial functions of the single independent variable $\psi_B(r)$. Two approaches to evaluate these two reference activities as well as K_{AB} seem pertinent.

A model-independent approach entails the separate elucidation of $z_A(r_F)$ and $z_B(r_F)$ from each experiment by writing Eqs. (21) and (22) in the forms

$$\bar{C}_A(r_F)/[\psi_B(r)]^p = z_A(r_F) + \bar{b}_{AB}z_A(r_F)z_B(r_F)\psi_B(r) + 2\bar{b}_{AA}[z_A(r_F)]^2[\psi_B(r)]^p + \cdots \quad (23)$$

$$\bar{C}_B(r_F)/\psi_B(r) = z_B(r_F) + 2\bar{b}_{BB}[z_B(r_F)]^2\psi_B(r) + \bar{b}_{AB}z_A(r_F)z_B(r_F)[\psi_B(r)]^p + \cdots , \quad (24)$$

which allow estimation of $z_A(r_F)$ and $z_B(r_F)$ as the ordinate intercepts of the respective dependences of $\bar{C}_A(r)/[\psi_B(r)]^p$ and $\bar{C}_B(r)/\psi_B(r)$ upon $\psi_B(r)$. On the grounds that the application of Eq. (17) then allows the evaluation of $z_A(r)$ and $z_B(r)$ throughout the distributions in each experiment, the results may be amalgamated into a single determination of K_{AB} by their analysis according to the combination of Eqs. (21) and (22) written as

$$\bar{C}_A(r) + \bar{C}_B(r) = z_A(r) + z_B(r) - 2B_{AA}[z_A(r)]^2 - 2B_{BB}[z_B(r)]^2 + 2(K_{AB} - B_{AB})z_A(r)z_B(r) + \cdots . \quad (25)$$

On the basis of spherical geometry for all species, values of the virial coefficients B_{AA}, B_{BB} and B_{AB} are available from the expressions [13, 15]

$$2B_{ii} = 32\pi NR_i^3/3 + [Z_i^2/(2I)][(1 + 2\kappa R_i)/(1 + \kappa R_i)^2] \quad (26)$$

$$B_{ij} = 4\pi N(R_i + R_j)^3/3 + [Z_iZ_j/(2I)] \times \{(1 + \kappa R_i + \kappa R_j)/[(1 + \kappa R_i)(1 + \kappa R_j)]\} , \quad (27)$$

in which the molar ionic strength (I) may be used to calculate the inverse screening length (κ) as $3.27 \times 10^7 \sqrt{I}$ at 20 °C. R_i and Z_i denote the respective radius and net charge of solute i, and N is Avogadro's number. Irrespective of the overall reaction stoichiometry the binding constant for 1:1 interaction (K_{AB}) may thus be obtained as the only parameter of unknown magnitude in the dependence of the summed constituent concentrations, $\overline{C}_A(r) + \overline{C}_B(r)$, upon the determined values of $z_A(r)$ and $z_B(r)$.

Although this procedure has the merit of being model-independent, the values of $z_A(r)$ and $z_B(r)$ used in the curve-fitting of data to Eq. (25) have been obtained without the advantage of any input from the equilibrium distribution for the other component in the extrapolations to find $z_A(r_F)$ and $z_B(r_F)$ – parameters upon which the calculated values of $z_A(r)$ and $z_B(r)$ via Eq. (17) are heavily reliant. The alternative is to introduce model-dependence from the outset in a procedure that accommodates concomitant analysis of both constituent distributions in terms of Eqs. (21) and (22) to find the best-fit set of values for K_{AB}, $z_A(r_F)$ and $z_B(r_F)$ in a given sedimentation equilibrium experiment; however, the experimental validity of such values depends upon the adequacy of the 1:1 interaction model upon which the analysis is based. Extension of this model-dependent analysis to include the two constituent distributions from several sedimentation equilibrium experiments has the potential to return a global best-fit value of K_{AB} as well as the corresponding best-fit estimates of $z_A(r_F)$ and $z_B(r_F)$ for each experiment. This approach, which appears to be the more tractable for a 1:1 interaction, is given further consideration after a description of the simulation procedures used to generate data for its illustrative application.

Simulation of nonideal sedimentation equilibrium distributions for 1:1 heterogeneous association

Consider a series of sedimentation equilibrium experiments in which neither reactant self-associates and in which the constitutive molar concentrations, $\overline{C}_A$ and $\overline{C}_B$, of interacting components A and B may both be determined as a function of radial distance. We shall confine attention to the situation where chemical interaction is restricted to 1:1 stoichiometry and where nearest-neighbour interactions (second virial coefficient terms) suffice to describe effects of thermodynamic nonideality. Data have been simulated for 1:1 interaction between dissimilar reactants with the molecular size and charge characteristics of ovalbumin and cytochrome c, but which is characterized by a binding constant that is some 20-fold smaller than the value reported from an experimental study at pH 6.3, $I = 0.03$ [5].

Sedimentation equilibrium distributions have been simulated for a nonideal heterogeneous association entailing 1:1 complex formation between reactants A and B with buoyant molecular weights, $M_i(1 - \overline{v}_i \rho_s)$, of 11,340 and 3,645, respectively. Second virial coefficients were assigned magnitudes on the basis of spherical geometry for all species and the above expressions for $2B_{ii}$ and B_{ij}, an ionic strength of 0.1 M being used for calculation of the inverse screening length. The radii (R_i) and net charges (Z_i) of the species were as follows: $R_A = 2.92$ nm, $Z_A = -12$; $R_B = 1.90$ nm, $Z_B = +12$; $R_C = 3.20$ nm, $Z_C = 0$. Such calculations led to the following magnitudes for the various second virial coefficients: $2B_{AA} = 814$ l/mol, $2B_{BB} = 544$ l/mol, $2B_{CC} = 661$ l/mol, $B_{AB} = -85$ l/mol, $B_{AC} = 572$ l/mol, and $B_{BC} = 334$ l/mol.

Sedimentation equilibrium distributions were simulated for experiments at 20 °C and a rotor speed of either 15,000 or 20,000 rpm. Radial extremities of the liquid column were 6.90 and 7.20 cm, and the reference radial position (r_F) was taken as 7.05 cm, the column midpoint. Each simulation was initiated by assigning magnitudes to $z_A(r_F)$ and $z_B(r_F)$, whereupon the thermodynamic activities of the two reactants throughout the distribution were generated by means of Eq. 17. Equations (21) and (22) were then used to obtain sedimentation equilibrium distributions for the A and B constituents, respectively, for a system with $K_{AB} = 3,000$ M^{-1}. These simulated distributions were accorded a greater semblance of experimental realism by the incorporation of Gaussian noise with a standard deviation of 0.2 μM – a value in keeping with experimental distributions recorded by the XL-A analytical ultracentrifuge.

Because Eqs. (21) and (22) form the basis of the analysis, it was deemed appropriate to also test the analytical procedure with simulated data that had been generated by other means. This entailed expressions of molar concentration of each species as its thermodynamic activity divided by a composition-dependent activity coefficient given by the expressions [13, 15]

$$\gamma_A(r) \approx \exp[2B_{AA}C_A(r) + B_{AB}C_B(r) + B_{AC}C_C(r)] \quad (28)$$

$$\gamma_B(r) \approx \exp[2B_{BB}C_B(r) + B_{AB}C_A(r) + B_{BC}C_C(r)] \quad (29)$$

$$\gamma_C(r) \approx \exp[2B_{CC}C_C(r) + B_{AC}C_A(r) + B_{BC}C_B(r)] . \quad (30)$$

The simulation was again initiated by assigning magnitudes to $z_A(r_F)$ and $z_B(r_F)$, but $z_C(r_F)$ was also calculated as $K_{AB}z_A(r_F)z_B(r_F)$ to allow generation, via Eq. (17), of distributions in terms of the thermodynamic activities of all three species. These activities were then converted to molar concentrations by an iterative procedure in which the activity coefficients were calculated via Eqs. (28)–(30) with the relevant activity, $z_i(r)$, being used as the initial estimate of the corresponding concentration: four iterations sufficed. Sedimentation equilibrium distributions reflecting partial contributions from the third and fourth virial coefficient terms were then obtained on the basis that $\overline{C}_A(r) = C_A(r) + C_C(r)$ and $\overline{C}_B(r) = C_B(r) + C_C(r)$.

Global analysis of simulated data

Having developed quantitative expressions, Eqs. (21) and (22), that make realistic allowance for the effects of thermodynamic nonideality on sedimentation equilibrium distributions for heterogeneously associating systems, we now need to demonstrate their utility for the evaluation of K_{AB} from sedimentation equilibrium distributions reflecting the consequences of such nonideality. For this purpose simulated data have the advantage of being free from ambiguities that may arise from systematic error (such as a baseline shift) that can plague experimental results. The first aim of this section is therefore to ascertain the extent to which the analysis of simulated data in terms of Eqs. (21) and (22) returns the input parameters, $z_A(r_F)$, $z_B(r_F)$, and K_{AB}.

The dependence of either $\overline{C}_A(r)$ or $\overline{C}_B(r)$ upon $\psi_B(r)$ in any given sedimentation equilibrium experiment could, in principle, provide an opportunity for nonlinear regression analysis in terms of Eqs. (21) and (22) to obtain K_{AB}, $z_A(r_F)$ and $z_B(r_F)$ as curve-fitting parameters; however, such analysis is likely to require unattainable accuracy in the experimentally determined distribution of a single component, and neglects the obligatory requirement that the same magnitude of K_{AB} must pertain to the descriptions of $\overline{C}_A(r)$ and $\overline{C}_B(r)$ in terms of Eqs. (21) and (22), respectively. Joint analysis of the $\overline{C}_A(r)$ and $\overline{C}_B(r)$ distributions needs to be effected by the simultaneous application of Eqs. (21) and (22) based on the minimization of the combined sum-of-squares residual for both distributions.

This joint analysis of distributions for the two constituents is readily adapted to the global analysis of results emanating from several sedimentation equilibrium runs. Under such circumstances, estimates of $z_A(r_F)$ and $z_B(r_F)$ are curve-fitting parameters pertaining to individual experiments, but the equilibrium constant

pertains to all experiments simultaneously. Thus, on the basis of an assigned initial estimate of K_{AB}, the joint sum-of-squares residual can be minimized for each experiment separately to find the best estimates of pairs of $[z_A(r_F), z_B(r_F)]$ parameters, and their global sum-of-squares residual can be minimized to find a new estimate of K_{AB} by holding constant all other parameters pertaining to individual experiments. By iteration of this process the globally best-fit value of K_{AB} can be determined along with commensurate estimates of the reference activities pertaining to each separate experiment.

Application of this global analysis of simulated data by means of Matlab software and a Marquardt–Levinthal subroutine for the nonlinear least-squares minimizations is summarized in Table 1 for a 1:1 interaction between reactants with the size and charge characteristics of ovalbumin and cytochrome c that is governed by a binding constant of 3,000 M^{-1} in buffer with an ionic strength of 0.1 M. For the first series of simulations (runs 1–3) the returned values of $z_A(r_F)$ and $z_B(r_F)$ are very satisfactory estimates of the input parameters – a situation manifested in the return of a best-fit K_{AB} that differs by about 1% from the value used in the simulation; however, because the simulated data were generated by the same expressions that were used for their global fitting, the results of this analysis reflect only the effect of the added "noise" and almost certainly overstate the likely accuracy attainable by the procedure. The second series of simulations (runs 4–6) provides a sterner test in the sense that the calculated nonideal distributions reflect some contribution from third and fourth virial coefficient terms. On this occasion the returned equilibrium constant again has trivial uncertainty in its statistical precision, but the deviation from the input value is approximately 16%. Inspection of Table 1 shows that the comparison of input and best-fit values of the activities at the reference radial position is still extremely satisfactory (within 1%), but that all of the returned values are overestimates of the input activities. In other words the relatively small contributions to the effects of thermodynamic nonideality arising

Table 1 Application of Eq. (17) for the global analysis of simulated sedimentation equilibrium experiments (15,000 rpm, 20 °C) for 1:1 complex formation between reactants A and B governed by a binding constant of 3,000 M^{-1}

Run no.[a]	Input activities (μM)		Output activities (μM)[b]		K_{AB} (M^{-1})
	$z_A(r_F)$	$z_B(r_F)$	$z_A(r_F)$	$z_B(r_F)$	
Run 1	25.00	25.00	24.75 ($\pm$0.02)	24.98 ($\pm$0.03)	
Run 2	25.00	50.00	25.71 ($\pm$0.02)	49.76 ($\pm$0.03)	3043 ($\pm$4)
Run 3	25.00	100.00	26.27 ($\pm$0.02)	99.18 ($\pm$0.03)	
Run 4	25.00	25.00	25.19 ($\pm$0.02)	25.21 ($\pm$0.03)	
Run 5	25.00	50.00	25.27 ($\pm$0.02)	50.36 ($\pm$0.03)	2607 ($\pm$4)
Run 6	25.00	100.00	25.11 ($\pm$0.02)	100.70 ($\pm$0.03)	

[a] Runs 1–3 refer to distributions simulated on the basis of Eqs. (21) and (22), whereas runs 4–6 refer to distributions simulated on the basis of composition-dependent activity coefficients
[b] Numbers in parentheses denote the uncertainty ($\pm$2 SD) of the returned estimates

from terms of higher order than those employed for analysis using the truncated versions of Eqs. (21) and (22) are countered in the global curve-fitting by an increase in the thermodynamic activities of the reactants, and hence by a concomitantly diminished estimate of K_{AB}. In that regard the situation would be improved markedly by the inclusion of third-order terms of Eqs. (21) and (22) for the global curve-fitting procedure; however, as already noted, such endeavours must await expressions for B_{AAB} and B_{ABB}. Meanwhile, the current global analysis in terms of expressions restricted to effects of thermodynamic nonideality arising from second-order virial terms should prove adequate for most experimental purposes.

Global analysis of solute self-association

At present there are no examples of experimental investigations in which the chemical interaction between dissimilar reactants has been sufficiently weak to warrant the use of solute concentration ranges commensurate with the existence of a significant effect of thermodynamic nonideality due to nonchemical forces on the form of the sedimentation equilibrium distribution. We therefore revert to published sedimentation equilibrium results [6, 16] reflecting the self-association of lysozyme in order to illustrate application of the global approach to the analysis of experimental data.

Adaptation of the Hill and Chen approach [10] to the problem of characterizing solute self-association equilibria (1 = monomer, 2 = dimer, 3 = trimer, etc.) has led to the expression [6]

$$\bar{c}(r)/M_1 = z_1(r_F)\psi_1(r) + 2\bar{b}_2[z_1(r_F)\psi_1(r)]^2$$
$$+ 3\bar{b}_3[z_1(r_F)\psi_1(r)]^3 + \cdots , \qquad (31)$$

where

$$\bar{b}_2 = K_2 - B_{11} \qquad (32)$$

$$\bar{b}_3 = K_3 - K_2 B_{12} + 2B_{11}^2 - B_{111}/2 \qquad (33)$$

$$\bar{b}_4 = K_4 + B_{22}K_2^2 - B_{13}K_2 + (1/2)[4B_{12}B_{11} + B_{12}^2 - B_{112}]K_2$$
$$- (1/3)[B_{1111} - 9B_{11}B_{111} + 16B_{11}^3] . \qquad (34)$$

Solute concentration is expressed in base-molar terms by dividing the weight concentration, $\bar{c}(r)$, by the molar mass of monomer (M_1), and the thermodynamic activity of monomer, $z_1(r)$, is given by Eq. (17) with i identified as monomer ($i = 1$). On the basis that the magnitudes will already have been ascribed to B_{11}, B_{12} and B_{111} [16], the dimerization (K_2) and trimerization (K_3) constants are thus the two invariant parameters to be evaluated by global fitting of several dependences of $\bar{c}(r)/M_1$ upon $\psi_1(r)$, whereas $z_1(r_F)$ is a local fitting parameter for each sedimentation equilibrium distribution.

Results from four sedimentation equilibrium experiments on lysozyme [6, 16] are summarized in Table 2, which presents experimental details and the $z_1(r_F)$ values returned for each run by global analysis in terms of Eq. (31): values of 504 M^{-1} and 4.34×10^4 M^{-2} were returned for K_2 and K_3, respectively. The results and best-fit descriptions are also illustrated diagrammatically in Fig. 1. Although the values of the equilibrium constants differ significantly from those of 585 M^{-1} and 2.04×10^5 M^{-2} obtained previously [6], the current analysis illustrates the feasibility of elucidating a general pattern of self-association without matching up overlapping data sets – the procedure used previously but a circumstance which may not always be afforded by the available experimental information. In that regard we note that extension of Eq. (31) in an attempt to take into account the possible formation of tetramers (the $\bar{b}_4$ term) led to divergence of the iterative fitting procedure, the corresponding finding in the previous study being that inclusion of the additional term in Eq. (31) could not be justified on statistical grounds [6]. It therefore appears that the general iterative procedure is capable of leading

Table 2 Global analysis of sedimentation equilibrium distributions [6, 16] reflecting lysozyme self-association (pH 8.0, $I = 0.15$, 15 °C)

Expt. no.	Speed (rpm)	r (cm)	$\bar{c}(r)/M_1$ (mM)	$z_1(r_F)$ (mM)[a]
1	20,000	6.924–7.147	0.097–0.338	0.1806
2	15,000	6.918–7.140	0.328–0.724	0.4237
3	11,000	6.908–7.113	0.614–0.919	0.5911
4	8,000	6.928–7.155	1.040–1.368	0.7837

[a] Thermodynamic activity of monomer at the reference radial position, taken as 7.050 cm for each experiment: the standard error of the estimate is 0.0003 mM in each case

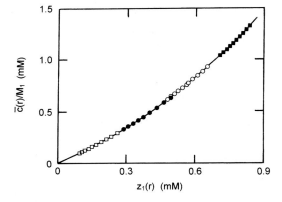

Fig. 1 Best-fit description (——) of the self-association of lysozyme obtained by global analysis of four sedimentation equilibrium experiments (Table 2) in terms of Eq. (31) with r_F taken as 7.050 cm in each experiment. For purposes of clarity only every fifth data point is shown

to sound conclusions, despite the fact that least-squares minimization is not constrained by the demand for ranges of common concentration between experiments. This finding augurs well for the corresponding global characterization of interactions between dissimilar reactants, for which the choice of r_F on the basis of common values of $\overline{C}_A(r_F)$ and $\overline{C}_B(r_F)$ in the various sedimentation equilibrium experiments is not an option.

Concluding remarks

This investigation has served two major purposes. First, it has led to quantitative expressions whereby rigorous allowance may be made for the effects of thermodynamic nonideality on sedimentation equilibrium distributions reflecting interaction between dissimilar reactants. Second, those quantitative expressions have been used to illustrate an analysis which yields global estimates of the equilibrium constants and the corresponding reference thermodynamic activities of the reactants pertaining to several sedimentation runs. In that regard the present analysis bears obvious similarities with existing procedures for the study of heterogeneous association by sedimentation equilibrium [2–4], but differs therefrom by virtue of its ability to take rigorous account of the effects of thermodynamic nonideality.

We conclude this investigation by noting that four decades have elapsed since Adams and Fujita [17] drew attention to the need for account to be taken of the effects of thermodynamic nonideality in the analysis of sedimentation equilibrium distributions reflecting macromolecular equilibria. Despite obvious inadequacies in that initial approach, progress towards a more realistic quantitative assessment of such nonideality effects has been undeniably slow. At last that challenge has been met by this development of quantitative expressions which take rigorous account of the effects of thermodynamic nonideality in interacting systems on the statistical-mechanical basis of excluded volume and charge–charge interactions.

References

1. Johnson ML, Correia JJ, Halvorson HR, Yphantis DA (1981) Biophys J 36:575–588
2. Laue TM, Senear DF, Eaton S, Ross AJB (1993) Biochemistry 32:2469–2472
3. Kim T, Tsukiyama T, Lewis MS, Wu C (1994) Protein Sci 3:1040–1051
4. Bailey MF, Davidson BE, Minton AP, Sawyer WH, Howlett GJ (1996) J Mol Biol 263:671–684
5. Winzor DJ, Jacobsen MP, Wills PR (1998) Biochemistry 37:2226–2233
6. Wills PR, Jacobsen MP, Winzor DJ (1996) Biopolymers 38:119–130
7. Jacobsen MP, Wills PR, Winzor DJ (1996) Biochemistry 35:13173–13179
8. Wills PR, Jacobsen MP, Winzor DJ (1997) Prog Colloid Polym Sci 107:1–10
9. McMillan WG, Mayer JE (1945) J Chem Phys 13:276–305
10. Hill TL, Chen YD (1973) Biopolymers 12:1285–1312
11. Haschemeyer RH, Bowers WF (1970) Biochemistry 9:435–445
12. Milthorpe BK, Jeffrey PD, Nichol LW (1975) Biophys Chem 3:169–176
13. Wills PR, Winzor DJ (1992) In: Harding SE, Rowe AJ, Horton JC (eds) Analytical ultracentrifugation in biochemistry and polymer science. Royal Society of Chemistry, Cambridge, pp 311–330
14. Wills PR, Comper WD, Winzor DJ (1993) Arch Biochem Biophys 300:206–212
15. Winzor DJ, Wills PR (1995) In: Gregory RB (ed) Protein–solvent interactions. Dekker, New York, pp 483–520
16. Wills PR, Nichol LW, Siezen RJ (1981) Biophys Chem 11:71–82
17. Adams ET, Fujita H (1959) In: Williams JW (ed) Ultracentrifugal analysis in theory and experiment. Academic Press, New York, pp 119–128

Progr Colloid Polym Sci (1999) 113 : 76–80
© Springer-Verlag 1999

G.M. Pavlov
S.E. Harding
A.J. Rowe

Normalized scaling relations as a natural classification of linear macromolecules according to size

This paper is dedicated to the memory of
professor Sergey Yu Frenkel

G.M. Pavlov (⊠)
Institute of Physics, University
Ulianovskaya str. 1
198904 St. Petersburg, Russia
e-mail: gpolym@onti.niif.spb.su
Tel.: +7-812-4284365
Fax: +7-812-4287240

G.M. Pavlov · S.E. Harding · A.J. Rowe
National Centre for Macromolecular
Hydrodynamics, University of Nottingham
Sutton Bonington LE12 5RD, UK

Abstract The scaling relationships (Mark–Kuhn–Houwink–Sakurada type) are considered for the following hydrodynamic values: intrinsic viscosity, velocity sedimentation coefficient and translational diffusion coefficient and the concentration sedimentation coefficient (Gralen coefficient). By also taking into account the mass per unit length we can obtain "normalized scaling plots" which provide a convenient way of representing the rigidity of linear polymers.

Key words Hydrodynamic values · Equilibrium rigidity · Scaling relationships

Introduction

Molecular hydrodynamics provides a powerful means for studying polymers and macromolecular systems. The principal experimental values we can measure are the coefficients of translational diffusion (D) and velocity sedimentation (s), which both manifest translational motion, and the intrinsic viscosity ($[\eta]$), which manifests the rotational motion. Fundamental relations exist which connect these values with the molecular weight (M) and the size of a polymeric chain (in terms of either $\langle h^2 \rangle$, the mean square end-to-end distance, or $\langle s^2 \rangle$, the mean square radius of gyration) are the well-known relations of Einstein–Stokes, Kirkwood–Riseman, Svedberg and Flory [1–5] which are, respectively

$$D = kT/f \tag{1}$$

$$f = P_0 \eta_0 \langle h^2 \rangle^{1/2} \tag{2}$$

$$s = (1 - v\rho_0)M/N_A f \tag{3}$$

$$[\eta] = \Phi_0 \langle h^2 \rangle^{3/2}/M \ . \tag{4}$$

In these relations, f is the translation frictional coefficient, η_0 the viscosity of solvent, T the temperature (K), N_A Avogadro's number, k the Boltzmann constant, and Φ_0 and P_0 are dimensionless hydrodynamic parameters. These parameters (Φ_0 and P_0) are dependent on the relative contour length (L/A, where L is the contour length and A is the Kuhn segment length, see later), the relative transverse dimension of the chain (d/A, where d is hydrodynamic diameter of the chain) and also on the thermodynamic quality of the solvent [6].

The key molecular parameters are the molecular weight and the size of the macromolecule. The molecular weight may be characterized by $M = M_L L$, where M_L is the molecular weight per unit length. This parameter M_L is distinctive of a particular type of macromolecule.

The size is determined by the equilibrium rigidity, the diameter of the chain and the excluded-volume effect. The equilibrium rigidity may be defined for sufficiently long linear chains in the θ condition (Gaussian statistic) as the ratio of the mean square end-to-end distance to the contour length $A = \langle h^2 \rangle/L$ [2, 6, 7]. This ratio is called the Kuhn segment length. Another characteristic of equilibrium rigidity is the persistence length, a, which is a half the Kuhn segment length: $A = 2a$ [2, 6, 7].

Molecular information is also contained in the first coefficients of the concentration expansion for these experimental values. For instance, the most widely known is that for the intrinsic viscosity $[\eta] \equiv k_1$, that is the first concentration coefficient in the expansion of the dynamic viscosity of a solution

$$\eta = \eta_0(1 + k_1 c + k_2 c^2 + \cdots) \tag{5}$$

$$k_1 \equiv [\eta] \equiv \lim(\eta - \eta_0)/\eta_0 c \tag{6}$$

with the concentration, c, in grams per cubic centimetre, and the intrinsic viscosity, $[\eta]$, in cubic centimetres per gram.

In some cases molecular information can also be obtained from comparison of s_0 with the concentration-dependence "Gralen" coefficient (k_s), [8–10] from the relation

$$s^{-1} = s_0^{-1}(1 + k_s c + \cdots) \quad , \tag{7}$$

with k_s also in cubic centimetres per gram. In terms of molecular parameters

$$k_s = B\langle h^2 \rangle^{3/2}/M \quad , \tag{8}$$

where B is a dimensionless parameter.

The values of $[\eta]$, s_0 and D_0 are also directly related to the molecular weight by the well known Mark–Kuhn–Houwink–Sakurada (MKHS) relationships [1, 2, 7]:

$$[\eta] = K_\eta M^{b_\eta} \tag{9}$$

$$s_0 = K_s M^{b_s} \tag{10}$$

$$D_0 = K_d M^{b_d} \tag{11}$$

These are often referred to as hydrodynamic "scaling relations". The additional scaling relations may be obtained for any pair of experimental values ($[\eta] \sim s_0^{b_{\eta s}}$; $s_0 \sim D_0^{b_{sD}}$; $k_s \sim s_0^{b_{ks}}$, etc). Particularly informative is the relation of k_s to s_0 ($k_s \sim s_0^{b_{ks}}$) since this relation can be obtained in a single series of sedimentation velocity experiments [9, 11].

Discussion

Hydrodynamic theory of a wormlike chain
with excluded-volume effect

A more complete interpretation of experimental hydrodynamic values (s_0 or D_0) and $[\eta]$ can be derived by applying the theories of the translational-friction coefficient [12] and intrinsic viscosity [13] for wormlike chains, after taking into account excluded-volume effects. In these theories [12, 13] the Porod statistic is applied for the neighbouring (adjacent) segments and the excluded-volume effects take into account the remote

segments by the parameter ε in the relations $\langle h^2 \rangle \sim N^{1+\varepsilon} \sim M^{1+\varepsilon}$ [14].

The analytical expression may be obtained [15, 16] only for translational friction on the basis of the theory [12]:

$$[s]PN_A = M[D]Pk^{-1} = Mf^{-1}\eta_0 P$$
$$= [3/(1-\varepsilon)(3-\varepsilon)](M_L^{(1+\varepsilon)/2}/A^{(1-\varepsilon)/2})M^{(1-\varepsilon)/2}$$
$$+ (M_L P/3\pi)[\ln A/d - d/3A - \varphi(\varepsilon)] \quad . \tag{12}$$

The asymptotic limit ($M \to \infty$) for the purely nondraining case corresponds to

$$f = [\eta_0 P_0(1-\varepsilon)(3-\varepsilon)/3]A^{(1-\varepsilon)/2}M_L^{-(1+\varepsilon)/2}M^{(1+\varepsilon)/2} \quad . \tag{13}$$

For the intrinsic viscosity only the asymptotic limit ($M \to \infty$) is currently known theoretically [13, 15, 17]:

$$[\eta] = \Phi(\varepsilon)A^{(3-3\varepsilon)/2}M_L^{-(3+3\varepsilon)/2}$$
$$\times [1 + (5/6)\varepsilon + (1/6)\varepsilon^2]^{-1}M^{(1+3\varepsilon)/2} \quad . \tag{14}$$

In the case of $\varepsilon = 0$ these relations transform into the well-known Flory (intrinsic viscosity) and the Kirkwood–Riseman (translational friction) relations for nondraining Gaussian coils. Since $(1 + 2\varepsilon)/2 \equiv b_\eta$ and $|b_d| \equiv (1+\varepsilon)/2$ [14] Eqs. (13) and (14) reveal the physical sense of the MKHS parameters (K_η, K_d and K_s) and their correlations with the scaling indices become clear [15].

The analysis of these relations (Eqs. 12–14) allows us to enumerate the molecular parameters (except M) which underpin $[\eta]$, s_0 and other experimental hydrodynamic values such as

1. The Kuhn segment length (which characterizes the equilibrium rigidity of the chain).
2. M_L – the mass per unit length.
3. d – the hydrodynamic diameter of the chain.
4. ε – the thermodynamic quality of the polymer–solvent system.

MKHS relationships and fractal concept

The fulfilment of the MKHS relationships reflects the fundamental principle of scale invariance for polymeric molecules [18]. This principle applies not only to linear polymers, but also to branched polymers. The scale indices b_η, b_s and b_d are simply related to the fractal (scale) dimensionality of these particles (i.e. in this case, the macromolecules).

The "fractal" concept was introduced by Mandelbrot [19] and is now widely applied in physics [20, 21]. The physical fractal may be formed by the connection of the separate particles into a single loose integer, called a "cluster", which has, as a rule, a noninteger fractal dimension. This dimension may be determined from the dependence of the number (N) of single particle clusters on the distance r on which the number is calculated:

$$N \sim r^{d_f} \quad . \tag{15}$$

Since for identical particles N is directly proportional to the mass (weight) M, it is possible to give the weight distribution of clusters in terms of size:

$$M \sim r^{d_f} \quad . \tag{16}$$

An individual macromolecule is regarded as a fractal object (cluster of connected repeat units, connected monomer cluster), which may be represented by the actual fractal dimension [19]. The relations (Eqs. 1–4, 8) and (Eqs. 9–11) provide us with the possibility of relating the fractal dimension to the scaling indices of MKHS relationships:

$$d_f = |b_d|^{-1} = (1 - b_s)^{-1}$$
$$= 3/(1 + b_\eta) = (b_{k_s} + 3)/(b_{k_s} + 1) \quad . \tag{17}$$

This use of the fractal concept provides us with the possibility of relating completely different objects with the same fractal dimension or same scale invariance. From the point of view of molecular physics and molecular hydrodynamics the fractal concept is a more general one than is the simple use of particular properties of individual objects. Molecular hydrodynamics allows us in fact to estimate not only the scaling indicies (fractal dimension), but also to interpret at the molecular level parameters such as Φ_0, P_0, K_η, K_d and K_s [1–6].

Normalized scaling relations

It is possible to eliminate from our consideration the effects of different M_L values by using the corresponding normalized scaling relations.

We will consider the dependence of the terms $[\eta]M_L$, $[s]/M_L$ and $[D]$ on the contour length of the molecules $L = M/M_L$ and $k_s M_L$ on $[s]/M_L$ on a double-logarithmic scale, where $[s] \equiv s_0\eta_0/(1-\upsilon\rho_0)$ and $[D] \equiv D_0\eta_0/T$. In these cases the terms $[\eta]M_L$, $[s]/M_L$, $[D]$ and $k_s M_L$ following Eqs. (1)–(4) and Eq. (8) will depend first of all on the size of the coil in the solution. The normalized MKHS plot is shown in Fig. 1. In this plot all possible conformations of linear macromolecules are repres-ented.

We have now found that the normalized (reduced) scaling relations allow the classification of polymers according to the size of the molecules, which will depend in the main on the equilibrium rigidity of the chain. This preliminary classification of the macromolecules on the basis of their rigidity (extra rigid, rigid, semiflexible, flexible) is important for the choice of the corresponding theory for the interpretation of hydrodynamic data (with or without the excluded-volume effects).

The same principles underlie the plot of $k_s M_L$ versus $[s]/M_L$ (Fig. 2) as discussed by us earlier [22]. It is clear

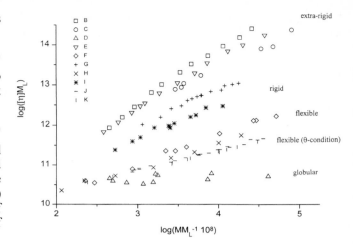

Fig. 1 Normalized double-logarithmic plot of $[\eta]M_L$ against MM_L^{-1} for the following polymer–solvent systems: B – schizophylan in water [26], C – DNA in aqueous buffer [27–30], D – globular proteins in aqueous buffer [1, 31], E – xanthan in 0.1 M NaCl [32], F – poly(1-vinyl-2-pyrrolidone) in 0.1 M sodium acetate [15], G – cellulose nitrate in ethylacetate [33], H – pullulan in water [34–38], I – methylcellulose in water [39], J – poly-α-methylstyrene in cyclohexane (θ condition) [40–43] and in *trans*-decalin (θ condition) [43], K – polystyrene in cyclohexane (θ condition) [42, 44]

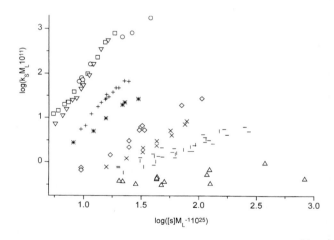

Fig. 2 Normalized double-logarithmic plot of $k_s M_L$ against $[s]M_L^{-1}$ for the same polymer–solvent systems as shown in Fig. 1

evidence of the usefulness of a direct comparison of the values for k_s and s_0, determined from a single series of experiments.

It is worth noting that the behaviour represented in Figs. 1 and 2 is similar in several important ways. First of all the terms $[\eta]M_L$ and $k_s M_L$ and the slopes of their dependencies are greater for more-rigid polymers. This is because the terms $[\eta]M_L \sim k_s M_L \sim \langle h^2 \rangle^{3/2}/L \sim V/L$ follow from Eqs. (4) and (8), where V is the volume occupied by macromolecule and L is its contour length. This ratio characterises the volume occupied per unit length of macromolecule (it is easy to choose the length

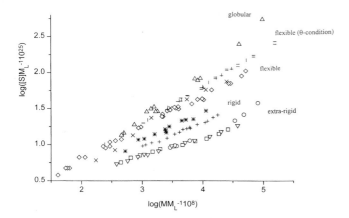

Fig. 3 Normalized double-logarithmic plot of $[s]M_L^{-1}$ against MM_L^{-1} for the same polymer–solvent systems as shown in Fig. 1

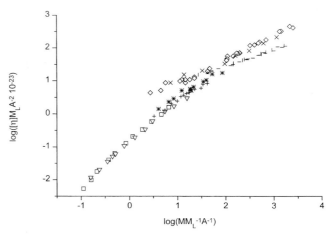

Fig. 4 Double normalized double-logarithmic plot of $[\eta]M_L/A^2$ against $MM_L^{-1}A^{-1}$ (master curve) for the same polymer systems as shown in Fig. 1

of a repeat unit in this case). Obviously these values are greater for rigid polymers, and small for the compact molecules (e.g. globular proteins or dendrimers [23, 24]). By contrast the dependence of $[s]M_L^{-1}$ versus MM_L^{-1} (Fig. 3) reveals the opposite tendency because $[s]/M_L \sim L/\langle h^2 \rangle^{1/2}$, and this ratio characterises the degree of coiling or the contraction of a macromolecule; this ratio is obviously greater for small and compact molecules and smaller for rigid ones. In this way the sedimentation coefficient is more sensitive to changes in molecular weight for compact molecules and is smaller for rigid ones [25]. Similar arguments may be applied to the $[D] \sim MM_L^{-1}$ plot.

We can now see how it is possible, using the equilibrium rigidity parameter, to develop the concept of a "master curve", bringing together all types of polymer conformation. The plot required is one in which $[\eta]M_L/A^2$ is plotted as a function of M/M_LA (Fig. 4). This dependence, which follows directly from Eq. 4, shows that polymers whose equilibrium rigidities vary

by up to about 200 fold [from flexible chains such as pullulan, poly(vinyl pyrrolidine) and poly(styrene) to extra-rigid rods such as shizophylan and xanthan)] all follow the same trend when represented in this way. There are small deviations from this curve, which in the upper region may be attributed to the excluded-volume effect, and in the lower region to the influence of the diameter of the chain, decreasing the draining of the macromolecules. In this new approach we have developed (and intend in the future to develop further) a formalism which allows the well-known MKHS relationships to be alternatively described in terms of the fractal dimension associated with the polymer, in particular with the equilibrium rigidity parameter of the linear polymer chain. It is clear that this newly derived relationship enables us to give, over a very wide range of solution types and parameters, an integrated description of polymer hydrodynamic behaviour.

References

1. Tanford C (1961) Physical chemistry of macromolecules. Wiley, New York
2. Tsvetkov VN, Eskin VE, Frenkel SYa (1970) Structure of macromolecules in solution. Butterworths, London
3. Kirkwood J, Riseman J (1948) J Chem Phys 16:565
4. Svedberg T, Pedersen KO (1940) The ultracentrifuge. Oxford University Press, Oxford
5. Flory P (1953) Principles of polymer chemistry. Cornell University Press Ithaca
6. Yamakawa H (1971) Modern theory of polymer solutions. Harper and Row, New York
7. Fujita H (1990) Polymer solutions. Elsevier, Amsterdam
8. Wales M, van Holde K (1954) J Polym Sci 14:81
9. Rowe A (1977) Biopolymers 16:2595
10. Pavlov G, Frenkel S (1995) Prog Colloid Polym Sci 99:101
11. Pavlov G (1997) Eur Biophys J 25:385
12. Gray H, Bloomfield V, Hearst J (1967) J Chem Phys 46:1493
13. Sharp P, Bloomfield V (1968) J Chem Phys 48:2149
14. Ptitsyn OB, Eizner Yu E (1959) Zh Tekh Fiz 29:1105
15. Pavlov G, Panarin E, Korneeva E, Kurochkin C, Baikov V, Ushakova V (1990) Makromol Chem 191:2889
16. Bushin SV, Astapenko EP (1986) Vysokomol Soedin 28:1499
17. Pavlov GM, Korneeva EV, Michailova NA, Ivanova NP, Panarin EF (1993) Vysokomol Soedin 35:1647
18. de Gennes P (1979) Scaling concepts in polymer physics. Cornell University Press, Ithaca
19. Mandelbrot B (1982) The fractal geometry of nature. Freeman San Fransisco
20. Pietronero L, Tosatti E (eds) (1986) Fractal in physics. Elsevier, Amsterdam
21. Grosberg A, Khochlov A (1989) Physics in the world of polymers. Nauka, Moscow

22. Pavlov G, Rowe A, Harding S (1997) Trends Anal Chem 16:401
23. Pavlov G, Korneeva E, Harding S, Jumel K, Meijer E, Nepogodiev S, Peerling H, Stoddart J (1999) Carbohydr Polym 38:195
24. Pavlov G, Korneeva E, Roy R, Michailova N, Cejas Ortega P, Alamino Perez M (1999) Progr Colloid Polym Sci PCPS 109
25. Tsvetkov VN (1989) Rigid-chain polymers. Consultants Bureau, New York
26. Yanaki T, Norisuye T, Fujita H (1980) Macromolecules 13:345
27. Geiduschek EP, Holtzer A (1958) Adv Biol Med Phys 6:431
28. Crothers DM, Zimm BH (1965) J Mol Biol 12:525
29. Aten JBT, Cohen JA (1965) J Mol Biol 12:537
30. Eigner J, Doty P (1965) J Mol Biol 12:549
31. Creeth J, Knight C (1965) Biochim Biophys Acta 102:549
32. Sato T, Norisuye T, Fujita H (1984) Macromolecules 17:2696
33. Pavlov GM, Kozlov AN, Martchenko GN, Tsvetkov VN (1982) Vysokomol Soedin 24B:284
34. Kawahara K, Ohta K, Miyamoto H, Nakamura S (1984) Carbohydr Polym 4:335
35. Kato T, Katsuki T, Takahashi A (1984) Macromolecules 17:1726
36. Buliga GS, Brant DA (1987) Int J Biol Macromol 9:71
37. Nishinari K, Kohyama K, Williams PA, Phillips GO, Burchard W, Ogino K (1991) Macromolecules 24:5590
38. Pavlov GM, Korneeva EV, Yevlampieva NP (1994) Int J Biol Macromol 16:318
39. Pavlov GM, Michailova NA, Tarabukina EB, Korneeva EV (1995) Prog Colloid Polym Sci 99:109
40. Noda J, Saito S, Fujimoto T, Nagasawa M (1967) J Phys Chem 71:4048
41. Abe M, Sakato K, Kageyama T, Fukatsu M, Kurata M (1968) Bull Chem Soc Jpn 41:2330
42. Kotera A, Saito T, Hamada T (1972) Polym J 3:421
43. Noda J, Mizutani K, Kato T (1977) Macromolecules 10:618
44. Peeters FAH, Smits HJE (1981) Bull Soc Chim Belg 90:111

Progr Colloid Polym Sci (1999) 113 : 81–86
© Springer-Verlag 1999

J. Garcia de la Torre
B. Carrasco

Universal size-independent quantities for the conformational characterization of rigid and flexible macromolecules

J. Garcia de la Torre (✉) · B. Carrasco
Departamento de Quimica Fisica
Universidad de Murcia
E-30071 Murcia, Spain
Fax: +34-968-364148
e-mail: jgt@fcu.um.es

Abstract Solution properties of macromolecules can be combined to form dimensionless quantities that are universal in the sense of depending on the shape or conformation of the macromolecular solute, being independent of its size. A number of such quantities have been formulated at different times by different authors, and as a consequence they differ widely from one to another, not only in their notation and the order of magnitude of their values, but also in the way in which the primary solution properties enter in them. In this work we propose a new set of universal size-independent quantities in a systematic and consistent way. First, solution proper-

ties are expressed in the form of radii of an equivalent sphere. Then the equivalent radii are combined to give ratios of radii. We propose the use of the ratios of radii as indicators of macromolecular conformation. Examples of their values are given for three macromolecular structures: (1) ellipsoids, with application to globular proteins, (2) oligomeric arrays of subunits, with seed globulins as examples and (3) flexible-chain macromolecules, illustrated by polystyrene in two solvents.

Key words Rigid and flexible macromolecules · Conformation · Globular proteins · Seed globulins · Polystyrene

Introduction

Hydrodynamic properties, such as the sedimentation and diffusion coefficients, s and D_t, the intrinsic viscosity, $[\eta]$, and equilibrium solution properties such as the radius of gyration, R_g, can be combined to construct dimensionless quantities that are universal in the sense of being independent of the size of the macromolecular particle, while they depend more or less sensitively on its shape or conformation. Typical examples are the Scheraga-Mandelkern parameter [1], β, given by

$$\beta = \frac{M^{1/3}[\eta]^{1/3}\eta_0}{100^{1/3}f} \tag{1}$$

The friction coefficient f is derived from the measured sedimentation coefficient, s, as $f = M(1 - \bar{v}\rho)/(N_A s)$, or from the diffusion coefficient as $f = kT/D_t$ where k is the Boltzmann constant and T the absolute temperature. In

Eq. (1), M and $\bar{v}$ are, respectively, the molecular weight and the partial specific volume of the macromolecule, and η_0 the solvent viscosity. Other classical size-independent combinations are the Flory parameters [2], that combine the intrinsic viscosity, $[\eta]$, and the radius of gyration, R_g:

$$\Phi_0 = \frac{[\eta]M}{6^{3/2}R_g^3} \tag{2}$$

and another combining the friction coefficient with the radius of gyration:

$$P_0 = \frac{f}{6\eta R_g} \tag{3}$$

These quantities have been proposed along the years, at different times and by different eminent scientists, after whom they are named. As a consequence of the diversity in their origin, the set of classical universal size-independent quantities suffers some inconveniences. Two of them, unimportant but somehow cumbersome, are related to the diversity not only in the symbols employed to represent them, but mainly in the disparity of their numerical values and the order of magnitude for typical cases; thus, β takes the values of 2.112×10^6 and about 2.3×10^6 for a sphere and a random coil, respectively, while the values for these two structures in the case of the Φ_0 are 9.23×10^{23} and 2.6×10^{23}.

A more important aspect is their different sensitivity to macromolecular conformation, as illustrated by the numerical examples that we have given in the above paragraph. It is notorious that the Scheraga-Mandelkern parameter is very insensitive to the specificities of the macromolecular shape, while the Flory parameters changes much more pronouncedly from one conformation to the other. In principle this is a consequence of the combination involved, $[\eta]$-f in one case, and $[\eta]$-R_g in the other, but the sensitivity also depends on the way that the compound quantity has been defined: the intrinsic viscosity enters as $[\eta]^{1/3}$ in β and as $[\eta]$ in Φ_0. If the Scheraga-Mandelkern parameter had been defined as $\beta' = \beta^3 = M[\eta]\eta_0/(100 f^3)$, involving the ratio $[\eta]/f^3$ rather than $[\eta]^{1/3}/f$, the relative difference between its numerical values would be more noticeable.

Anyhow, the classical and other, more recently derived, size-independent functions have been very successfully applied as indicators of macromolecular shape. This has been done for rigid macromolecules, modelled as ellipsoidal shapes [3] or as arbitrarily shaped bead models [4], and also for flexible chain polymers [5]. In the present work, we propose the use of a set of universal, size-independent quantities that are defined in a rather consistent way, so that the above-commented inconveniences of the classical functions are removed. We present the numerical values for some commonly employed models of rigid and flexible macromolecules, and we show some examples that illustrate their applicability for the characterization of the macromolecular conformation.

Theory

Ratios of equivalent radii

As a preliminary step for the subsequent definition of the new universal shape-dependent quantities, we first define the equivalent radii for the various solution properties. For some property, X, the equivalent radius a_x is defined as the geometrical radius of a spherical particle that would have the same value X experimentally obtained for the particle. The solution properties that we employ are: the friction coefficient, f; the intrinsic viscosity, $[\eta]$; the radius of gyration, R_g; and the particle's volume, V. Other solution properties could be included in this analysis, but then the number of ratios of radii (see below) would be much larger.

Expressions for the various equivalent radii a_T, a_I, a_G, and a_V are collected in Table 1. We note that, on the contrary to the primary properties, that display different sensitivities to size and shape, all the equivalent radii depend proportionally on a linear dimension of the particle; if the particle is uniformly expanded by a given factor, all the radii are multiplied by the same factor.

The universal size-independent quantities are defined as ratios of the equivalent radii. With the properties considered here, the following combinations can be formulated:

$$IT \equiv \frac{a_I}{a_T} = \left(\frac{3[\eta]M}{10\pi N_A}\right)^{1/3} \frac{6\pi\eta_0}{f} \tag{4}$$

$$TV \equiv \frac{a_T}{a_V} = \frac{f}{6\pi\eta_0(3V/(4\pi))^{1/3}} \tag{5}$$

$$GT \equiv \frac{a_G}{a_T} = \frac{6\pi\eta_0}{f} \frac{R_g}{(3/5)^{1/2}} \tag{6}$$

$$IV \equiv \frac{a_I}{a_V} = \left(\frac{2[\eta]M}{5VN_A}\right)^{1/3} \tag{7}$$

$$GI \equiv \frac{a_G}{a_I} = \left(\frac{10\pi}{3[\eta]M}\right)^{1/3} \left(\frac{5}{3}\right)^{1/2} R_g \tag{8}$$

$$GV \equiv \frac{a_G}{a_V} = \left(\frac{4\pi}{3V}\right)^{1/3} \left(\frac{5}{3}\right)^{1/2} R_g \tag{9}$$

All these ratios of radii have a remarkable property: their values for a spherical particle are equal to unity, and take values larger than unity for any other conformation, but always with the same order of magnitude. As illustrated in next section, typical values for most macromolecular structures are between 1 and, say, 5 (not very elongated).

Table 1 Expression for the equivalent radii for various solution properties

Property	Symbol	Acronym	Equivalent radii
Translational friction coefficient	f	T	$a_T = f/6\pi\eta_0$
Intrinsic viscosity	$[\eta]$	I	$a_I = (3/10\pi N_A)^{1/3}([\eta]M)^{1/3}$
Volume	V	V	$a_V = (3/4\pi)^{1/3}V^{1/3}$
Radius of gyration	R_g	G	$a_G = \sqrt{(5/3)}R_g$

Hydration

The application of universal shape functions, either the classical ones or the new ratios of radii, requires the consideration of an unclear problem: hydration. In the definition of shape functions, either the classical ones or the ratios of radii, it is implicitly assumed that the particle "seen" by the various solution properties is the same. Then, for compatibility with hydrodynamic properties, the particle volume used for the calculation of a_V must be the hydrated volume, i.e., V_{hyd}. From the molecular weight and partial specific volume, M and $\bar{v}$, we can readily calculate the anhydrous volume, $V_{anh} = \bar{v}M/N_A$. For large particles, the thickness of the hydration layer will be small in comparison with the size of the macromolecule and both volumes will be approximately equal. However, for other macromolecules, particularly small or medium sized proteins, this approximation is not valid. It is commonly assumed that the hydration effect causes a uniform expansion of the macromolecule (in terms of the ellipsoidal models employed below, p is the same for the hydrated and the "dry" particle). We must take

$$V = V_{hyd} = h^3 V_{anh} \tag{10}$$

where h is the hydration expansion factor. The quantity usually employed to express hydration of proteins is the ratio δ, of the grams of water per gram of macromolecule. Then the expansion factor is

$$h = (1 + \delta/\bar{v}\rho)^{1/3} \tag{11}$$

The most delicate aspect is the treatment of hydration effects in the radius of gyration. There is some controversy regarding the amount of hydration that is detected by scattering; some authors [6] have argued that such amount is smaller than for hydrodynamic properties but not negligible. We have proposed to treat this situation [7] by means of another expansion factor, t, that relates the experimental radius of gyration and the anhydrous value

$$R_g(\exp) = t R_g(anh) \tag{12}$$

Note that $1 < t < h$, where the extreme cases are that scattering sees no hydration at all ($t = 1$) or the full hydration ($t = h$). The radius of gyration to be used in the calculation of the equivalent radii must be the one that we would have for the full hydrated particle, and can be estimated suppressing the t factor and applying next the h factor:

$$R_g = R_g(hyd) = (h/t) R_g(\exp) \tag{13}$$

For the purpose of the applications to be presented later, we have made a compromise, taking for t the midpoint between the two extremes, $t = (h + 1)/2$, so that $h/t = 2h/(h + 1)$.

Models and applications

In this section we describe the values of the ratios of radii for some simple but very useful models: ellipsoids, oligomeric structures, and flexible chains, and we apply this technique to typical macromolecules that are represented by these models.

Ellipsoidal models; application to a globular protein

Ellipsoidal models are useful to describe primarily the non-spherical shape of rigid, compact macromolecules such as globular proteins, and also to give estimations of the behavior of rigid elongated structures. Indeed, the analysis of macromolecular size and shape in terms of ellipsoidal models has been a motivation for the definition and usage of universal shape functions, and for this purpose the classical models have been used by some authors [3, 8]. Other workers have made the shape analysis of proteins using ellipsoidal models in terms of equivalent radii and their ratios [9, 10].

For the illustrative purposes of this paper, we consider only prolate ellipsoids of revolution; the case of oblates is trivially similar and the study could be extended with some work but without much difficulty to triaxial ellipsoids.

For ellipsoids with semiaxes a, a, and b, and axial ratio $p = b/a$, the friction coefficient and the intrinsic viscosity can be calculated from the equations of Perrin [11] and Simha [12]. The radius of gyration and the particle volume are given by $R_g = [(2a^2 + b^2)/5]^{1/2}$ and $V = 4\pi a^2 b/3$. The universal size-independent quantities are functions of the axial ratio, p. Numerical values of the ratio of radii are displayed in Fig. 1.

A common application of ellipsoidal models is the determination of the overall size and shape of globular proteins. We have chosen a well-studied protein, lysozyme, to illustrate the applicability of the ratios of radii to estimate the shape. Experimental data for lysozyme are: $M = 14211$ (from chemical composition), $\bar{v} = 0.703\,cm^3/g$ [13], $D_t = 11.1 \times 10^{-7}\,cm^2/s$ (a consensus of various data, from sedimentation, diffusion, and scattering measurements [14]), $[\eta] = 3.0\,cm^3/g$ [15, 16]. Regarding the experimental value of the radius of gyration, there is some scatter in the literature data, that range from $R_g = 15.2$ Å [13, 16], to $R_g = 13.8$ Å [17]. We have adopted a consensus value, $R_g = 14.5$ Å, which surely is within ± 1 Å of the true value. With these data, and assigning a value to δ [and $t = (h + 1)/2$], the effective volume and effective radius of gyration are evaluated, and from them we calculate a set of equivalent radii and a set of ratios. On the other hand, calculated values of the ratios can be obtained for ellipsoids with varying p. The deviation between the experimental and calculated sets can be measured by the

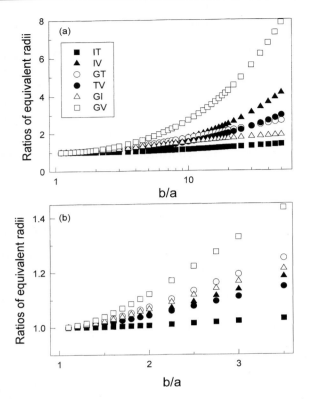

Fig. 1 Variation of the ratios of radii with the axial ratio, p, of prolate ellipsoids, in two ranges of p

sums of square differences, $\Sigma_Z (Z_{cal} - Z_{exp})^2$, where Z runs over six available ratios of radii. This sum is minimized with p and δ as adjustable parameters, and the least deviation is found for $p = 1.8$ and $\delta = 0.4$ g/g and the corresponding theoretical ratios are listed in Table 2. The analysis is more sensitive to δ than to p; therefore, the former quantity is statistically more reliable than the latter. This value for hydration is in the range that is presently accepted. The overall dimensions of lysozyme in the crystal are about $45 \times 30 \times 30$ Å, so that we would expect $p \approx 1.5$.

The shape analysis of ellipsoidal shapes presents a difficulty caused by the low sensitivity of some universal size-independent functions to p. For the illustrative purposes of this paper, we have restricted our analysis to the prolate shape, but it can be remarked that some universal functions for oblate ellipsoids are even less sensitive to shape than for prolates. An extreme case is the *IT* combination for oblates [9]. Anyhow, the results calculated with the crystallographic estimate, $p = 1.5$ (shown in Table 2), do not differ greatly from those of the best fitted for $p = 1.8$. A more elaborate analysis, in which the error bars of the properties would be considered, probably would not differentiate between the two values of p and, at any rate, this effort could be of scarce significance, because the primary model, i.e., the prolate ellipsoid, is not a very good description of the shape of lysozyme.

Oligomeric structures; application to seed globulins

For multisubunit proteins, composed of nearly spherical subunits arranged with a polygonal or polyhedral geometry, oligomeric arrays of spheres are useful to describe their solution properties. Values of oligomer-to-monomer ratios of a given property have been given in the literature [18, 19]. The results can be easily recast in the form of ratios of equivalent radii for pairs of properties. As illustration of such results, we present in Table 3 values for hexameric structures, composed by six spherical subunits.

We again note the different sensitivity to shape of the various ratios. Again, the least sensitive is *IT*, where we note incidentally that the values for two structures are slightly under unity, which is due to residual deficiencies in the hydrodynamic treatment (the cubic substitution) used in their calculation [20]. The largest sensitivity, detected as largest departures of the ratios from unity, is found for the *GV* ratio.

As an example of this kind of structure, we present an analysis of the structure of seed globulins. In a previous paper [21], this aim was pursued using the classical size-independent quantities, and now we repeat the analysis with our ratios of radii. From literature data for oilseed and rapeseed globulins [22–26] compiled by Carrasco et al. [21], we compute the equivalent radii. (Parenthetically we note some discrepancy in the literature values of the sedimentation coefficient: while a value of 11.8 S is given for sunflower globulin [26], other authors [22]

Table 2 Ratios of radii for lysozyme

	Experimental $\delta = 0.4$	Ellipsoid $p = 1.8$	Ellipsoid $p = 1.5$
IT	0.994	1.005	1.002
TV	1.049	1.032	1.015
GT	1.098	1.054	1.025
IV	1.043	1.037	1.017
GI	1.105	1.049	1.022
GV	1.152	1.088	1.040

Table 3 Ratios of radii calculated for arrays of six spherical subunits

Structure	*GT*	*IT*	*TV*	*GI*	*IV*	*GV*
Sphere	1.000	1.000	1.000	1.000	1.000	1.000
Hexagonal ring	1.221	1.000	1.248	1.221	1.248	1.524
Trigonal prism	1.055	0.987	1.153	1.070	1.138	1.217
Octahedron	1.018	0.991	1.126	1.027	1.115	1.146
Linear	1.755	1.089	1.418	1.611	1.544	2.488

report 12.8 S, which is the value also given in a recent review [27].)

Regarding hydration, we consider three possible values for δ, for which the corresponding hydration expansion factor is evaluated according to Eq. (11). The effective radius of gyration is calculated from Eq. (13) with $h/t = 2h/(h + 1)$, as in the above study of lysozyme. Then, the ratios of radii are calculated, and their values are listed in Table 4. These values are to be compared with those in Table 3 for the various structures, looking for the structure that gives the best match. This is done in Table 4 for the various ratios of radii for the two globulins, and for the various possibilities for hydration. In most cases, the best fitting structure is an octahedron, in agreement with the conclusion of our previous work [21].

Flexible-chain polymers; application to polystyrene in different solvents

In contrast to the case of the rigid structures described above, we have also considered the applicability of our ratios of radii to flexible chain macromolecules. Such is the case for most synthetic polymers in solution, and for biopolymers such as polysaccharides and denatured proteins. These macromolecules present the typical random coil conformation. This structure is not compact so that the volume V cannot be defined. Experimental values are usually available for f (from s or D_t), $[\eta]$, and R_g. It was indeed for the random coil conformation that Flory proposed the parameters Φ_0 and P_0 [2], although they can also be applied to rigid

structures. A key aspect in the conformational behavior of flexible-chain macromolecules is solvent quality; in theta solvent/temperature conditions the coil obeys the statistics of the so-called phantom chain, but in a good solvent the coil is expanded due to excluded volume. The different conformation must be reflected in the universal quantities, whose numerical values have been the subject of successive revisions as the hydrodynamic theory has been improved. Good estimates of Φ_0 and P_0 have been obtained by Monte Carlo simulation: $\Phi_0 = 2.53 \times 10^{23}$ and $P_0 = 6.0$ for theta conditions [28, 29], and $\Phi_0 = 1.9 \times 10^{23}$ and $P_0 = 5.3$ in good solvent conditions [30]. These results can be recast in the form of ratios of radii, for which the values are given in Table 5.

To illustrate the use of the ratios of radii for flexible polymers, we have chosen the well-documented properties of polystyrene in various solvents and temperatures. In the literature, we find solution properties, D_t or s, $[\eta]$, and R_g of polystyrene, expressed in the form of property versus molecular weight relationships. Results are avaliable in the Θ solvent cyclohexane at 35 °C [31–33] and in the good solvents ethylbenzene at 25 °C [34, 35] and toluene at 20 °C [36, 37]. Interpolating in the property-M relationships for a sufficiently large value of M, of the order of 10^5, numerical values of the properties can be estimated, and from them we calculate the equivalent radii and the ratios of radii that are presented in Table 5. We see that the agreement between experimental and calculated ratios of radii is excellent.

Usually, solvent quality is determined from the exponent in the property-M relationships, which requires the determination of properties for various samples with different M. Looking at Table 5, we note that the ratios of radii take appreciably different values in good and theta solvents, and therefore they can be used to ascertain solvent quality with data for a single sample.

Computer programs

We have written computer programs to evaluate ratios of equivalent radii; these are publicly available and can

Table 4 Ratios of radii for seed globulins, obtained from experimental data and the indicated hydration parameters: [a] for $[\eta] = 4.2$ ml/g; [b] for $[\eta] = 3.6$ ml/g. The attachment letters mean that the value is best fitted by an octahedron (o) or a trigonal prism (p)

Hydration, δ	Low (0.20 g/g)	Medium (0.35 g/g)	High (0.50 g/g)
Sunflower seed			
GT	0.93 (o)	0.95 (o)	0.97 (o)
IT[a]	1.03 (o)	1.03 (o)	1.03 (o)
IT[b]	0.98 (p)	0.98 (p)	0.98 (p)
TV	1.18 (p)	1.12 (o)	1.08 (o)
GI[a]	0.90 (o)	0.93 (o)	0.94 (o)
GI[b]	0.95 (o)	0.97 (o)	0.99 (o)
GV	1.15 (o)	1.15 (o)	1.15 (o)
IV[a]	1.22 (p)	1.16 (p)	1.11 (o)
IV[b]	1.16 (p)	1.10 (o)	1.05 (p)
Rape seed			
GT	0.97 (o)	0.99 (o)	1.01 (o)
IT	1.00 (o)	1.00 (o)	1.00 (o)
TV	1.18 (p)	1.12 (o)	1.07 (o)
GI	0.97 (o)	0.99 (o)	1.01 (o)
GV	1.19 (p)	1.19 (p)	1.19 (p)
IV	1.18 (p)	1.12 (o)	1.07 (o)

Table 5 Ratios of radii theoretically calculated for flexible polymer chains, in Θ and good-solvent conditions, and obtained from experimental data for polystyrene in a Θ solvent and two good solvents

	Θ conditions		Good-solvent conditions		
	Theoretical	Cyclohexane 35 °C	Theoretical	Ethylbenzene 25 °C	Toluene 20 °C
GI	1.53	1.53	1.69	1.77	1.74
GT	1.65	1.68	1.87	2.07	1.88
IT	1.08	1.10	1.11	1.17	1.08

86

be downloaded from our web site, http://leon-ardo.fcu.um.es/macromol, where the computer programs HYDRO [38] and SOLPRO [4, 7] can also be found. For bead models, the calculation of the ratios is within the latest version of SOLPRO (file solpro-x.f, where x is a version number), while for revolution ellipsoids we provide a separate subroutine (file ufelcy-x.for).

Acknowledgements We acknowledge support by grant PB96-1106 from the Direccion General de Enseñanza Superior. B.C. is the recipient of a predoctoral fellowship from the same source. Further support was provided by grant 01758/CV/98 from Fundacion Seneca, Comunidad Autonoma, Region de Murcia.

References

1. Scheraga HA, Mandelkern L (1953) J Am Chem Soc 75:179
2. Flory PJ (1953) Principles of polymer chemistry. Cornell University Press, New York
3. Harding SE (1995) Biophys Chem 55:69
4. García de la Torre J, Carrasco B, Harding SE (1997) Eur Biophys J 25:361
5. Yamakawa H (1971) Modern theory of polymer solutions. Harper & Row, New York
6. Kumosinski T, Pessen H (1985) Methods Enzymol 117:154
7. García de la Torre J, Harding SE, Carrasco B (1999) Eur Biophys J 28:119
8. Harding SE, Cölfen H (1995) Anal Biochem 228:131
9. Durchschlag H, Zipper P (1997) J Appl Crystallogr 30:1112
10. Durchschlag H, Zipper P (1997) Prog Colloid Polym Sci 107:43
11. Perrin F (1936) J Phys Radium 7:1
12. Simha R (1940) J Phys Chem 44:25
13. Van Holde KE (1971) Physical biochemistry. Prentice Hall, Englewood Cliffs
14. Kuntz ID, Kauzmann W (1974) Adv Protein Chem 28:239
15. Sophianopoulus AJ, Van Holde KE (1964) J Biol Chem 239:2516
16. Luzatti V, Witz J, Nicolaieff A (1961) J Mol Biol 3:367
17. Stuhrmann HB, Fuess H (1976) Acta Crystallogr Sect A 32:67
18. García Bernal JM, García de la Torre J (1981) Biopolymers 20:129
19. García de la Torre J (1989) In: Harding SE, Rowe AJ (eds) Dynamic properties of macromolecular assemblies. Royal Society of Chemistry, Cambridge, pp 3–31
20. Carrasco B, García de la Torre J (1999) Biophys J 76:3044
21. Carrasco B, Harding SE, García de la Torre J (1998) Biophys Chem 74:127
22. Schwenke KD, Pahtz W, Linow KJ, Schultz M (1979) Nahrung 23:241
23. Schwenke KD, Schultz M, Linow KJ, Gast K, Zirwer D (1980) Int J Pept Protein Res 16:12
24. Harding SE, Kyles P, West G, Norton G (1987) Biochem Soc Trans 15:684
25. Prakash V (1994) J Sci Ind Res 53:684
26. Plietz P, Damaschun G, Muller JJ, Schwenke KD (1983) Eur J Biochem 130:315
27. Prakash V (1992) In: Harding SE, Rowe AJ, Horton JC (eds) Analytical ultracentrifugation in biochemistry and polymer science. Royal Society of Chemistry, Cambridge, pp 445–469
28. García de la Torre J, López Martínez MC, Tirado MM, Freire JJ (1984) Macromolecules 17:2715
29. Freire JJ, Rey A, García de la Torre J (1986) Macromolecules 19:457
30. García Bernal JM, Tirado MM, García de la Torre J (1991) Macromolecules 24:693
31. Roovers J (1985) J Polym Sci Polym Phys Ed 23:1117
32. Bohdanecky M, Petrus V, Porsch B, Sundelof LO (1983) Makromol Chem 184:309
33. Huber K, Bantle S, Lutz P, Burchard W (1985) Macromolecules 18:1461
34. Jamieson AM, Venkataswamy K (1984) Polym Bull 12:275
35. Venkataswamy K, Jamieson AM, Petschek RG (1986) Macromolecules 19:124
36. Raczek J (1980) PhD Dissertation, Mainz
37. Huber K, Burchard W, Akcasu AZ (1985) Macromolecules 18:2743
38. García de la Torre J, Navarro S, López Martínez MC, Díaz FG, López Cascales JJ (1994) Biophys J 67:530

Progr Colloid Polym Sci (1999) 113:87–105
© Springer-Verlag 1999

H. Durchschlag
P. Zipper

Calculation of structural parameters from hydrodynamic data

H. Durchschlag (✉)
Institute of Biophysics
and Physical Biochemistry
University of Regensburg
Universitätsstrasse 31
D-93040 Regensburg, Germany
e-mail: helmut.durchschlag@biologie.
uni-regensburg.de
Tel.: +49-941-9433041
Fax: +49-941-9432813

P. Zipper
Institute of Physical Chemistry
University of Graz
Heinrichstrasse 28
A-8010 Graz, Austria

Abstract Biopolymers of simple overall shape (nonconjugated and conjugated proteins, viruses, ribonucleic acids) can be modeled by whole-body approaches, approximating their shape by prolate/oblate ellipsoids of revolution or spheres. A sophisticated rearrangement of the theoretical formalism already applied for the prediction of hydrodynamic data from solution scattering or crystal data allows the inverse procedure, the prediction of structural parameters on the basis of hydrodynamic data. Sedimentation and diffusion coefficients, in addition to molar masses, partial specific volumes and values for hydration, are used to predict structural parameters typical of small-angle X-ray scattering studies (radii of gyration, volumes, surface-to-volume ratios and surface areas), together with estimates of axial ratios. For particles of simple shape such as globular proteins, fair agreement between observed and predicted values was achieved. Far-reaching conformity between experiments and calculations was also obtained for the prediction of subtle ligand-induced shape changes. A critical assessment of errors reveals the validity of anticipations of structural parameters from hydrodynamic data. The accuracy to be obtained, however, turns out to be less than for the reverse procedure. Use of viscosity data for parameter predictions cannot be recommended.

Key words Biopolymers · Analytical ultracentrifugation · Viscometry · Small-angle X-ray scattering · Parameter predictions

Introduction

To understand and compare the information provided by different techniques of structural analysis in solution and in crystals, quantitative correlations between the parameters obtained by various physico-chemical techniques have to be established. In this context, the results from solution scattering [light scattering, small-angle X-ray scattering (SAXS) and neutron scattering] and hydrodynamic techniques (analytical ultracentrifugation, viscometry, densimetry, size-exclusion chromatography) have to be considered, together with data from high-resolution methods such as X-ray crystallography, NMR spectroscopy or electron microscopy.

For correlating structural and hydrodynamic parameters, biopolymers can be modeled by simple geometric structures such as spheres, prolate or oblate ellipsoids of revolution, triaxial ellipsoids, rods, etc. Such whole-body approaches to obtain structural parameters from hydrodynamic data and to predict hydrodynamic parameters from scattering data have a long tradition (see Refs. [1–25] and references therein). More recent applications apply bead modeling to simulate the structure of complex molecules [26–37], including predictions based on high-resolution 3D structures [28, 31, 33, 37].

In general, hydrodynamacists use sedimentation and diffusion coefficients and intrinsic viscosity as principal hydrodynamic probes, in addition to the second virial coefficient, several decay constants and relaxation times. Establishing hydration-independent shape functions turned out to be particularly successful [7, 12, 19]; however, besides the high sensitivity of some functions to experimental errors, the poor sensitivity to shape is a serious drawback in the case of relatively isometric molecules (e.g., globular proteins with axial ratios less than 2). In connection with the interpretation of scattering data, the so-called G function [19] should be addressed, since the radius of gyration R_G (obtained either from light or small-angle scattering) is involved in the calculations. For recent reviews on the hydrodynamic analysis of macromolecules see Refs. [38–40].

For the comparison of X-ray data (SAXS, X-ray diffraction) with the results from hydrodynamics, SAXS investigators usually prefer another type of whole-body approach. As shown by Kumosinski and Pessen [14, 41, 42], use of simple geometric forms (spheres, prolate or oblate ellipsoids of revolution) is straightforward and successful for modeling hydrated globular proteins and spherical viruses in solution to predict hydrodynamic properties. Later on, Müller et al. [18] succeeded in modeling small ribonucleic acids. Recently, we applied similar approaches to nonstandard proteins (conjugated proteins, anisometric and multisubunit proteins and high-molecular-weight ribonucleic acids) and extended the existing approaches and correlations considerably, including the repertoire of possible input and output parameters [43–47]. Since SAXS parameters [48, 49] implicitly contain hydration, no assumptions concerning hydration have to be made. In particular, sedimentation and diffusion coefficients could be predicted reliably from molar masses, partial specific volumes and X-ray data (radii of gyration, hydrated volumes and/or surface-to-volume ratios). The prediction of intrinsic viscosities turned out to be less successful. If atomic coordinates were used for the predictions, appropriate hydration contributions had to be considered.

Following the guidelines in our above-mentioned approaches [43–47], in this study we convert the theoretical formalism used previously, in order to predict structural parameters from hydrodynamic data. On the basis of known sedimentation and diffusion coefficients as well as intrinsic viscosities, and applying molar masses and partial specific volumes as auxiliary data and, if necessary, reasonable assumptions for hydration, structural parameters such as axial ratios, radii of gyration, hydrated volumes, surface-to-volume-ratios, and surfaces of the hydrated particles are predicted. However, predictions of structural parameters from the results of hydrodynamics are more problematic than the opposite procedures, partly caused by the low accuracy of hydrodynamic data used as input parameters (e.g., viscosity). To compare the results from both types of prediction, for the present calculations the same dataset of molecules as used previously [45–47] was applied.

Theory

Correlations between hydrodynamic and scattering data

For modeling homogeneous particles of roughly globular shape in solution, whole-body approaches are applied, using spheres or prolate (axial ratio $p > 1$) or oblate ($p < 1$) ellipsoids of revolution. For the prediction of hydrodynamic parameters from X-ray data, the theoretical formalism underlying our presentation has been outlined in detail [9, 14, 41–43, 45, 46]. For the inverse approach, i.e., the prediction of structural parameters from hydrodynamic data, already-existing equations had to be rearranged and completed appropriately. In the following, the theoretical formalism required for these types of calculations is summarized in brief, using the same nomenclature as in our previous papers [45, 46].

Calculation of structural parameters
from hydrodynamic data

For the prediction of structural parameters, sedimentation and diffusion coefficients, s and D, and the intrinsic viscosity, $[\eta]$, are used. In addition, two further auxiliary quantities, molar mass, M, and partial specific volume, $\bar{v}$, must be known. Both quantities may be taken from databases or may be determined experimentally [45, 46]. For some cases, the amount of particle hydration, δ_1, or at least reasonable assumptions for this quantity are necessary. In the case of simple proteins, a hydration of 0.35 g g^{-1} may be used as a reasonable default value [39, 40, 43, 44, 49–52].

Provided all mentioned input parameters (s, D, $[\eta]$, M, $\bar{v}$, δ_1) are known, two main procedures are possible.

The first approach uses both ultracentrifugal and viscosity data, commonly combining s or D with M and $[\eta]$[1]. The assumption of a definite value for the hydration is not required. Combining two different hydrodynamic techniques allows the estimation of the amount of hydration involved. In principle, the calculative combination of s, D and $[\eta]$ is also possible, without direct knowledge of M. For the calculation of the axial ratio, p, the amount of hydration need not be known.

The second approach considers only ultracentrifugal data, neglecting less accurate results from viscometric studies. For this procedure, however, realistic hydra-

---

[1] Diffusion coefficients may also be obtained by dynamic light scattering

tional contributions (δ_1) have to be taken into account explicitly as input values. Results stem from combinations of two of the three quantities, s, D and M, with δ_1.

In principle, all approaches should yield the same results; however, one has to be aware of the fact that parameters calculated according to different approaches represent different averages.

1. Combining s, M and $[\eta]$:

If s and M are known, these two parameters and the frictional ratio of the hydrated particle, f/f_0, are related by a modified form of the Svedberg equation

$$A = \frac{M^{2/3}(1 - \bar{v}\rho)}{6\pi\eta Ns\left(\frac{3}{4\pi N}\right)^{1/3}} = \frac{f}{f_0}\,\bar{v}_h^{1/3} \ , \tag{1}$$

where M and $\bar{v}$ are the anhydrous molar mass and the partial specific volume of the biopolymer, respectively, $\bar{v}_h$ is the hydrated partial specific volume, ρ and η are the density and viscosity of water at 293 K, respectively, and N symbolizes Avogadro's number; f_0 is the frictional coefficient of a sphere with the same volume as the hydrated particle [24].

It should be stressed that in order to consider the effects of hydration correctly $\bar{v}_h$ has to be used instead of $\bar{v}$ on the right-hand side of Eq. (1). The hydrated partial specific volume, $\bar{v}_h$, is given by

$$\bar{v}_h = \bar{v}_2 + \delta_1\bar{v}_1 \ , \tag{2}$$

where $\bar{v}_2$ and $\bar{v}_1$ denote the partial specific volumes of the macromolecule and water, respectively, and δ_1 stands for macromolecule hydration (e.g., grams of water per gram of protein).

The hydrated particle volume, V, is related to $\bar{v}_h$:

$$V = \frac{M\bar{v}_h}{N} \ . \tag{3}$$

From V a radius, R_o, can be derived, which represents the radius of a sphere with a volume identical to V:

$$R_o = \left(\frac{3V}{4\pi}\right)^{1/3} \ . \tag{4}$$

The following equation relates this fictive radius R_o to the Stokes (hydrodynamic) radius, R_{D^2}.

$$R_D = R_o\,(f/f_0) \ . \tag{5}$$

The frictional ratio, f/f_0, of the hydrated particle is given by Perrin's formulae for prolate or oblate ellipsoids of revolution [1, 2]:

$$\frac{f}{f_0} = \frac{(p^2 - 1)^{1/2}}{p^{1/3}\ln\left[p + (p^2 - 1)^{1/2}\right]} \quad (p > 1, \text{ prolate}) \ , \tag{6}$$

$$\frac{f}{f_0} = \frac{(1 - p^2)^{1/2}}{p^{1/3}\tan^{-1}\left[(1 - p^2)^{1/2}/p\right]} \quad (p < 1, \text{ oblate}) \ , \tag{7}$$

where the axial ratio $p = a/b$ is defined as the ratio of the semiaxis a of revolution to the equatorial semiaxis b of the ellipsoid.

The hydrated partial specific volume, $\bar{v}_h$, and the hydrated particle volume, V, are related to the intrinsic viscosity, $[\eta]$, via the Simha factor, v, which is a function of the axial ratio, p [3, 6, 8]:

$$[\eta] = \frac{VN}{M}\quad v = \bar{v}_h\,v \ . \tag{8}$$

The combination of Eqs. (1) and (8) yields

$$\frac{A}{[\eta]^{1/3}} = \left(\frac{f}{f_0}v^{-1/3}\right) \tag{9}$$

from which the axial ratio, p, of prolate and oblate ellipsoids can be derived using tabulated values for $(f/f_0)v^{-1/3}$.

Since now the axial ratio, p, and hence the hydrated volume, V, are known, the radius of gyration, R_G, of the particle under consideration may be derived from the relation

$$R_G = \left(\frac{3V}{4\pi p}\right)^{1/3}\left(\frac{p^2 + 2}{5}\right)^{1/2} \ . \tag{10}$$

Equation (10) applies to both prolate and oblate ellipsoids of revolution.

The fact that relations between surface-to-volume-ratios, S/V, and radius of gyration, R_G, on the one hand, and axial ratios, p, on the other, have been established for the case of prolate and oblate ellipsoids [9, 14], respectively, allows derivation of equations for the estimation of S/V, and consequently, by use of V, of the surface area, S:

$$\frac{S}{V} = \frac{3}{2pR_G}\left[1 + \frac{p^2}{(p^2 - 1)^{1/2}}\sin^{-1}\frac{(p^2 - 1)^{1/2}}{p}\right]$$

$$\times \left(\frac{p^2 + 2}{5}\right)^{1/2} \quad (p > 1) \tag{11}$$

or

$$\frac{S}{V} = \frac{3}{2pR_G}\left[1 + \frac{p^2}{(1 - p^2)^{1/2}}\tanh^{-1}(1 - p^2)^{1/2}\right]$$

$$\times \left(\frac{p^2 + 2}{5}\right)^{1/2} \quad (p < 1) \tag{12}$$

$$S = \left(\frac{S}{V}\right)V \ . \tag{13}$$

---

[2] For nomenclature see Ref. [45]

In this context it has to be noted that the values for structural parameters determined by such approaches refer only to the properties of model bodies and do not necessarily reflect the properties of particles such as biopolymers. This especially holds for the values of S/V and S which may differ significantly from true surface peculiarities of real particles, due to the considerable surface roughness of most biopolymers.

2. Combining D, M and $[\eta]$:

The combination of the values for D, M and $[\eta]$ leads to similar expressions as outlined above for combining s, M and $[\eta]$:

$$B = \frac{kT}{6\pi\eta D M^{1/3}\left(\frac{3}{4\pi N}\right)^{1/3}} = \frac{f}{f_0}\,\bar{v}_h^{1/3}\;, \qquad (14)$$

where k is the Boltzmann constant, T the temperature, and the term B is related to $[\eta]$ by

$$\frac{B}{[\eta]^{1/3}} = \left(\frac{f}{f_0}v^{-1/3}\right)\;. \qquad (15)$$

Applying the term B in Eqs. (14) and (15) and the formalism outlined in (1), this again yields predicted values for $\bar{v}_h$, δ_1, V, p, R_G, S/V and S, both for prolate and oblate ellipsoidal models.

3. Combining s, D and $[\eta]$:

Concomitant use of s, D and $[\eta]$ yields an expression from which M and $\bar{v}_h$ have been eliminated

$$C = \left(\frac{4\pi k^2 T^2}{3}\right)^{1/3}\frac{1}{6\pi\eta}\left(\frac{(1-\bar{v}\rho)}{sD^2[\eta]}\right)^{1/3} = \frac{f}{f_0}v^{-1/3}\;. \qquad (16)$$

Application of Eq. (16) allows calculation of the axial ratio, p, and implies that for its calculation explicit knowledge of the particle mass or hydration is not required. On the basis of p, the Simha factor, v, can be obtained, thus allowing calculation of $\bar{v}_h$ from $[\eta]$. Combining s and D yields M, which may be used for estimating the hydrated volume, V, and R_G. Finally S/V and S may be derived.

4. Combining s, M and δ_1:

Use of s, M and δ_1 yields the frictional ratio, f/f_0:

$$\frac{f}{f_0} = \frac{M^{2/3}(1-\bar{v}\rho)}{\bar{v}_h^{1/3}N^{2/3}6\pi\eta s}\left(\frac{4\pi}{3}\right)^{1/3}\;. \qquad (17)$$

The corresponding values for p follow from f/f_0, applying Eqs. (6) and (7) for the cases of prolate and oblate ellipsoids, respectively. Use of a definite value for δ_1 (e.g., 0.35 g g^{-1}) yields $\bar{v}_h$ and V when applying Eqs. (2) and (3), and finally R_G, S/V and S, both for the prolate and oblate models.

5. Combining D, M and δ_1

The frictional ratio may also be obtained by means of D, M and δ_1,

$$\frac{f}{f_0} = \frac{kT}{6\pi\eta DM^{1/3}\bar{v}_h^{1/3}}\left(\frac{4\pi N}{3}\right)^{1/3}\;, \qquad (18)$$

and, again, leads to the above-mentioned structural parameters, provided appropriate hydration contributions are considered.

6. Combining s, D and δ_1:

Similar to (4) and (5), structural data may be predicted on the basis of f/f_0, derived from s and D, in connection with reasonable estimates for δ_1 required for the calculation of $\bar{v}_h$:

$$\frac{f}{f_0} = \left(\frac{4\pi k^2 T^2}{3}\right)^{1/3}\frac{1}{6\pi\eta}\left(\frac{(1-\bar{v}\rho)}{sD^2\bar{v}_h}\right)^{1/3}\;. \qquad (19)$$

Results and discussion

Choice of biopolymers

For the prediction of hydrodynamic parameters from scattering data, more than 50 globular biopolymers (simple and conjugated proteins, viruses, ribonucleic acids) of different molar mass and shape as well as particles undergoing conformational changes (apo- and holoproteins) were used [45, 46]; therefore, for the prediction of structural data from the results of hydrodynamics, the same dataset was used (Table 1). This allows a comparison of the accuracy to be achieved on a rational basis. To illustrate the behavior of simple proteins, 21 nonconjugated apoproteins ($M = 14$–220 kg mol^{-1}, $p = 0.2$–5) were chosen (Figs. 1–3). This corresponds exactly to the dataset used previously for illustrating several correlations between X-ray and hydrodynamic data [47]. Citrate synthase was chosen as a representative example in order to outline some features of our approaches and the accuracy achievable (Table 2). This example, together with possible errors involved in the calculations, was also discussed in detail when using the reverse approach, i.e., when predicting hydrodynamic data from X-ray results [45].

Survey of experimental data

An inspection of the experimental values found for various biopolymers (Table 1) indicates some scatter, if the parameters are considered as a function of the molar masses, M. In Fig. 1 this behavior is demonstrated more clearly for the selected simple proteins.

Table 1 Predictions of structural parameters from hydrodynamic data. Molecules are arranged in compositional classes according to increasing molar mass. The following models were used: spheres (S); prolate ellipsoids of revolution (PE); oblate ellipsoids of revolution (OE). Reference values from small-angle X-ray scattering (SAXS) are listed

Biopolymers (source) Input parameters: $M, \bar{v}, s, D, [\eta]$[a]	Parameter[c]	Reference values[b] SAXS	$V, [\eta]$	Predicted values[b] $s, M, [\eta]$ PE	OE	$D, M, [\eta]$ PE	OE	$s, D, [\eta]$ PE	OE	$s, M, \delta_1 = 0.35$[d] PE	OE	$D, M, \delta_1 = 0.35$[d] PE	OE	$s, D, \delta_1 = 0.35$[d] PE	OE
1. Globular apoproteins:															
Ribonuclease (bovine pancreas)	δ_1	0.27								1.05	0.29		0.30		0.30
$M = 13.69$	$\bar{v}_h$	0.97								23.9	1.74		1.72		1.73
$\bar{v} = 0.702$	V	22.0													
$s = 1.78$	p (PE)	1.87, 3.69	2.68							3.28		3.14		3.19	
$D = 10.68$	R_G	1.48								1.92	2.23	1.88	2.19	1.89	2.21
$[\eta] = 3.30$	S/V	2.9								2.03	53.4	2.01	52.4	2.02	52.7
	S	63.8								48.6		48.0		48.2	
α-Lactalbumin (bovine milk)	δ_1	0.36								1.05	0.42		0.31		0.34
$M = 14.18$	$\bar{v}_h$	1.07								24.8	1.60		1.74		1.69
$\bar{v} = 0.704$	V	25.1													
$s = 1.92$	p (PE)	1.43, 2.81	1.87							2.34		3.11		2.85	
$D = 10.57$	R_G	1.45								1.67	1.92	1.89	2.15	1.82	2.07
$[\eta] = 3.01$	S/V	2.4								1.85	47.6	1.98	53.4	1.93	51.4
	S	60.2								45.8		49.1		48.0	
Lysozyme (chicken egg white)	δ_1	0.32								1.05	0.37		0.45		0.42
$M = 14.31$	$\bar{v}_h$	1.02								25.0	1.65		1.57		1.60
$\bar{v} = 0.702$	V	24.2													
$s = 1.91$	p (PE)	1.42, 2.92	2.06							2.63		2.17		2.32	
$D = 11.2$	R_G	1.43								1.76	2.00	1.62	1.86	1.67	1.91
$[\eta] = 3.0$	S/V	2.5								1.89	50.0	1.81	46.6	1.84	47.7
	S	60.5								47.3		45.2		45.9	
Myoglobin (sperm whale)	δ_1	0.41		-0.29		-0.14		-0.19					0.73		0.91
$M = 17.0$	$\bar{v}_h$	1.16		0.46		0.61		0.55		1.10			1.54		1.51
$\bar{v} = 0.748$	V	32.7		13.1		17.1		16.0		31.0					
$s = 1.97$	p (PE)	2.22, 2.96	1.91	6.02		4.69		5.12				1.37		1.10	
$D = 10.9$	R_G	1.79		2.22		2.09		2.15				1.54	1.54	1.51	1.51
$[\eta] = 3.3$	S/V	2.02		2.97		2.51		2.64				1.57	48.6	1.54	47.8
	S	65.9		38.8		43.1		42.2				48.5		47.8	
α-Chymotrypsin (bovine pancreas)	δ_1	0.28								1.09	0.71				
$M = 22.0$	$\bar{v}_h$	1.02								39.7	1.68				
$\bar{v} = 0.736$	V	37.2													
$s = 2.4$	p (PE)	2.00, 2.02								1.41					
$D = 10.2$	R_G	1.80								1.69	1.45				
	S/V	1.57								1.45	57.5				
	S	58.4								57.4					

Table 1 (*Contd.*)

Biopolymers (source) / Input parameters: M, $\bar{v}$, s, D, $[\eta]^a$	Parameter^c	SAXS	V, $[\eta]$	$s, M, [\eta]$ PE	$s, M, [\eta]$ OE	$D, M, [\eta]$ PE	$D, M, [\eta]$ OE	$s, D, [\eta]$ PE	$s, D, [\eta]$ OE	$s, M, \delta_1 = 0.35^d$ PE	$s, M, \delta_1 = 0.35^d$ OE	$D, M, \delta_1 = 0.35^d$ PE	$D, M, \delta_1 = 0.35^d$ OE	$s, D, \delta_1 = 0.35^d$ PE	$s, D, \delta_1 = 0.35^d$ OE
Chymotrypsinogen A (bovine pancreas) $M = 25.67$, $\bar{v} = 0.736$, $s = 2.58$, $D = 9.5$, $[\eta] = 2.5$	δ_1	0.15													
	$\bar{v}_h$	0.89								1.09					
	V	37.8	1.86							46.3					
	p (PE)	2.00, 2.12								1.97	0.50	1.46	0.68	1.64	0.60
	R_G	1.81								1.93	1.88	1.78	1.78	1.83	1.81
	S/V	1.6								1.45	1.48	1.38	1.38	1.40	1.41
	S	60.5								66.9	68.3	63.8	64.1	64.9	65.4
Riboflavin-binding protein, pH 3.7 (chicken egg white) $M = 32.5$, $\bar{v} = 0.720$, $s = 2.70$	δ_1	0.51													
	$\bar{v}_h$	1.23								1.07					
	V	66.5								57.7					
	p (PE)	1.63, 3.58								4.88	0.19				
	R_G	2.06								3.21	2.67				
	S/V	2.03								1.70	2.07				
	S	135.0								97.9	120				
Pepsin (not specified) $M = 34.2$, $\bar{v} = 0.725$, $s = 3.20$, $D = 8.71$, $[\eta] = 3.35$	δ_1	0.24		0.33	−0.29	−0.20		−0.08							
	$\bar{v}_h$	0.97		1.06	0.43	0.52		0.65		1.08					
	V	54.9	2.75	60.0	24.6	29.7		34.8		61.1					
	p (PE)	2.00, 4.76		2.38	0.11	5.49		4.46		2.29	0.43	1.32	0.75	1.70	0.58
	R_G	2.05		2.25	2.43	2.76		2.57		2.23	2.14	1.93	1.92	2.02	2.00
	S/V	2.60		1.38	3.86	2.20		1.95		1.36	1.41	1.24	1.25	1.28	1.30
	S	142.7		82.8	95.0	65.1		68.1		83.0	86.0	76.0	76.1	78.4	79.2
β-Lactoglobulin A, dimer (bovine milk) $M = 36.73$, $\bar{v} = 0.750$, $s = 2.87$, $D = 7.82$, $[\eta] = 3.4$	δ_1	0.24													
	$\bar{v}_h$	0.99								1.10					
	V	60.3	2.71							67.1					
	p (PE)	2.13, 2.93								3.13	0.31	2.64	0.37	2.80	0.34
	R_G	2.16								2.65	2.42	2.44	2.30	2.51	2.34
	S/V	1.66								1.42	1.55	1.36	1.44	1.38	1.48
	S	100.0								95.5	104.0	91.4	96.6	92.7	99.0
Albumin (bovine serum) $M = 66.30$, $\bar{v} = 0.735$, $s = 4.5$, $D = 5.9$, $[\eta] = 4.1$	δ_1	0.55		0.49	−0.63										
	$\bar{v}_h$	1.29		1.22	0.11					1.09					
	V	142.0	2.39	134	11.7					119					
	p (PE)	2.49, 3.88		2.62	1.81					3.24	0.30	4.09	0.23	3.80	0.25
	R_G	3.06		3.07	3.39					3.26	2.97	3.70	3.20	3.55	3.12
	S/V	1.46		1.08	15.5					1.18	1.30	1.26	1.47	1.24	1.41
	S	207.3		145	181					141	155	151	175	148	168
Citrate synthase (pig heart) $M = 97.94$, $\bar{v} = 0.735$	δ_1	0.33		−0.24		−0.12		−0.16							
	$\bar{v}_h$	1.07		0.51		0.62		0.58		1.09					
	V	174.4		82.2		101		96.0		177					

v̄ = 0.740, s = 6.2, D = 5.8, [η] = 3.95

	1	2	3	4	5	6	7	8	9	10
δ_l		3.00								
p (PE)	1.78, 1.65	6.51	5.43	5.78	1.96	0.50	2.25	0.44	2.15	0.46
R_G	2.91	4.30	4.13	4.21	3.01	2.94	3.16	3.04	3.11	3.01
S/V	0.88	1.65	1.45	1.51	0.92	0.94	0.95	0.98	0.94	0.97
S	154.2	135	147	145	164	167	168	174	167	172

Cellobiose dehydrogenase (*P. chrysosporium*)
M = 107.2, v̄ = 0.741, s = 5.6

	1	5	6
δ_l	0.19	1.09	
$\bar{v}_h$	0.93		
V	165	194	
p (PE)	4.50	4.55	0.20
R_G	4.35	4.62	3.91
S/V		1.11	1.33
S		215	257

Histone core complex (calf thymus)
M = 110, v̄ = 0.73, s = 6.6, D = 5.4

	1	5	6	7	8	9	10
δ_l	0.30	1.08					
$\bar{v}_h$	1.03						
V	188.6	197					
p (OE)	0.29, 0.21	2.83	0.34	2.81	0.34	2.82	0.34
R_G	3.48	3.61	3.37	3.60	3.36	3.60	3.36
S/V	1.43	0.97	1.04	0.97	1.03	0.97	1.03
S	268.9	191	204	190	203	191	204

7S Seed globulin (*Phaseolus vulgaris*)
M = 137.5, v̄ = 0.729, s = 7.1, D = 4.53

	1	5	6	7	8	9	10
δ_l	0.58	1.08					
$\bar{v}_h$	1.31						
V	300	246					
p (OE)	0.29, 0.22	4.14	0.23	4.50	0.21	4.38	0.21
R_G	4.05	4.74	4.10	4.97	4.21	4.89	4.17
S/V	1.22	1.00	1.16	1.02	1.22	1.01	1.20
S	365	246	286	252	300	250	295

Glyceraldehyde-3-phosphate dehydrogenase (baker's yeast)
M = 142.9, v̄ = 0.737, s = 7.6, D = 5.0, [η] = 3.45

	1	4	5	6	7	8	9	10
δ_l	0.38	0.41	1.09					
$\bar{v}_h$	1.11							
V	264.2		258					
p (OE)	0.64		2.91	0.33	2.61	0.37	2.71	0.36
R_G	3.21		4.00	3.71	3.81	3.59	3.87	3.63
S/V			0.89	0.96	0.87	0.92	0.88	0.93
S			230	247	224	236	226	240

Tryptophan synthase (*E. coli*)
M = 143.15, v̄ = 0.755, s = 6.4, D = 4.83

	1	5	6	7	8	9	10
δ_l	0.38	1.11					
$\bar{v}_h$	1.14						
V	270	263					
p (OE)	0.27	4.54	0.20	3.06	0.31	3.53	0.27
R_G	4.01	5.10	4.32	4.13	3.79	4.44	3.97
S/V		1.00	1.20	0.90	0.97	0.93	1.04
S		263	315	236	256	245	274

Table 1 (*Contd.*)

Biopolymers (source) / Input parameters: M, $\bar{v}$, s, D, $[\eta]$[a]	Parameter[c]	SAXS	V, $[\eta]$	$s, M, [\eta]$ PE	$s, M, [\eta]$ OE	$D, M, [\eta]$ PE	$D, M, [\eta]$ OE	$s, D, [\eta]$ PE	$s, D, [\eta]$ OE	$s, M, \delta_1 = 0.35$[d] PE	$s, M, \delta_1 = 0.35$[d] OE	$D, M, \delta_1 = 0.35$[d] PE	$D, M, \delta_1 = 0.35$[d] OE	$s, D, \delta_1 = 0.35$[d] PE	$s, D, \delta_1 = 0.35$[d] OE
Lactate dehydrogenase (dogfish) $M = 145.17$, $\bar{v} = 0.741$, $s = 7.54$, $D = 5.05$, $[\eta] = 3.8$	δ_1	0.31		0.57	0.25	0.03		0.16							
	$\bar{v}_h$	1.05		1.31	0.99	0.77		0.90		1.09					
	V	253.3		316	238	185		209		263					
	$p(OE)$	0.41, 0.41	0.30	1.99	0.27	4.27		3.57		2.95	0.33	2.36	0.41	2.55	0.38
	R_G	3.47		3.67	3.67	4.38		4.13		4.05	3.75	3.67	3.51	3.80	3.59
	S/V	0.89		0.76	1.07	1.11		1.01		0.89	0.96	0.84	0.88	0.86	0.90
	S	226.2		241	255	205		211		234	252	222	230	226	237
β-Lactoglobulin A, octamer (bovine milk) $M = 146.94$, $\bar{v} = 0.750$, $s = 7.38$	δ_1	0.13													
	$\bar{v}_h$	0.88								1.10					
	V	215.0								268					
	$p(OE)$	0.35, 0.26								2.82	0.34				
	R_G	3.44								3.99	3.72				
	S/V	1.25								0.87	0.93				
	S	268.8								234	250				
Malate synthase (baker's yeast) $M = 187.0$, $\bar{v} = 0.745$, $s = 8.6$, $D = 4.4$	δ_1	0.34													
	$\bar{v}_h$	1.09								1.10					
	V	338								340					
	$p(OE)$	0.36, 0.40								3.27	0.29	3.18	0.30	3.21	0.30
	R_G	3.96								4.65	4.22	4.58	4.18	4.60	4.19
	S/V	0.80								0.84	0.92	0.83	0.91	0.83	0.91
	S	270								285	313	283	309	283	310
Hemoglobin, dodecameric subunit (*Lumbricus terrestris*) $M = 190$, $\bar{v} = 0.733$, $s = 9.1$	δ_1	0.08													
	$\bar{v}_h$	0.81								1.08					
	V	255.0								342					
	$p(OE)$	0.32								3.33	0.29				
	R_G	3.74								4.70	4.25				
	S/V									0.84	0.93				
	S									287	317				
Pyruvate kinase (brewer's yeast) $M = 219.5$, $\bar{v} = 0.754$	δ_1	0.36													
	$\bar{v}_h$	1.11								1.10					
	V	406.0								402					
	$p(OE)$	0.32, 0.30								4.23	0.22			3.41	0.28

Parameter	Col 1	Col 2	Col 3	Col 4	Col 5	Col 6	Col 7
(continued) $s = 8.7$, $D = 4.2$							
R_G	4.35	5.65	4.86	4.73	4.35	5.02	4.52
S/V	0.88	0.85	1.00	0.78	0.84	0.80	0.89
S	356.9	343	401	312	338	322	358
Catalase (bovine liver) $M = 230.34$, $\bar{v} = 0.730$, $s = 11.3$, $D = 4.1$, $[\eta] = 3.9$							
δ_1	0.37	−0.16	0.32				
$\bar{v}_h$	1.10	0.57	1.05				
V	420.0	219	430				
p (OE)	0.47, 0.43 0.31	5.80	3.04				
R_G	3.98	5.55	4.85				
S/V	0.75	1.15	0.76				
S	315.8	251	327				
α-Globulin (sesame) $M = 270$, $\bar{v} = 0.730$, $s = 12.8$, $D = 3.95$							
δ_1	0.11						
$\bar{v}_h$	0.84	1.08					
V	375	484					
p (OE)	0.36, 0.22	1.83	0.54	3.02	0.32	2.62	0.37
R_G	4.1	4.12	4.05	5.02	4.63	4.71	4.44
S/V	1.18	0.65	0.66	0.73	0.79	0.70	0.74
S	442	316	320	353	382	341	360
11S Seed globulin (rape seed) $M = 300$, $\bar{v} = 0.729$, $s = 12.7$, $D = 3.78$							
δ_1	0.17						
$\bar{v}_h$	0.90	1.08					
V	450	538					
p (OE)	0.46, 0.31	3.12	0.31	3.16	0.30	3.15	0.30
R_G	4.1	5.29	4.84	5.32	4.86	5.31	4.85
S/V	0.91	0.71	0.77	0.71	0.78	0.71	0.78
S	410	382	416	383	418	383	418
Legumin (Vicia faba) $M = 312.48$, $\bar{v} = 0.729$, $s = 13.0$, $D = 3.38$							
δ_1	0.59						
$\bar{v}_h$	1.32	1.08					
V	685	560					
p (OE)	0.60, 0.50	3.18	0.30	4.85	0.19	4.27	0.22
R_G	4.45	5.42	4.94	6.83	5.69	6.34	5.44
S/V	0.62	0.70	0.77	0.79	0.97	0.77	0.90
S	426	394	431	445	542	428	502
Leucine aminopeptidase (bovine) $M = 317.22$, $\bar{v} = 0.751$, $s = 12.6$, $D = 3.75$							
δ_1	0.17						
$\bar{v}_h$	0.92	1.10					
V	485	580					
p (OE)	0.37, 0.21	2.41	0.41	2.89	0.33	2.73	0.35
R_G	4.45	4.82	4.60	5.22	4.85	5.09	4.77
S/V	1.14	0.65	0.68	0.68	0.73	0.67	0.71
S	553	377	393	394	422	388	412
11S Seed globulin (sunflower) $M = 321.63$, $\bar{v} = 0.730$, $s = 11.8$, $D = 3.78$							
δ_1	0.04						
$\bar{v}_h$	0.77	1.08					
V	410	577					
p (OE)	0.47, 0.35	5.14	0.18	2.79	0.35	3.52	0.27
R_G	3.95	7.14	5.87	5.13	4.79	5.76	5.15
S/V	0.87	0.80	0.99	0.67	0.72	0.72	0.80
S	356	462	573	389	415	413	462

Table 1 (*Contd.*)

Biopolymers (source) / Input parameters: M, $\bar{v}$, s, D, $[\eta]$[a]	Reference values[b] Parameter[c]	SAXS	V, $[\eta]$	s, M, $[\eta]$ PE	s, M, $[\eta]$ OE	D, M, $[\eta]$ PE	D, M, $[\eta]$ OE	s, D, $[\eta]$ PE	s, D, $[\eta]$ OE	s, M, $\delta_1 = 0.35$[d] PE	s, M, $\delta_1 = 0.35$[d] OE	D, M, $\delta_1 = 0.35$[d] PE	D, M, $\delta_1 = 0.35$[d] OE	s, D, $\delta_1 = 0.35$[d] PE	s, D, $\delta_1 = 0.35$[d] OE
Glutamate dehydrogenase (bovine liver) $M = 333.37$, $\bar{v} = 0.749$, $s = 11.4$, $D = 3.5$, $[\eta] = 3.2$															
δ_1	0.46										0.19		0.25		0.23
$\bar{v}_h$	1.21									1.10					
V	668	1.55								608					
p (PE)	1.98, 2.30									4.77		3.75		4.07	
R_G	4.70									6.94	5.81	6.06	5.35	6.35	5.50
S/V	0.65									0.77	0.93	0.72	0.81	0.73	0.85
S	433									467	567	436	494	446	518
Apoferritin (horse spleen) $M = 466.9$, $\bar{v} = 0.723$, $s = 17.60$, $D = 3.61$, $[\eta] = 5.16$															
δ_1	0.69		−0.25	0.48	−0.55	0.17	−0.48	0.24		0.32		0.64		0.48	
$\bar{v}_h$	1.42								1.07						
V	1098	2.95	370		132		171		832						
p (S)	1.00		8.43		17.2		13.7		3.00		1.55		2.07		
R_G	5.33		8.36		9.43		8.84		6.00	5.53	4.73	4.70	5.13	4.99	
S/V			1.08		1.93		1.64		0.61	0.66	0.53	0.53	0.56	0.57	
S			401		255		280		506	546	441	444	464	475	
2. Globular holoproteins:															
Riboflavin-binding protein, pH 7.0 (chicken egg white) $M = 32.5$, $\bar{v} = 0.720$, $s = 2.92$															
δ_1	0.31										0.27				
$\bar{v}_h$	1.03								1.07						
V	55.6								57.7						
p (PE)	1.76, 3.62								3.50						
R_G	1.98								2.67	2.39					
S/V	2.13								1.54	1.72					
S	118.4								89.0	99.4					
Citrate synthase + oxaloacetate (pig heart) $M = 98.20$, $\bar{v} = 0.739$, $s = 6.4$															
δ_1	0.27										0.66				
$\bar{v}_h$	1.01								1.09						
V	164.0								178						
p (PE)	1.67, 1.28								1.51						
R_G	2.80								2.81	2.79					
S/V	0.85								0.88	0.89					
S	139.6								157	158					
Glyceraldehyde-3-phosphate dehydrogenase + NAD$^+$ (baker's yeast) $M = 145.5$, $\bar{v} = 0.735$															
δ_1	0.30										0.40		0.44		0.43
$\bar{v}_h$	1.03								1.09						
V	250.0								262						
p (OE)	0.61								2.43		2.22		2.29		

Protein / descriptor	Parameter							
s = 8.0, D = 5.1	R_G	3.17	3.71	3.54	3.58	3.46	3.63	3.48
	S/V		0.85	0.89	0.83	0.86	0.84	0.87
	S		222	232	218	225	220	227
Malate synthase + glyoxylate (baker's yeast) M = 187.2, $\bar{v}$ = 0.745, s = 8.7	δ_I	0.35						
	$\bar{v}_h$	1.10	1.10					
	V	341	340					
	p (OE)	0.38		0.31				
	R_G	3.91	3.10	4.15				
	S/V		4.53					
	S		0.83	0.90				
			281	306				
Pyruvate kinase + fructose diphosphate (brewer's yeast) M = 220.8, $\bar{v}$ = 0.754, s = 8.81	δ_I	0.35						
	$\bar{v}_h$	1.11	1.10					
	V	406.0	405					
	p (OE)	0.35, 0.32	0.23					
	R_G	4.25	4.81					
	S/V	0.86	0.98					
	S	347.1	340	395				
Pyruvate decarboxylase + thiamine pyrophosphate + Mg^{2+} (brewer's yeast) M = 248.4, $\bar{v}$ = 0.751, s = 9.95, D = 5.55	δ_I	0.26						
	$\bar{v}_h$	1.02	1.10					
	V	419	454					
	p (OE)	0.33, 0.27	3.57	5.35				
	R_G	4.38	4.78					
	S/V	0.93	0.87					
	S	391	354	397				
3. Glycoproteins: Ceruloplasmin (human) M = 122.2, $\bar{v}$ = 0.714, s = 7.2, D = 5.30	δ_I	0.58	1.06					
	$\bar{\delta}_h / \bar{v}_h$	1.29	1.06					
	V	262	216					
	p (PE)	2.00, 2.01	3.56	4.18	2.63	2.93	0.33	0.37
	R_G	3.45	3.73	3.60	3.79	3.50	3.39	
	S/V	0.82	1.00	0.92	0.95	1.02	0.97	1.12
	S	213.9	215	241	199	205	220	210
Immunoglobulin IgG1 (human) M = 150, $\bar{v}$ = 0.733, s = 6.81, $[\eta]$ = 6.20	δ_I	0.59		1.10				
	$\bar{v}_h$	1.32	1.84	1.08				
	V	329	457	270				
	p (PE)	4.03	2.64	5.72	5.91			
	R_G	5.03	4.63	4.73				
	S/V	5.84	0.72	1.07	1.37			
	S		329	287	369			

Table 1 (*Contd.*)

Biopolymers (source) / Input parameters: $M, \bar{v}, s, D, [\eta]^a$	Parameter^c	Reference SAXS^b	Reference $V, [\eta]$	$s, M, [\eta]$ PE	$s, M, [\eta]$ OE	$D, M, [\eta]$ PE	$D, M, [\eta]$ OE	$s, D, [\eta]$ PE	$s, D, [\eta]$ OE	$s, M, \delta_1 = 0.35^d$ PE	$s, M, \delta_1 = 0.35^d$ OE	$D, M, \delta_1 = 0.35^d$ PE	$D, M, \delta_1 = 0.35^d$ OE	$s, D, \delta_1 = 0.35^d$ PE	$s, D, \delta_1 = 0.35^d$ OE
Fibronectin (human)	δ_1	0.64								1.07					
$M = 510$	$\bar{v}_h$	1.36													
$\bar{v} = 0.72$	V	1149								906					
$s = 13.25$	p (OE)	0.10	0.11							10.8	0.07	10.6	0.08	10.7	0.08
$D = 2.27$	R_G	8.75								13.2	9.05	13.0	8.96	13.1	8.99
$[\eta] = 10.2$	S/V									0.87	1.44	0.87	1.41	0.87	1.42
	S									790	1310	784	1280	786	1290
Immunoglobulin IgM$_{ser}$ (bovine)	δ_1	0.42								1.07					
$M = 950$	$\bar{v}_h$	1.14													
$\bar{v} = 0.724$	V	1800								1690					
$s = 17.7$	p (OE)	0.07	0.07							14.7	0.05	12.6	0.06	13.3	0.06
$D = 1.73$	R_G	11.5								19.9	12.6	18.0	11.9	18.7	12.1
	S/V	1.28								0.78	1.48	0.74	1.32	0.76	1.37
	S	2300								1320	2510	1260	2230	1280	2320

4. Nucleoproteins, spherical viruses:

Biopolymers (source) / Input parameters: $M, \bar{v}, s, D, [\eta]^a$	Parameter^c	Reference SAXS^b	Reference $V, [\eta]$	$s, M, [\eta]$ PE	$s, M, [\eta]$ OE	$D, M, [\eta]$ PE	$D, M, [\eta]$ OE	$s, D, [\eta]$ PE	$s, D, [\eta]$ OE	$s, M, \delta_1 = 0.35^d$ PE	$s, M, \delta_1 = 0.35^d$ OE	$D, M, \delta_1 = 0.35^d$ PE	$D, M, \delta_1 = 0.35^d$ OE	$s, D, \delta_1 = 0.35^d$ PE	$s, D, \delta_1 = 0.35^d$ OE
Bacteriophage fr (E. coli)	δ_1	0.91		0.41						1.02					
$M = 3620$	$\bar{v}_h$	1.58		1.09											
$\bar{v} = 0.673$	V	9500		6530						6150					
$s = 79$	p (S)	1.00	0.54	3.40						3.73	0.25	6.53	0.14	5.51	0.16
$D = 1.4$	R_G	10.58		12.7						13.1	11.5	18.2	14.1	16.4	13.2
$[\eta] = 4.4$	S/V			0.32						0.33	0.38	0.39	0.53	0.37	0.47
	S			2060						2030	2300	2410	3240	2280	2900
Southern bean mosaic virus	δ_1	0.46		3.12		−0.12		−0.07		1.04					
$M = 6690$	$\bar{v}_h$	1.15		0.73		0.58		0.62							
$\bar{v} = 0.694$	V	12770		8060		6430		6780		11600					
$s = 115$	p (S)	1.00	0.28	5.21		6.36		5.97		3.09	0.31	2.79	0.35	2.89	0.33
$D = 1.39$	R_G	12.3		17.3		18.1		17.8		14.7	13.4	14.0	13.0	14.2	13.2
$[\eta] = 4.4$	S/V			0.33		0.38		0.37		0.26	0.28	0.25	0.26	0.25	0.27
	S			2690		2460		2500		2950	3210	2880	3070	2900	3110
Tomato bushy stunt virus	δ_1	0.99								1.05					
$M = 8700$	$\bar{v}_h$	1.69													
$\bar{v} = 0.700$	V	24429								15200					
$s = 132$	p (S)	1.00								3.34	0.28	4.46	0.21	4.07	0.23
$D = 1.15$	R_G	13.2								16.7	15.1	19.5	16.6	18.5	16.1
$[\eta] = 3.44$	S/V									0.24	0.26	0.26	0.31	0.25	0.29
	S									3600	3980	3910	4650	3810	4420

5. Ribonucleic acids:

tRNA^phe (yeast) — $M = 25.0$, $\bar{v} = 0.54$, $D = 7.8$

δ_l	0.34
$\bar{v}_h$	0.88 · 0.89
V	36.6 · 36.9
p (OE)	0.18, 0.17 · 0.14
R_G	2.31 · 6.25 · 2.52
S/V	2.66 · 3.22 · 2.12 · 2.82
S	97.5 · 78.5 · 104

5S rRNA (E. coli) — $M = 44.0$, $\bar{v} = 0.54$, $s = 5.3$, $D = 6.2$

δ_l	0.23
$\bar{v}_h$	0.77 · 0.89
V	56.0 · 65.0
p (OE)	0.10, 0.10 · 0.13 · 0.12 · 0.13
R_G	3.27 · 6.56 · 3.10 · 7.18 · 3.20 · 6.97 · 3.17
S/V	2.88 · 4.00 · 1.79 · 2.41 · 4.23 · 1.84 · 2.56 · 4.15 · 1.82 · 2.51
S	161.0 · 116 · 157 · 119 · 167 · 118 · 163

5S rRNA (rat liver) — $M = 45.0$, $\bar{v} = 0.54$, $s = 4.9$, $D = 5.9$

δ_l	0.24
$\bar{v}_h$	0.78 · 0.89
V	58.0 · 66.5
p (OE)	0.10, 0.10 · 0.09 · 0.10 · 0.10
R_G	3.31 · 8.85 · 3.50 · 8.21 · 3.40 · 8.42 · 3.43
S/V	2.92 · 4.87 · 1.95 · 2.96 · 4.64 · 1.90 · 2.80 · 4.72 · 1.92 · 2.85
S	169.5 · 130 · 197 · 127 · 186 · 128 · 190

MS2 RNA (bacteriophage MS2) — $M = 1088$, $\bar{v} = 0.457$, $s = 26.6$, $[\eta] = 44.0$

δ_l	11.28 · 15.3, 13.3 · 0.81
$\bar{v}_h$	11.73 · 15.8, 13.8
V	21200 · 28500, 24900 · 1460
p (OE)	0.28 · 0.29 · 1.80, 0.39 · 0.01
R_G	18.1 · 16.0, 16.3 · 43.9 · 19.3
S/V	39.2 · 0.17, 0.20 · 1.18 · 4.01
S	1720 · 4770, 4910 · 5840

ᵃ Experimental input parameters are given in the following units. M: kg mol⁻¹; $\bar{v}$: cm³ g⁻¹; s: 10⁻¹³ s; D: 10⁻⁷ cm² s⁻¹; $[\eta]$: cm³ g⁻¹. Experimental values were taken from previous compilations [45, 46]; the OE approach for catalase was adopted from Ref. [33]; for $[\eta]$ of apoferritin the value given in Refs. [15, 28] was used

ᵇ Reference values and predicted structural parameters are given in the following units. δ_l: g g⁻¹; $\bar{v}_h$: cm³ g⁻¹; V: nm³; R_G: nm; S/V: nm⁻¹; S: nm²

ᶜ In the case of the p values derived from SAXS, the first value refers to the result from SAXS, the second one to the approach which combines R_G and V. For hydrodynamic modeling, the particle was assumed to be a solid sphere ($p = 1.00$) in all cases where the particle shape was known to be approximately spherical, whether or not it contained hollow spaces (e.g., apoferritin, spherical viruses). In a few cases, special calculation procedures were applied; for details see Ref. [45]. The p value derived from viscosity data combines $[\eta]$ with the hydrated volume V obtained from SAXS

ᵈ The approaches which combine s, D, M with δ_l estimates yield the same values for $\bar{v}_h$ and V. These values are located on the left of the "s, M, $\delta_l = 0.35$" column

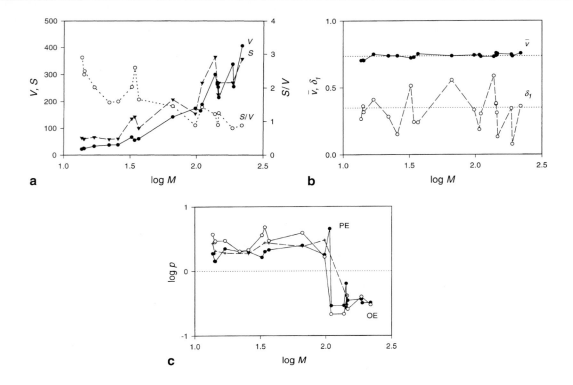

Fig. 1a–c Observed structural parameters of selected globular proteins as a function of log M. **a** Experimental values for hydrated volume, V (nm³), surface, S (nm²), and surface-to-volume ratio, S/V (nm⁻¹), obtained from small-angle X-ray scattering (*SAXS*). **b** Experimental values for partial specific volume, $\bar{v}$ (cm³ g⁻¹), usually obtained from densimetry, and particle hydration, δ_1 (g g⁻¹), obtained from the hydrated SAXS volume. Mean values of observed $\bar{v}$ and δ_1 values: (0.734 ± 0.016) cm³ g⁻¹ and (0.32 ± 0.13) g g⁻¹, respectively. The *dotted lines* correspond to 0.735 cm³ g⁻¹ and 0.35 g g⁻¹, taken as reasonable default values for $\bar{v}$ and δ_1, if necessary. **c** Experimental values for the logarithm of axial ratio, p, obtained from SAXS [(●) and (○): (R_G,V) and ($R_G, S/V$) approaches, respectively] or from a combination of SAXS with viscosity data [(+): (V, [η]) approach via the Simha factor]. The *dotted line* corresponds to $p = 1$, indicating the switch from prolate ellipsoids of revolution (*PE*) to oblate ones (*OE*)

Volume, V, and surface area, S, increase as a function of log M, whereas the surface-to-volume ratio, S/V, decreases slightly (Fig. 1a). The resulting curves are not smooth due to differing particle shapes, arrangement of subunits and peculiarities of the particle surface. While the partial specific volume, $\bar{v}$, remains nearly constant with increasing mass, the experimentally observed values for particle hydration, δ_1, show pronounced deviations from a constant value (Fig. 1b). This is obviously due to the specific characteristics of proteins (e.g., localization of different amounts of hydrophilic amino acids on the protein surface) and/or experimental deficiencies in determining the hydrated volume required for estimates of hydration values [49]. As may be expected, the axial ratios, p, differ for the proteins under consideration (Fig. 1c). An interesting fact is the observation that small proteins (M less than about 100 kg mol⁻¹) are of prolate shape, whereas large proteins have oblate shapes. This phenomenon is presumably caused by the fact that small proteins are composed of one or two subunits, thus yielding elongated particles, while proteins composed of three or more subunits lead to oblate shapes, and assemblies of many protein subunits may form roughly spherical shapes. This finding may be used for hydrodynamic modeling, to differentiate between traditional ambiguities of particles of prolate or oblate shape. The behavior of experimental R_G, s, D and [η] values has already been presented (Fig. 1a in Ref. [47]).

Documentation of data

Table 1 lists the available input parameters (M, $\bar{v}$, s, D, [η]) of selected examples, together with reference values obtained from SAXS (δ_1, $\bar{v}_h$, V, p, R_G, S/V, S), and the structural parameters as revealed from the parameter predictions mentioned. Calculations were performed on the basis of both prolate and oblate ellipsoidal models. Results may be compared with the reference data column. Obviously there are some gaps in the middle portion of the table. This is due to two facts: in many cases viscosity data are not available, and the approaches applying [η] frequently fail because of mathematical inconsistencies (especially in the case of

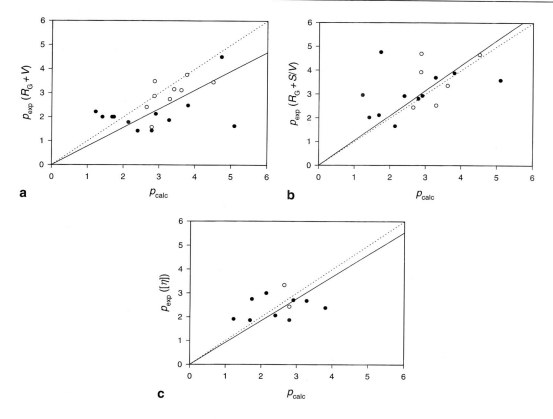

Fig. 2a–c Observed and predicted axial ratios, p, of selected globular proteins. For better comparison with ellipsoids of prolate shape ($p > 1$), axial ratios of oblate ellipsoids ($p < 1$) have been transformed to $1/p$ (thus giving values greater than 1). Experimental values from SAXS and viscometric studies are plotted against the values predicted on the basis of hydrodynamic studies. Experimental data: **a** SAXS: (R_G, V) approach; **b** SAXS: (R_G, S/V) approach; **c** SAXS and viscosity: (V, $[\eta]$) approach. The predicted axial ratios represent the average values of the various approaches combining (s, M, δ_1), (D, M, δ_1), and (s, D, δ_1), for both prolate (●) and oblate (○) ellipsoids of revolution. The *solid line* is the regression line, starting from the origin of the coordinate system. The *dotted line* represents the median

the oblate ellipsoid approaches). Erroneous/unrealistic viscosities may lead to negative hydrational contributions.

An inspection of Table 1 reveals that in the majority of cases the combinations applying s, D, M and δ_1 are successful, whereas the results predicted on the basis of viscosity data, if feasible, are rather ambiguous. The best results are obtained for predictions of R_G and V, while the results for S/V and S are influenced by the deficiencies of the smooth models used. True molecules exhibit a more or less pronounced surface roughness. The values for p show the greatest deviations, even if we differentiate between prolate and oblate shape models. In this context, however, we have to remember that the axial ratios given by the models do not necessarily reflect physical reality. The accuracy of the results obtained may be improved by averaging the six

approaches which combine s, D, M and δ_1 (prolate ellipsoid and oblate ellipsoid approaches), but neglect the viscosity data.

For the case of selected simple proteins, the graphs shown in Figs. 2 and 3 allow a direct comparison of observed and predicted values. There is only restricted agreement between experimental and calculated axial ratios (Fig. 2), irrespective of the calculational approach used for the experimental determination of p. In contrast, the predictions for R_G and V (Fig. 3a, b) show excellent coincidence, both for prolate and oblate modeling approaches. The agreement is somewhat less well defined for S/V and S, because of obvious deficiencies in modeling the true protein surface. The predicted surface is smaller than the true one.

The accordance between observed and predicted values is optimum for simple (nonconjugated) globular proteins (Table 1, Figs. 2, 3). Comparing the results for conjugated proteins or ribonucleic acids, however, reveals only rough correspondence. Serious discrepancies between experimental and calculated values occur in those cases where special assumptions have already been made for the prediction of hydrodynamic parameters from scattering data [45], i.e., for molecules containing pronounced inhomogeneities (e.g., fibronectin, spherical viruses, MS2 RNA). Modeling the latter molecules obviously fails, if δ_1 values of 0.35 g g^{-1}, typical of simple proteins, are used for the modeling approaches.

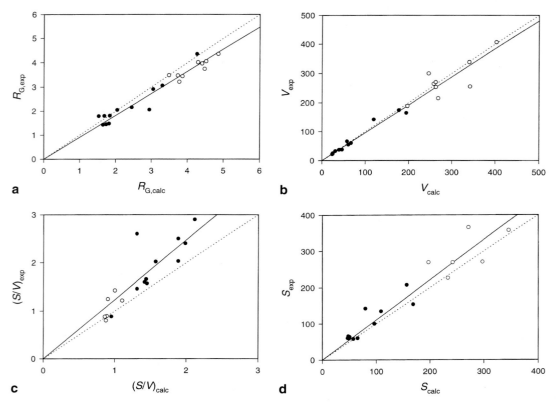

Fig. 3a–d Observed and predicted structural parameters of selected globular proteins of prolate (●) or oblate (○) shape. Experimental and calculated values for **a** radius of gyration, R_G (nm), **b** hydrated volume, V (nm^3), **c** surface-to-volume ratio, S/V (nm^{-1}), and **d** surface area, S (nm^2). The *solid line* is the regression line, starting from the origin of the coordinate system. The *dotted line* represents the median

On the other hand, combined use of ultracentrifugal and viscosity data may lead to rough parameter estimates (see, e.g., the predicted values for MS2 RNA).

Citrate synthase: a representative example

The dimeric apoenzyme (for which a hydration of $\delta_1 = 0.33$ g g^{-1} was found by a SAXS analysis) can be modeled correctly, applying s, M and/or D and assuming a default hydration of $\delta_1 = 0.35$ g g^{-1} (Table 1). The predicted (averaged) values for $\bar{v}_h$, V, p (prolate ellipsoid), R_G, S/V and S (1.09 cm^3 g^{-1}, 177 nm^3, 2.12, 3.05 nm, 0.95 nm^{-1} and 169 nm^2, respectively), are in reasonable accordance with the observed values (1.07 cm^3 g^{-1}, 174.4 nm^3, 1.72, 2.91 nm, 0.88 nm^{-1}, 154.2 nm^2). Similar statements hold for calculated (1.09 cm^3 g^{-1}, 178 nm^3, 1.51, 2.80 nm, 0.89 nm^{-1}, 158 nm^2) and experimental (1.01 cm^3 g^{-1}, 164.0 nm^3, 1.48, 2.80 nm, 0.85 nm^{-1}, 139.6 nm^2) values of the holoenzyme ($\delta_1 = 0.27$ g g^{-1} according to SAXS re-

sults). As shown for the apoenzyme, the combinations involving [η] are by no means satisfactory.

Prediction of small structural changes

A comparison between apo- and holoforms of the enzyme citrate synthase proves that the occurrence and the nature of subtle structural changes may also be anticipated correctly. The radius of gyration, R_G, the most important structural parameter in connection with the registration of structural changes [53], decreases with both observed (2.91 nm → 2.80 nm) and predicted (3.05 nm → 2.80 nm) values (Table 1). The prediction of such effects is not trivial because of changes in M, $\bar{v}$, δ_1, V and p upon ligand binding.

The correct prediction of the occurrence of structural changes (decrease in R_G) can also be found in the case of all other proteins mentioned (apo- and holoforms of riboflavin-binding protein, glyceraldehyde-3-phosphate dehydrogenase, malate synthase and pyruvate kinase).

Accuracy of parameter predictions

The prediction of structural parameters, and particularly the anticipation of the occurrence of conformational changes, rests on the reliability of the experimental data

Table 2 Error propagation from hydrodynamic data to predicted structural parameters, for citrate synthase (apoenzyme) and typical error bars for the input parameters[a]

Input parameters Deviation (%)			Structural parameters				$s, M, \delta_1 = 0.35$[c]		$D, M, \delta_1 = 0.35$[c]		$s, D, \delta_1 = 0.35$[c]	
s	D	$[\eta]$	Parameter	$s, M, [\eta]$ PE[b]	$D, M, [\eta]$ PE[b]	$s, D, [\eta]$ PE[b]	PE	OE	PE	OE	PE	OE
+1/−1	0	0	δ_1	−23/27	0	−14/15						
			$\bar{v}_h$	−11/12	0	−4/4						
			V	−11/12	0	−3/3						
			p	10/−10	0	3/−3	−8/8	10/−8			−2/2	3/−3
			R_G	2/−2	0	1/−1	−3/3	−2/2			−0.9/0.9	−0.6/0.6
			S/V	7/−7	0	2/−2	−2/2	−2/2			−0.5/0.5	−0.8/0.8
			S	−5/5	0	0/0	−2/2	−2/2			−0.5/0.5	−0.8/0.8
0	+2/−2	0	δ_1	0	−110/149	−52/63						
			$\bar{v}_h$	0	−21/28	−14/17						
			V	0	−21/28	−16/20						
			p	0	22/−21	14/−14			−14/14	17/−13	−10/10	12/−9
			R_G	0	5/−6	2/−3			−5/6	−4/4	−4/4	−2/2
			S/V	0	15/−14	10/−10			−3/3	−4/5	−2/2	−3/3
			S	0	−9/10	−9/10			−3/3	−4/5	−2/2	−3/3
0	0	+5/−5	δ_1	−28/35	−71/90	−48/60						
			$\bar{v}_h$	−13/16	−13/17	−13/17						
			V	−13/16	−13/17	−13/17						
			p	16/−16	18/−18	17/−17						
			R_G	5/−6	6/−6	5/−6						
			S/V	10/−10	10/−11	10/−10						
			S	−4/5	−4/5	−4/5						
+1/−1	+2/−2	+5/−5	δ_1	−48/67	−164/276	−102/160						
			$\bar{v}_h$	−22/31	−31/52	−28/44						
			V	−22/31	−31/52	−29/45						
			p	27/−25	42/−39	37/−34	−8/8	10/−8	−14/14	17/−13	−12/12	15/−11
			R_G	7/−8	10/−13	8/−11	−3/3	−2/2	−5/6	−4/4	−4/5	−3/3
			S/V	17/−16	26/−25	24/−22	−2/2	−2/2	−3/3	−4/5	−3/3	−4/4
			S	−9/10	−13/14	−13/14	−2/2	−2/2	−3/3	−4/5	−3/3	−4/4

[a] Observed and predicted values for citrate synthase are given in Table 1. Representative errors including the maximum deviations are included; all other permutations of input parameters (not shown) lead to deviations smaller than the cited maximum errors

[b] The table includes only the values for the PE approach; with the given input parameters, the OE approach was not successful

[c] A given value for the hydration, δ_1, yields definite values for $\bar{v}_h$ and V, i.e., no deviations of these parameters are observed

4

used. Error propagations, based on reasonable assumptions for the most important input parameters ($s \pm 1\%$, $D \pm 2\%$, $[\eta] \pm 5\%$), have been performed for various biopolymers. A few results for citrate synthase are given in Table 2. They disclose the possibility to predict structural parameters and structural changes correctly, provided the predictions are based on accurate values of s and/or D. They also show why use of erroneous input parameters can lead to unreliable results (very large errors and even negative values for some parameters; for example, the -110% deviation of hydration δ_1, resulting from $\Delta D = +2\%$). Under similar environmental conditions, of course, some of the experimental errors cancel out to some extent.

Conclusions

The values listed in Table 1 may be used for various purposes: a data source for many problems dealing with scattering and hydrodynamic parameters, establishment of correlations between these parameters, compatibility tests, assessment of the reliability of structural predictions on the basis of hydrodynamic data, etc.

Summarizing our results we may state that the radius of gyration and the hydrated volume of the majority of the simple globular proteins investigated were predicted correctly, including the anticipation of small structural changes. Less accurate predictions were obtained in the case of the surface-to-volume ratio and the surface area, presumably caused by differences in the surface roughness of models, on the one hand, and true particles, on the other. Only poor agreement, however, was achieved for the axial ratio, mainly due to insufficiencies of the modeling approaches by ellipsoids.

Predictions for glyco- and nucleoproteins or ribonucleic acids were less successful. Particles with extraordinary features (mass inhomogeneities, unusual water binding behavior) cannot be modeled correctly by simple whole-body approaches, if a hydration of 0.35 g g^{-1}, characteristic of simple proteins, is applied. This implies that use of a correct value for the hydrodynamically effective hydration is of utmost importance.

Differentiation between prolate and oblate particles seems to be of secondary importance. Small nonconjugated proteins as well as glycoproteins should preferably be modeled by prolate models, whereas for large proteins and ribonucleic acids the oblate shape is superior.

Use of viscosity data for modeling purposes cannot be recommended because of the rather low accuracy of viscosity data and the fact that for particles of roughly spherical shape the values for intrinsic viscosity do not differ very much. By contrast, use of thermodynamically more rigid data such as sedimentation and diffusion coefficients yields reliable results for parameter predictions.

References

1. Perrin F (1934) J Phys Radium Série VII 5:497–511
2. Perrin F (1936) J Phys Radium Série VII 7:1–11
3. Simha R (1940) J Phys Chem 44:25–34
4. Mehl JW, Oncley JL, Simha R (1940) Science 92:132–133
5. Oncley JL (1941) Ann NY Acad Sci 41:121–150
6. Saito N (1951) J Phys Soc Jpn 6:297–301
7. Scheraga HA, Mandelkern L (1953) J Am Chem Soc 75:179–184
8. Scheraga HA (1955) J Chem Phys 23:1526–1532
9. Luzzati V, Witz J, Nicolaieff A (1961) J Mol Biol 3:367–378
10. Luzzati V, Witz J, Nicolaieff A (1961) J Mol Biol 3:379–392
11. Slegers H, Clauwaert J, Fiers W (1973) Biopolymers 12:2033–2044
12. Rowe AJ (1977) Biopolymers 16:2595–2611
13. Squire PG, Himmel ME (1979) Arch Biochem Biophys 196:165–177
14. Kumosinski TF, Pessen H (1982) Arch Biochem Biophys 219:89–100
15. Harding SE, Rowe AJ (1982) Int J Biol Macromol 4:160–164
16. Harding SE, Rowe AJ (1982) Int J Biol Macromol 4:357–361
17. Harding SE, Rowe AJ (1983) Biopolymers 22:1813–1829 and Erratum (1984) 23:843
18. Müller JJ, Damaschun H, Damaschun G, Gast K, Plietz P, Zirwer D (1984) Stud Biophys 102:171–175
19. Harding SE (1987) Biophys J 51:673–680
20. Harding SE, Cölfen H (1995) Anal Biochem 228:131–142
21. Harding SE, Horton JC, Cölfen H (1997) Eur Biophys J 25:347–359
22. Behlke J (1997) Eur Biophys J 25:319–323
23. Harding SE, Horton JC, Winzor DJ (1998) Biochem Soc Trans 26:737–741
24. Tanford C (1961) Physical chemistry of macromolecules. Wiley, New York
25. Cantor CR, Schimmel PR (1980) Biophysical chemistry, parts I–III. Freeman, San Francisco
26. García de la Torre J (1989) In: Harding SE, Rowe AJ (eds) Dynamic properties of biomolecular assemblies. Royal Society of Chemistry, Cambridge, UK, pp 3–31
27. García de la Torre J, Navarro S, Lopez Martinez MC, Diaz FG, Lopez Cascales JJ (1994) Biophys J 67:530–531
28. Byron O (1997) Biophys J 72:408–415
29. García de la Torre J, Carrasco B, Harding SE (1997) Eur Biophys J 25:361–372
30. Spotorno B, Piccinini L, Tassara G, Ruggiero C, Nardini M, Molina F, Rocco M (1997) Eur Biophys J 25:373–384
31. Zipper P, Durchschlag H (1997) Prog Colloid Polym Sci 107:58–71
32. García de la Torre J, Harding SE, Carrasco B (1998) Biochem Soc Trans 26:716–721
33. Zipper P, Durchschlag H (1998) Biochem Soc Trans 26:726–731
34. García de la Torre J, Carrasco B (1998) Eur Biophys J 27:549–557
35. García de la Torre J, Harding SE, Carrasco B (1999) Eur Biophys J 28:119–132

36. Carrasco P, García de la Torre J, Zipper P (1999) Eur Biophys J (in press)
37. Zipper P, Durchschlag H (1999) In: SAS99, Abstracts of the 11th International Conference on Small-Angle Scattering. Brookhaven National Laboratory, Upton, New York, p 261
38. Harding SE (1989): In: Harding SE, Rowe AJ (eds) Dynamic properties of biomolecular assemblies. Royal Society of Chemistry, Cambridge, UK, pp 32–56
39. Harding SE (1995) Biophys Chem 55:69–93
40. Harding SE (1997) Prog Biophys Mol Biol 68:207–262
41. Kumosinski TF, Pessen H (1985) Methods Enzymol 117:154–182
42. Pessen H, Kumosinski, TF (1993) In: Baianu IC, Pessen H, Kumosinski TF (eds) Physical chemistry of food processes, vol 2: advanced techniques, structures, and applications. Van Nostrand Reinhold, New York, pp 274–306
43. Durchschlag H, Zipper P, Purr G, Jaenicke R (1996) Colloid Polym Sci 274:117–137
44. Durchschlag H, Zipper P (1996) J Mol Struct 383:223–229
45. Durchschlag H, Zipper P (1997) J Appl Crystallogr 30:1112–1124
46. Durchschlag H, Zipper P (1997) Prog Colloid Polym Sci 107:43–57
47. Durchschlag H, Zipper P (1998) Biochem Soc Trans 26:731–736
48. Glatter O, Kratky O (eds) (1982) Small angle X-ray scattering. Academic Press, London
49. Durchschlag H (1993) In: Baianu IC, Pessen H, Kumosinski TF (eds) Physical chemistry of food processes, vol 2: advanced techniques, structures, and applications. Van Nostrand Reinhold, New York, pp 18–117
50. Kratky O (1971) In: Broda E, Locker A, Springer-Lederer H (eds) Proceedings of the 1st European Biophysics Congress, vol VI: theoretical molecular biology, biomechanics, biomathematics, environmental biophysics, techinques, education. Wiener Medizinische Akademie, Vienna, pp 373–396
51. Pessen H, Kumosinski TF (1985) Methods Enzymol 117:219–255
52. Zhou H-X (1995) Biophys J 69:2298–2303
53. Durchschlag H, Zipper P, Wilfing R, Purr G (1991) J Appl Crystallogr 24:822–831

Progr Colloid Polym Sci (1999) 113 : 106–113
© Springer-Verlag 1999

P. Zipper
H. Durchschlag

Prediction of hydrodynamic parameters from 3D structures

P. Zipper (✉)
Institute of Physical Chemistry
University of Graz, Heinrichstrasse 28
A-8010 Graz, Austria
Tel.: +43-316-380 5415
Fax: +43-316-380 9850
e-mail: peter.zipper@kfunigraz.ac.at

H. Durchschlag
Institute of Biophysics and
Physical Biochemistry
University of Regensburg
Universitätsstrasse 31
D-93040 Regensburg, Germany

Abstract The performance of different expressions for the hydrodynamic interaction tensor in the prediction of translational and rotational friction coefficients and intrinsic viscosities of a broad spectrum of different models consisting of two unequal spheres, varying from separated to completely overlapping, was tested in a systematic manner. The emphasis of the investigation was laid on checking the efficiency of different averages of the bead radii, when used in combination with the interaction tensor of Rotne and Prager for overlapping equal beads, as an ad hoc expression in treating unequal spheres. It was found that only averages based on the arithmetic means of bead volumes or bead surfaces perform well in all cases studied. An analysis of our results favors the use of the average based on the bead surfaces. Computations carried out with various bead models of the enzyme aldolase demonstrate the success of our approach in impeding the occurrence of erratic results.

Key words Bead modeling · Hydrodynamic interaction tensor · Multibody approach · Overlapping unequal beads · Prediction of hydrodynamic parameters

Introduction

The prediction of hydrodynamic parameters from 3D structures can be performed by means of whole-body or multibody approaches. Whole-body predictions are restricted to 3D structures which can be modeled by spheres, ellipsoids of revolution, or triaxial ellipsoids [1–7]. Multibody predictions are possible for all 3D structures which can be modeled by an assembly of spherical elements (beads) [4, 8–12]. Though multibody approaches have a long history in hydrodynamics [13–15], widespread application became feasible only since appropriate computer programs have been made available to the scientific community. Among those programs, García de la Torre's program HYDRO [16] has found widest acceptance as a potent tool for predicting hydrodynamic parameters of oligomeric and polymeric structures.

In the prediction of hydrodynamic parameters from crystal or NMR structures of biopolymers, data reduction steps must precede the use of HYDRO, because the number of atomic coordinates in these structures (often several thousand) is much higher than the number of spherical elements that can be handled by the HYDRO program in a reasonable time[1] (typically about 300). Data reduction is crucial because the structural characteristics of the polymer particle (dimensions, mass distribution, shape, symmetry, surface, etc.) must be retained as much as possible. The performance of several reduction procedures has been investigated recently [10, 11, 17]. The result of data reduction is a low-resolution model consisting of small spheres, often of unequal size and some of them partially overlapping their neighbors. Until recently, parameter predictions for models containing overlapping unequal beads were hazardous and could lead to erratic and unphysical results [10, 11] due

---

[1] The prediction of hydrodynamic parameters for a model of N spherical elements requires the inversion of a $3N \times 3N$ supermatrix

to singularities [18] as described by Carrasco et al. [19], because a theoretically founded hydrodynamic interaction tensor of overlapping unequal spheres is lacking. We succeeded in handling the problem by introducing an ad hoc expression for this tensor [10, 11]. A recent systematic study of two-sphere problems convincingly demonstrated that our ad hoc expression removes the singularities efficiently and performs adequately [19]. This paper continues and completes the previous study by considering a broad spectrum of models and expressions for the interaction tensor.

Theory and methods

Hydrodynamic interaction tensors

Hydrodynamic interaction tensors play a central role in the calculation of hydrodynamic parameters for a given bead model (for the theory see [8, 9, 15, 19]). The following interaction tensors are implemented in the HYDRO program [16]:

1. The unmodified Oseen tensor [20], which is valid for widely separated spheres:

$$\boldsymbol{T}_{ij} = (8\pi\eta_0 R_{ij})^{-1}\left(\boldsymbol{I} + \frac{\boldsymbol{R}_{ij}\boldsymbol{R}_{ij}}{R_{ij}^2}\right) \tag{1}$$

Here η_0 is the viscosity of the solvent, $\boldsymbol{I}$ is the unit tensor, and R_{ij} designates the center-to-center distance of two beads.

2. The tensor by García de la Torre and Bloomfield (GTB tensor, [15]), a kind of modified Oseen tensor, which is applicable to non-overlapping spheres of equal or unequal size (expressed by the radii σ_i and σ_j):

$$\boldsymbol{T}_{ij} = (8\pi\eta_0 R_{ij})^{-1}\left(\boldsymbol{I} + \frac{\boldsymbol{R}_{ij}\boldsymbol{R}_{ij}}{R_{ij}^2} + \frac{\sigma_i^2 + \sigma_j^2}{R_{ij}^2}\left(\frac{1}{3}\boldsymbol{I} - \frac{\boldsymbol{R}_{ij}\boldsymbol{R}_{ij}}{R_{ij}^2}\right)\right) \tag{2}$$

3. The tensor by Rotne and Prager (RP tensor, [21]), which holds for overlapping spheres of equal size:

$$\boldsymbol{T}_{ij} = (6\pi\eta_0\sigma)^{-1}\left(\left(1 - \frac{9}{32}\frac{R_{ij}}{\sigma}\right)\boldsymbol{I} + \frac{3}{32}\frac{\boldsymbol{R}_{ij}\boldsymbol{R}_{ij}}{R_{ij}\sigma}\right) \tag{3}$$

In previous studies [10, 11] we have applied the RP tensor also to overlapping beads of non-equal size, replacing the bead radius σ in Eq. (3) with an averaged radius σ_{av}, preferably with the radius related to the arithmetic mean of the bead volumes:

$$\sigma_{av} = \sqrt[3]{(\sigma_1^3 + \sigma_2^3)/2} \tag{4}$$

Recently, Carrasco et al. [19] have tested the performance of this expression in the case of two-sphere models. The comparison with results obtained with σ_{av}, defined as the arithmetic mean of the bead radii:

$$\sigma_{av} = (\sigma_1 + \sigma_2)/2 \tag{5}$$

did not reveal any relevant differences, but the similarity of the radii σ_1 and σ_2 used in that test did not allow final conclusions about the proper choice of σ_{av}. In the present study, the comparison was checked in a systematic manner with models covering a broad spectrum of different radii (see below). In

addition, we tested further averages of bead radii: geometric mean, harmonic mean, and an average related to the arithmetic mean of the bead surfaces:

$$\sigma_{av} = \sqrt{(\sigma_1^2 + \sigma_2^2)/2} \tag{6}$$

Models

Two-sphere models

Following the ideas put forward in [19], we selected five models, each consisting of two spheres with constant radii σ_1 and σ_2 (obeying the condition $\sigma_1 + \sigma_2 = 1$) at varying distance R_{12}; σ_1 was chosen as 0.5, 0.6, 0.7, 0.8, and 0.9 (models A to E, illustrated in Fig. 1) and R_{12} was always varied between 1.5 and 0.01. The models thus covered the range from equal ($\sigma_1 = \sigma_2 = 0.5$) to extremely unequal beads ($\sigma_1 = 0.9$, $\sigma_2 = 0.1$) and, additionally, the whole transition from separated to touching beads ($R_{12} = 1$) and finally to completely overlapping beads. The range of complete overlapping is limited by the relation $R_{12} = \sigma_1 - \sigma_2$. Below this limiting distance the models presumably behave as if they were single spheres of radius $\sigma = \sigma_1$. The models A and B have already been used in the previous study [19].

Protein models: aldolase

To illustrate the efficiency of our ad hoc expression [Eq. (4)] in the prediction of hydrodynamic parameters from the crystal structure of biopolymers, the enzyme aldolase from rabbit muscle [22] was chosen as an example. Atomic coordinates were obtained from the

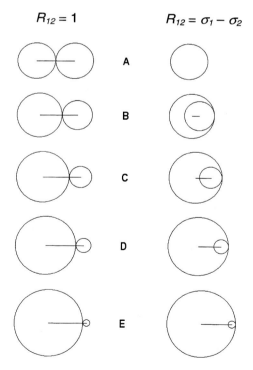

Fig. 1 Schematic drawing showing the two-sphere models *A–E*. The radii σ_1 and σ_2 obey the condition $\sigma_1 + \sigma_2 = 1$. The center-to-center distance R_{12} is variable. The models are presented for the two limits: touching spheres ($R_{12} = 1$; on the *left*), and completely overlapping spheres ($R_{12} = \sigma_1 - \sigma_2$; on the *right*)

Brookhaven Protein Data Bank [23] (accession code 1ADO). Application of the data reduction procedures of running means and cubic grids described in [10, 11] yielded various low-resolution models, based on either atomic or amino acid coordinates and containing 64–292 beads.

Computations by means of HYDRO

Computations of hydrodynamic parameters (sedimentation coefficient s, diffusion coefficient D, and intrinsic viscosity $[\eta]$) of protein models (e.g., aldolase) were performed using two versions of the program HYDRO: the version HYDROX_5 (downloaded from the www site of the U.S. Biophysical Society and upgraded to handle 300 beads), which treated overlapping unequal beads in an inadequate manner, and a modified version HYDRO-VZ [11], where our improved handling of overlapping [Eqs. (3) and (4)] had been implemented.

In the computations of hydrodynamic parameters of the two-sphere models the different interaction tensors were tested also outside their dedicated range of validity, in order to study their behavior and the continuity of results around the limiting distance $R_{12} = 1$, corresponding to touching spheres. The calculations were performed by means of special versions of HYDRO that forced the exclusive use of a selected interaction tensor and allowed the use of different kinds of averages σ_{av}. The volume corrections for intrinsic viscosity and rotational quantities [12] were applied, after the volume itself had been corrected, if necessary, for overlapping [19]. For convenience, the parameters obtained from these computations are presented in terms of reduced, dimensionless quantities (cf. [19]): f_t^* denotes the reduced translational friction coefficient, f_r^* the reduced rotation-

al friction coefficient for rotation around a perpendicular axis, and $[\eta]^*$ the reduced intrinsic viscosity.

Results

Two-sphere models

Equal spheres

The reduced hydrodynamic parameters f_t^*, f_r^* and $[\eta]^*$, as obtained for model A (equal beads, $\sigma_1 = \sigma_2 = 0.5$) by using different interaction tensors, are plotted versus the center-to-center distance R_{12} in Fig. 2. As Fig. 2a shows, both the unmodified Oseen tensor [Eq. (1)], the GTB tensor [(Eq. (2)], and the RP tensor [Eq. (3); here, of course, without averaging of radii] yield identical

Fig. 2 a–c Reduced translational friction coefficient, f_t^* (**a**), reduced rotational friction coefficient for rotation around a perpendicular axis, f_r^* (**b**), and reduced intrinsic viscosity, $[\eta]^*$ (**c**), predicted for the two-sphere model A (equal spheres, $\sigma_1 = \sigma_2 = 0.5$) as a function of the center-to-center distance R_{12}. Predictions were made by using the unmodified Oseen tensor (▲), the GTB tensor (○), and the RP tensor (*solid line*), respectively. The *vertical bars* (*dotted*) mark the case of touching spheres. The values drawn by the symbol ■ correspond to a single sphere with $\sigma = 0.5$

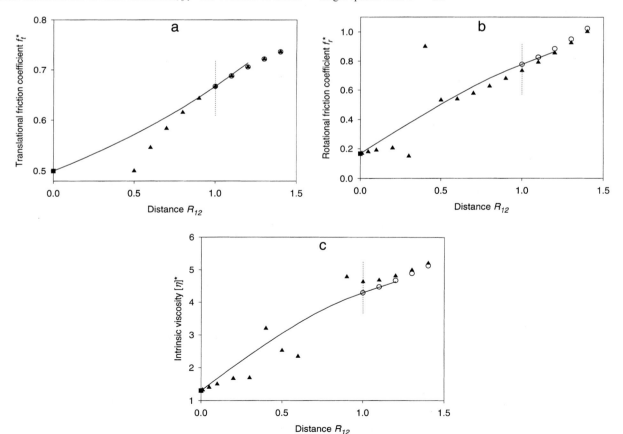

Table 1 Relative deviations Δf_t^*, Δf_r^* and $\Delta[\eta]^*$ of the predictions of hydrodynamic parameters by using the RP tensor, in connection with mean bead radii σ_{av} [defined by Eqs. (4)–(6)], from the predictions by means of the GTB tensor (at $R_{12} = 1.0$) or the values corresponding to single spheres of radius $\sigma = \sigma_1$ (at $R_{12} = \sigma_1 - \sigma_2$)

Model	R_{12}	Δf_t^* (%)			Δf_r^* (%)			$\Delta[\eta]^*$ (%)		
		Eq. (5)	Eq. (6)	Eq. (4)	Eq. (5)	Eq. (6)	Eq. (4)	Eq. (5)	Eq. (6)	Eq. (4)
B	1.0	0.01	0.01	0.03	−0.22	−0.03	0.13	0.13	0	−0.16
	0.2	2.25	1.22	0.63	22.5	20.5	19.0	10.2	9.1	8.3
C	1.0	0.14	0.12	0.23	−0.55	−0.22	−0.17	0.24	−0.06	−0.36
	0.4	1.70	0.47	0.13	21.1	19.2	18.0	9.2	8.2	7.6
D	1.0	0.31	0.25	0.43	−0.49	−0.36	−0.48	0.16	−0.08	−0.28
	0.6	0.85	0.15	0.14	14.6	13.8	13.4	6.2	5.8	5.5
E	1.0	0.27	0.20	0.31	−0.17	−0.17	−0.22	0.04	−0.04	−0.08
	0.8	0.30	0.09	0.16	7.4	7.2	7.2	3.1	2.9	2.9

values for the reduced translational friction coefficient f_t^* at $R_{12} = 1$ (i.e. for spheres just touching). While the coincidence of f_t^* values obtained by using the Oseen tensor and the GTB tensor evidently persists in the range of separated beads ($R_{12} > 1$), the figure reveals an increasing divergence of f_t^* values calculated with the Oseen tensor and the RP tensor in the range of partially overlapping spheres ($R_{12} < 1$). When R_{12} approaches the limit of complete overlapping, $R_{12} = 0$, the f_t^* values based on the RP tensor approach the value of 0.5, corresponding to a single sphere of radius $\sigma = 0.5$.

For touching spheres ($R_{12} = 1$), the GBT tensor and the RP tensor yield identical values also for the reduced rotational friction coefficient (for rotation around a perpendicular axis) f_r^* (Fig. 2b) and for the reduced intrinsic viscosity $[\eta]^*$ (Fig. 2c), whereas the results obtained with the Oseen tensor deviate significantly. The application of the Oseen tensor for calculating the quantities f_r^* and $[\eta]^*$ in the range of $R_{12} < 1$ obviously suffers from singularities which can be avoided if the RP tensor is used instead. At $R_{12} = 0$, the Oseen tensor and the RP tensor yield coincident values of f_r^* and $[\eta]^*$, respectively, which are identical to the values expected for a single sphere of radius $\sigma = 0.5$.

Unequal spheres

A first inspection of the results that were obtained for the models B–E (unequal beads) by using the RP tensor together with different kinds of averaged radii σ_{av} already revealed that only the averages σ_{av} defined by Eqs. (4)–(6) were able to remove singularities efficiently, while both the geometric mean and the harmonic mean of the bead radii led to severe singularities (data not shown). A closer inspection of the results showed that for the prediction of f_r^* and $[\eta]^*$ the choice of the average σ_{av} [Eqs. (4)–(6)] did not influence the results significantly, regardless of whether the bead radii were quite similar (model B: $\sigma_1 = 0.6$, $\sigma_2 = 0.4$) or extremely different (model E: $\sigma_1 = 0.9$, $\sigma_2 = 0.1$). For the pre-

diction of f_t^*, however, the dependence on the choice of the average σ_{av} was found to be enhanced with increasing dissimilarity of the bead radii σ_1 and σ_2.

In Table 1, predictions of f_t^*, f_r^*, and $[\eta]^*$, as obtained by using the RP tensor with the three different averages σ_{av} [Eqs. (4)–(6)], are compared, at $R_{12} = 1$, to the predictions obtained by means of the GTB tensor and, at $R_{12} = \sigma_1 - \sigma_2$, to the values calculated for a single sphere of radius $\sigma = \sigma_1$, respectively. The findings are illustrated by representative examples in Figs. 3–5, and are discussed in the following.

Figure 3 shows the predictions of the translational friction coefficient f_t^* for the models B–E. It is evident that in the range of separated beads ($R_{12} > 1$) the results obtained with the unmodified Oseen tensor and with the GTB tensor diverge. The discrepancies increase with decreasing separation and are more pronounced the greater the dissimilarity of the two beads. The f_t^* values obtained with the Oseen tensor in the range of partial overlap are strongly affected by singularities and fall outside the drawing below $R_{12} = 0.8$. As follows from Fig. 3a and from the tabulated relative differences Δf_t^* (Table 1: model B), the RP tensor and the GTB tensor yield virtually identical results at $R_{12} = 1$, regardless of the kind of average σ_{av}. In the adjacent range, the f_t^* values obtained with different averages σ_{av} are very similar, down to $R_{12} = 0.3$. At about this distance the graphs of f_t^* show a flat minimum, which is followed by an unphysical increase of f_t^* when R_{12} approaches zero[2]. The f_t^* values in the minimum nearly correspond to a single sphere of $\sigma = 0.6$ (identical with σ_1 of this model). It should be noted that this level of f_t^* is to be expected for the range of complete overlapping, i.e. below the limiting distance $R_{12} = 0.2$. At that distance, however, the predicted f_t^* values have already increased

---

[2] It should be borne in mind that, in principle, models with bead distances $R_{12} < \sigma_1 - \sigma_2$ are rather unphysical, because the smaller sphere is completely embedded in the larger one

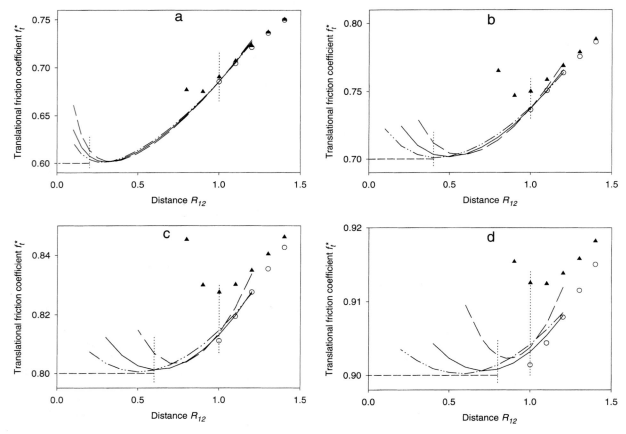

Fig. 3a–d Reduced translational friction coefficients, f_t^*, predicted for various two-sphere models with unequal spheres ($\sigma_1 + \sigma_2 = 1$) as a function of the center-to-center distance R_{12}. **a** Model B ($\sigma_1 = 0.6$), **b** model C ($\sigma_1 = 0.7$), **c** model D ($\sigma_1 = 0.8$), **d** model E ($\sigma_1 = 0.9$). Predictions were made by using the unmodified Oseen tensor (▲), the GTB tensor (○), or the RP tensor together with an averaged radius σ_{av} defined by the arithmetic mean of radii (*long dashes*), surfaces (*solid line*), and volumes (*dash-dotted line*) of the spheres, respectively. The *vertical bars* (*dotted*) mark the case of touching spheres and the limiting distance of complete overlapping. The *horizontal bars* (*short dashes*) represent the values corresponding to single spheres with $\sigma = \sigma_1$

by up to about 2% (cf. Table 1: model B, Δf_t^* at $R_{12} = 0.2$).

Figure 3b and, to a greater degree, Fig. 3c and d, exhibit considerable differences if compared to Fig. 3a. One difference concerns the behavior of the curves at $R_{12} = 1$, where the f_t^* values obtained with the RP tensor using different averages σ_{av} do not coincide with the value calculated by means of the GTB tensor, but are higher than that value. The discrepancy is always largest when Eq. (4) is used to calculate σ_{av} (cf. Table 1: models C–E, Δf_t^* values at $R_{12} = 1$). The other difference concerns the behavior of the f_t^* values obtained with the RP tensor at lower values of R_{12}. While in Fig. 3a the minima of f_t^* values occur at a similar distance of $R_{12} = 0.3$, the positions of the minima in Fig. 3b–d are

shifted towards larger R_{12} with increasing dissimilarity of the beads and at the same time spread in dependence on the average σ_{av} used. Certainly, the extreme behavior of f_t^* values based on the average σ_{av} according to Eq. (5) in the case of model E (Fig. 3d) is an argument against using the arithmetic mean of radii in connection with the RP tensor. The averages σ_{av} according to Eqs. (4) and (6), on the other hand, appear to be practicable, because both lead to minima of f_t^* near the limiting distance of complete overlapping and the values of f_t^* in the minima are similar to the value expected for a single sphere of radius $\sigma = \sigma_1$. The main differences between the application of Eqs. (4) and (6) are that by using Eq. (6) the rise to unphysical values of f_t^* on decreasing R_{12} occurs at somewhat larger separations R_{12} and from a somewhat higher level than by using Eq. (4). This behavior is also reflected by higher Δf_t^* values resulting upon application of Eq. (6) (Table 1: models C and D, $R_{12} = 0.4$ and 0.6, respectively). Model E is an exception because the f_t^* values predicted by means of Eq. (4) exceed the values obtained from Eq. (6) down to about $R_{12} = 0.7$ (Fig. 3d). Therefore the application of Eq. (4) results here in a higher Δf_t^* value if compared to Eq. (6) (Table 1: model E, $R_{12} = 0.8$).

Figure 4 presents predictions of the rotational friction coefficient f_r^*, and Fig. 5 of the intrinsic viscosity $[\eta]^*$.

Figure 4a, which shows the data obtained for model B (the model with the least dissimilar beads), reveals that the unmodified Oseen tensor leads to severe singularities in the range of overlapping, while the RP tensor, in connection with any of the three different averages σ_{av}, yields continuous and nearly identical results. The same statements hold also for the viscosity data of the same model in Fig. 5a. It should be noted that the values predicted with the RP tensor for f_r^* and $[\eta]^*$ at $R_{12} = 0.2$, the limiting distance of total overlapping, are significantly larger than the values calculated for a single sphere of radius σ_1 (see Table 1: model B, Δf_r^* at $R_{12} = 0.2$), whereas with the Oseen tensor the single sphere levels of f_r^* and $[\eta]^*$ values are reached at $R_{12} = 0.2$.

Predictions of f_r^* and $[\eta]^*$ for model E, which contains the most dissimilar beads, are presented in Figs. 4b and 5b. Also here the application of the Oseen tensor obviously leads to singularities in the range of overlapping; the effects are, however, less pronounced than in Figs. 4a and 5a. The application of the RP tensor with any of the averages σ_{av} again yields continuous curves of f_r^* and $[\eta]^*$, regardless of σ_{av} used. It is surprising that these curves are similar to the predictions from the Oseen tensor (particularly for f_r^*). A similarity between predictions of f_r^* and $[\eta]^*$, respectively, by means of the RP tensor, on one hand, and the Oseen tensor, on the other, was also observed with models of less dissimilar beads (curves not shown). In general, the similarity of predictions of f_r^* and $[\eta]^*$ by means of different tensors was found to be enhanced with increasing dissimilarity of the beads. It is obvious from Figs. 4b and 5b, and from the Δf_r^* and $\Delta[\eta]^*$ values for model E in Table 1, that at $R_{12} = 0.8$ none of the tensors leads to values of f_r^* and $[\eta]^*$ corresponding to a single sphere of radius $\sigma = \sigma_1 = 0.9$. The relative

deviations upon using the RP tensor (Table 1: model E), are, however, smaller than those found for model B (the tabulated Δf_r^* and $\Delta[\eta]^*$ values for models C and D are in good accordance with this finding).

Aldolase

Figure 6 summarizes the predictions of sedimentation coefficient s for a gamut of low-resolution models derived from the crystal structure of the enzyme aldolase. The filled symbols refer to the predictions obtained by means of the HYDROX_5 program version. These data scatter considerably, owing to the frequent occurrence of partial overlapping of neighboring unequal spheres. An analysis of the models revealed that 8–26% of the total volume of the models is involved in partial overlapping. The percentage of overlapping is constantly low (8–11%) for the "cubic grid" models, whereas it tends to increase with the decreasing number of beads for the models created by the "running means" procedure (in these models, even overlapping of more than two beads in the same space region occurs). Figure 6 clearly shows (cf. the lines and open symbols) that the application of the RP tensor together with Eq. (4) to all overlapping unequal spheres, as implemented in the program version HYDRO-VZ, removes the scatter and leads to continuous and consistent predictions of s for all models investigated. Analogous improvements were encountered in the predictions of diffusion coefficient D and intrinsic viscosity (data not shown).

Discussion and conclusions

The results obtained both for the various two-sphere models and for the models of the enzyme aldolase convincingly demonstrate that the prediction of hydrodynamic parameters for bead models containing overlapping unequal spheres can be essentially improved by

Fig. 4a, b Reduced rotational friction coefficients for rotation around a perpendicular axis, f_r^*, predicted for selected two-sphere models with unequal spheres ($\sigma_1 + \sigma_2 = 1$) as a function of the center-to-center distance R_{12}. **a** Model B ($\sigma_1 = 0.6$), **b** model E ($\sigma_1 = 0.9$). The meaning of symbols and lines is the same as in Fig. 3

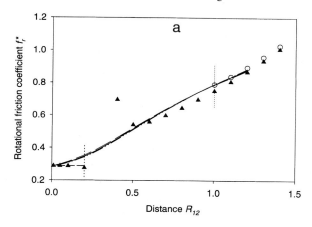

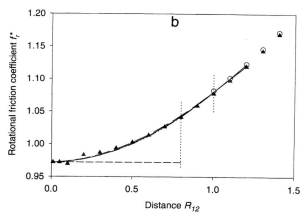

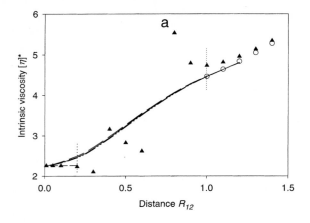
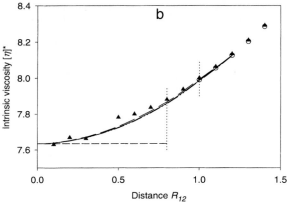

Fig. 5a, b Reduced intrinsic viscosities, $[\eta]^*$, predicted for selected two-sphere models with unequal spheres ($\sigma_1 + \sigma_2 = 1$) as a function of the center-to-center distance R_{12}. **a** Model B ($\sigma_1 = 0.6$), **b** model E ($\sigma_1 = 0.9$). The meaning of symbols and lines is the same as in Fig. 3

our approach of treating each pair of unequal beads in the calculation of the interaction tensor (and only in this context) like two equal beads of radius σ_{av}, thus the RP tensor [21] can be applied.

Of the various choices of σ_{av} investigated in this study, only two were found to be applicable in all cases of two-sphere models studied: the average σ_{av} based on the arithmetic mean of the bead volumes (Eq. (4), our previous proposal [10, 11]), and the average based on the arithmetic mean of the bead surfaces [Eq. (6)] which was studied here for the first time in a systematic manner. Both kinds of averages led to similar, often nearly identical, predictions of hydrodynamic parameters, and

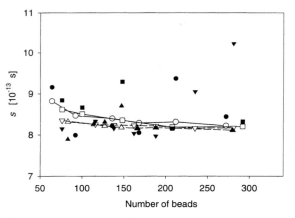

Fig. 6 Sedimentation coefficient s, predicted for various bead models of the enzyme aldolase. The models were derived from the crystal structure by using the data reduction procedures of running means ($\bullet$, $\blacksquare$) and cubic grids ($\blacktriangle$, $\blacktriangledown$). *Filled symbols* refer to the predictions made by means of the downloaded program version HYDROX_5. *Lines* and *empty symbols* show the results obtained with the special version HYDRO-VZ, applying the ad hoc expression [Eq. (4)] to overlapping unequal beads

the predicted parameters turned out to be continuous functions of the center-to-center distance of the beads.

The predictions of the translational friction coefficient were found to be excellent, as judged from comparisons (see the Δf_t^* values in Table 1) with results obtained by means of the GTB tensor [15], for touching spheres, and with values calculated for single spheres, at the limits of complete overlapping. From the tabulated data and from Fig. 3 we can conclude that Eq. (6) performs somewhat better than Eq. (4) for touching or moderately overlapping beads, whereas near or below the limit of complete overlapping, in general, Eq. (4) yields better results, except for very unequal beads (cf. Fig. 3d and Table 1: model E, Δf_t^*). The better performance of Eq. (6) for touching spheres is not unexpected, because the GTB tensor also utilizes the arithmetic mean of the surfaces [cf. Eq. (2)].

Also in predicting the rotational friction coefficient (for rotation around a perpendicular axis) or the intrinsic viscosity of touching or moderately overlapping beads, Eq. (6) generally yielded slightly better results than Eq. (4). The predictions of f_r^* and especially of $[\eta]^*$, made for the limit of complete overlapping, are comparatively poor. Possibly this is partly due to an overestimation because of the volume correction applied in the calculation of f_r^* and $[\eta]^*$ [12].

The analysis of the results obtained for the two-sphere models has proven that, on the whole, the averages σ_{av} based on the arithmetic means of bead volumes and bead surfaces perform similarly. The performance of Eq. (4) was found to decrease at high dissimilarity of the beads, presumably owing to an overestimation of the contribution of the larger bead. Therefore a slight preference for using the average σ_{av} based on the arithmetic mean of bead surfaces can be concluded from our data.

Overlapping of beads might be considered unphysical, and, in general, overlapping in bead models should be avoided when possible [19, 24]. This holds particularly for the extreme case of complete overlapping, which should never happen. Nevertheless, overlapping is of some importance in bead modeling of biopolymers.

Already in the crystal structure of a biopolymer a considerable amount of formal overlapping occurs, if each atom or atomic group is represented by a sphere with a radius corresponding to the volume of the atom or atomic group. In the crystal structure of tetrameric aldolase, for instance, about 39% of the total volume of a single chain, calculated as the sum of atomic volumes, is involved in binary or multiple overlaps; this is a higher percentage than in any of the reduced models derived from the crystal structure. Thus the occurrence of overlaps in the reduced models, used for computations by means of HYDRO, is not primarily an annoying byproduct of the data reduction procedure, but draws back to the crystal structure itself. Overlapping is required to make a bead model spacefilling and equivalent in volume with the particle to be modeled. Different data reduction procedures lead to different qualities of overlapping: by the approaches of running means, models can be created which consist of nearly equal but considerably overlapping beads, whereas with models generated by cubic grids approaches the amount of overlapping is less, but the beads are much more unequal. The computations performed for aldolase and other proteins [10, 11] have shown that our approach of treating overlaps is successful with both kinds of models. Therefore, it is not necessary any longer to construct bead models with exactly same-sized spheres in order to apply the RP tensor if overlapping occurs (cf. [17, 25]).

The latest version of the computer program HYDRO contains our ad hoc expression based on the arithmetic mean of bead volumes [Eq. (4)] for treating overlapping unequal spheres [19]. This program can be downloaded freely from García de la Torre's web site, http://leonardo.fcu.um.es/macromol.

References

1. Kumosinski TF, Pessen H (1982) Arch Biochem Biophys 219:89–100
2. Müller JJ (1991) Biopolymers 31:149–160
3. Harding SE (1989) In: Harding SE, Rowe AJ (eds) Dynamic properties of biomolecular assemblies. Royal Society of Chemistry, Cambridge, pp 32–56
4. Harding SE (1995) Biophys Chem 55:69–93
5. Durchschlag H, Zipper P, Purr G, Jaenicke R (1996) Colloid Polym Sci 274:117–137
6. Durchschlag H, Zipper P (1997) J Mol Struct 383:223–229
7. Durchschlag H, Zipper P (1998) Biochem Soc Trans 26:731–736
8. García de la Torre J, Bloomfield VA (1981) Q Rev Biophys 14:81–139
9. García de la Torre J (1989) In: Harding SE, Rowe AJ (eds) Dynamic properties of biomolecular assemblies. Royal Society of Chemistry, Cambridge, pp 3–31
10. Zipper P, Durchschlag H (1997) Prog Colloid Polym Sci 107:58–71
11. Zipper P, Durchschlag H (1998) Biochem Soc Trans 26:726–731
12. García de la Torre J, Carrasco B (1998) Eur Biophys J 27:549–557
13. Kirkwood JG, Riseman J (1948) J Chem Phys 16:565–573
14. Bloomfield V, Dalton WO, Van Holde KE (1967) Biopolymers 5:135–148
15. García de la Torre J, Bloomfield VA (1977) Biopolymers 16:1747–1763
16. García de la Torre J, Navarro S, López Martínez MC, Díaz FG, López Cascales JJ (1994) Biophys J 67:530–531
17. Byron O (1997) Biophys J 72:408–415
18. Zwanzig R, Kiefer J, Weiss GH (1968) Proc Natl Acad Sci USA 60:381–386
19. Carrasco B, García de la Torre J, Zipper P (1999) Eur Biophys J (in press)
20. Oseen CW (1927) Hydrodynamik. Akademie Verlag, Leipzig
21. Rotne J, Prager S (1969) J Chem Phys 50:4831–4837
22. Blom N, Sygusch J (1997) Nat Struct Biol 4:36–39
23. Sussman JL, Lin D, Jiang J, Manning NO, Prilusky J, Ritter O, Abola EE (1998) Acta Crystallogr Sect D 54:1078–1084
24. García de la Torre J, Harding SE, Carrasco B (1998) Biochem Soc Trans 26:716–721
25. Hellweg T, Eimer W, Krahn E, Schneider K, Müller A (1997) Biochim Biophys Acta 1337:311–318

Progr Colloid Polym Sci (1999) 113 : 114–120
© Springer-Verlag 1999

C. Tziatzios
H. Durchschlag
B. Sell
J.A. van den Broek
W. Mächtle
W. Haase
J.-M. Lehn
C.H. Weidl
C. Eschbaumer
D. Schubert
U.S. Schubert

Solution properties of supramolecular cobalt coordination arrays

C. Tziatzios (✉) · B. Sell · J.A. van den
Broek · D. Schubert
Institut für Biophysik, JWG-Universität
D-60590 Frankfurt am Main, Germany

H. Durchschlag
Institut für Biophysik und Physikalische
Biochemie der Universität
D-93040 Regensburg, Germany

W. Mächtle
Kunststofflaboratorium, BASF AG
D-67056 Ludwigshafen, Germany

W. Haase
MPI für Biophysik
D-60528 Frankfurt am Main, Germany

J.-M. Lehn
Laboratoire de Chimie Supramoleculaire
Université Louis Pasteur
F-67000 Strasbourg, France

C.H. Weidl · C. Eschbaumer
U.S. Schubert
Lehrstuhl für Makromolekulare Stoffe
TU München, D-85747 Garching
Germany

Abstract A number of chemically related gridlike Co coordination arrays were studied by UV/vis absorption spectroscopy, electron microscopy and, in particular, analytical ultracentrifugation and partial specific volume measurements, in order to determine their solubility, stability and association behavior in a variety of organic solvents. As judged by the naked eye, solubilization of the compounds occurred instantaneously or at least within minutes. In contrast, the UV/ vis absorbance of the samples distinctly changed for hours or even days, depending on the compound in question. In some cases, the spectral changes indicated dissociation events, probably involving dissolution of clusters or microcrystals. This was supported by ^{1}H NMR (on related Cd and Zn compounds) and electron microscopic observations at different time intervals after addition of the solvent. Under certain conditions, addition of 20–50 mM salt (necessary to obtain ultracentrifuge data not influenced by nonideal sedimentation behavior) again led to aggregation of the material. However, according to equilibrium sedimentation experiments in most solvents the solubilized Co coordination arrays finally were in the form of monomers, whereas in some solvents intermediate aggregates were predominant. Prolonged storage of the solubilized compounds at room temperature in most cases led to their decomposition or conversion. Reliably determining the partial specific volume, $\bar{v}$, of the compounds turned out to be the most difficult problem in our studies. Density measurements using a Paar density meter apparently suffered from disturbances (probably due to aggregation) at the relatively high compound concentrations required. $\bar{v}$ determinations applying the Edelstein–Schachman method to data collected in solvents of different density suffered from dependencies of $\bar{v}$ on the solvent. Combining measurements in nondeuterated and deuterated solvents (as in the original Edelstein–Schachman method) suffered from relatively low accuracy; in addition, it is applicable to a few solvents only. At present, weighted averages of the $\bar{v}$ values from the different ultracentrifuge methods seem to yield the most reliable figures.

Key words Cobalt coordination arrays · Solubility · Density · Self-association · Sedimentation equilibrium analysis

Introduction

The design, synthesis, and arrangement of special functional units with nanometer dimensions into defined molecular architectures are the main aims in polymer, supramolecular, and material science [1]. This requires precise control of the structures from molecular size to micrometers. One promising approach for the construction of such objects comes from supramolecular chemistry; it applies noncovalent interactions such as hydrogen-bonding [2] or metal–ligand interactions [3, 4] to assemble highly organized architectures.

Following the approach described, supramolecular coordination arrays with two-dimensional [2 × 2] grid-type architectures and interesting electronic, magnetic, and structural properties (such as electronic interactions between the metal centers and an antiferromagnetic transition at low temperatures [5, 6]) have been synthesized. They are formed by spontaneous self-assembly of 4,6-bis(6-(2,2-bipyridyl))pyrimidine ligands [5] (or functionalized derivatives [7]) and suitable metal ions such as Co(II) [6] (Fig. 1) and seem to be particularly suited to assemble into ordered and stable arrangements on surfaces or into ordered thin films. For such applications, knowing the solution properties of the compounds is of major importance. As a prerequisite, methods for characterizing these properties have to be provided; analytical ultracentrifugation seems to be the most powerful method to do this.

Our group has already used two members of the class of [2 × 2] grids to demonstrate the feasibility of analytical ultracentrifugation as a tool in studies on the state of association of supramolecular compounds in solution [8], and to illustrate the different types of suitable ultracentrifuge methods [9]. We have now made a broader study of the compounds' solution properties, which is based on five different but related [2 × 2] Co grids and, in addition, on a [1 × 1] Co grid representative of the organic ligands (Fig. 2). Again, the main technique was analytical ultracentrifugation, supplemented by determinations of solute density. In addition, UV/vis spectroscopy, ¹H NMR spectroscopy, and transmission

Organic grid ligands

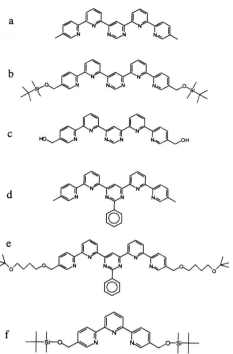

Fig. 2 The functionalized organic ligands which, together with Co ions, assemble into coordination arrays. Ligands a–e lead to the formation of [2 × 2] gridlike complexes (Fig. 1). Two molecules of ligand f, together with one Co ion, assemble into a [1 × 1] grid [10]

electron microscopy were applied. It will be shown that the solution properties of these compounds are distinctly more complex than could be anticipated.

Materials and methods

Materials

The Co coordination arrays were prepared by recently described methods [5–7, 10–12]. In brief, the organic ligands were synthesized by Stille-type cross-coupling procedures [5, 7, 10], and the metal coordination arrays were prepared and purified according to Refs.

Fig. 1 Schematic presentation of Co coordination array formation. Two hexafluorophosphate counterions are associated with each metal ion (not shown)

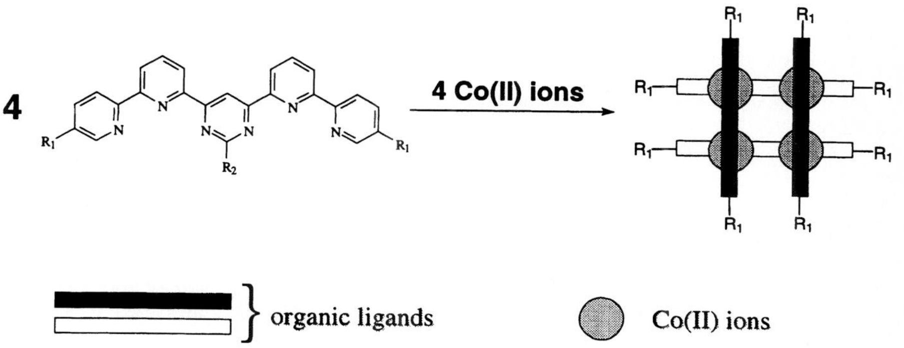

[6, 11, 12]. The solvents and salts used were purchased from Merck or Aldrich and were of analytical or spectroscopic grade. For most experiments, the concentration range was 1–100 mg/l, with emphasis on the lower part of this range. Much higher concentrations (4–12 g/l) were necessary for density measurements of the grid solutions. For these measurements, the compounds were solubilized at room temperature under permanent stirring and were used between 3 and 6 h after addition of the solvent. Following the measurement, the solvent was evaporated under vacuum, and the sample was used for another density measurement.

Methods

UV/vis spectra were measured with a Hitachi U-2000 spectrophotometer, in cells of 1 cm pathlength made from Suprasil (Hellma). ^{1}H NMR spectra were recorded on a Bruker AC 300 spectrometer. The chemical shifts were calibrated to the residual solvent peak or to tetramethylsilane. For electron microscopy, a Philips EM 208 electron microscope was used. Samples were prepared from solutions in acetone. They were either slowly dried in an acetone/air atmosphere, or frozen at approximately −200 °C and then freeze-dried at slowly increasing temperature (in all cases without additional staining).

Sedimentation equilibrium experiments were performed with a Beckman Optima XL-A ultracentrifuge in connection with an An-60 Ti or an An-50 Ti rotor, titanium double-sector centerpieces (pathlength 1.2 cm), and polyethylene gaskets (BASF). The rotor temperature was 20 °C. The absorbance-versus-radius profiles, $A(r)$, were recorded at a wavelength between 330 and 450 nm, depending on the sample concentration. The data were evaluated using the computer program DISCREEQ (sedimentation equilibrium) [13, 14]. In all cases, ideal sedimentation behavior of the samples was assumed [8]. In most fits, the position of the baseline was treated as a free parameter; however, the fits were accepted only if the calculated baseline position differed from zero by less than ±0.010 absorbance units. In a number of experiments, the baseline level was determined experimentally by prerunning the cell with solvent only, and afterwards exchanging solvent against sample in the sample sector.

In the determination of the partial specific volume, $\bar{v}$, of the solute, altogether five different procedures were applied and their results compared with each other:

1. Measurement of the density difference between (salt-free) sample solution and pure solvent [15].
2. Determination of $\bar{v}$ from the effective molar mass measured by sedimentation equilibrium, assuming the presence of either homogeneous solute monomers or a mixture of oligomers.
3. Application of the Edelstein–Schachman method for simultaneous determination of M and $\bar{v}$ [16], using two different solvents of distinctly different density.
4. Application of the original Edelstein–Schachman method, using nondeuterated and deuterated acetone.
5. Determination, by sedimentation equilibrium analysis, of the apparent effective mass, $M(1 - \bar{v}\rho_o)$, of the solute in solvent mixtures of different density, ρ_o, and extrapolation to the condition $M(1 - \bar{v}\rho_o) = 0$.

Solvent and solution densities were measured with a Paar DMA 02 density meter.

Results

The Co coordination arrays, the organic ligands of which are shown in Fig. 2, were studied by a variety of techniques.

Solubility and self-association

As judged by the naked eye, in most solvents and at moderate solute concentrations, solubilization of the Co coordination arrays seemed to occur virtually instantaneously or at least within minutes; however, other techniques revealed the process of solubilization to be much slower:

1. At compound concentrations around 1 g/l, the UV absorbance of the samples frequently increased severalfold within days or even weeks.
2. At similar compound concentrations, ^{1}H NMR spectra of the Cd or Zn analogues of compound d [6] initially contained broad lines which, within a few days, were transformed into narrow ones, suggesting the disappearance of intermolecular interactions.

Spectral changes within the same time interval were also observed in UV/vis spectra of the Co grids at solute concentrations below 100 mg/l, in most cases represented by shifts in the wavelength of the maxima of the curves but also by a decrease in peak height (Fig. 3). A delayed transformation of the solute into monomeric or oligomeric grids, after solvent addition, was also apparent from experiments in the analytical ultracentrifuge:

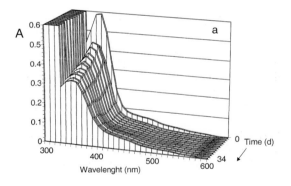

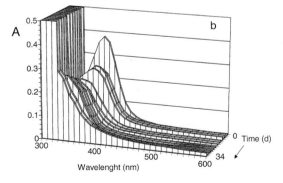

Fig. 3a, b Dependency of the UV/vis absorption spectra of two different [2 × 2] grids on time. **a** Grid c; **b** grid d. Solvent: acetone; compound concentrations: approximately 1.5 μM. The spectra were measured at 20 °C; between measurements, the samples were stored at 4 °C

Most of the compounds, when studied within a few hours after adding the solvent, sedimented to the bottom of the ultracentrifuge cell in the time interval required to reach a rotor speed of 40,000 rpm. On the other hand, after storage for a few days (at 4 °C), samples from the same tube were fully, or at least to a large percentage, monomeric (see later). Transmission electron microscopy of freeze-dried samples from the freshly prepared solutions showed both microcrystals and amorphous (cluster?) structures (maximum dimensions around 500 nm), which were much less abundant after prolonged incubation. Under some conditions, addition of 10–50 mM ammonium hexafluorophosphate accelerated the process of solubilization. Under other conditions, addition of the salt to the already solubilized compound led to reassociation into large aggregates, as indicated in ultracentrifuge measurements by the loss of up to more than 90% of the solute absorbance in the solution during the acceleration period of the rotor. This phenomenon was frequently accompanied by the formation of large crystals with dimensions of up to several millimeters, even at grid concentrations down to 100 mg/l (these crystals were not observed in the absence of the grids and thus indicate grid/salt cocrystallization). Tetrabutylammonium cations were much more effective than ammonium cations in producing this "salting out". It should be noted that, in the ultracentrifuge experiments, the presence of salt at concentrations above 10 mM is obligatory in order to suppress nonideal sedimentation behavior of the grids [8].

In appropriate solvents, the compounds studied were monomeric after incubation for a few days at 20 or 4 °C (see later). Prolonged storage of the solutions at room temperature in most cases, however, led to their partial degradation and/or structural changes. This was apparent from sedimentation equilibrium analysis, which revealed the appearance of components smaller than grid monomers, as well as the concomitant appearance of aggregated material, from ^{1}H NMR measurements on Cd or Zn grids [6], which showed changes in the shape and the intensity of the peaks as well as the appearance of new ones, and frequently by visual inspection, which revealed changes in the color of the solutions. However, the mechanism of conversion and the structure of the end products is not yet known.

In the time interval between incomplete solubilization and decomposition of the compounds (i.e., a few days until approximately 1–3 weeks after addition of the solvent), in appropriate solvents and at sufficiently low solute concentrations, all compounds studied were virtually homogeneous and monomeric. This was shown by sedimentation equilibrium analysis. The homogeneity is demonstrated, for solutions of the compounds in acetone, in Fig. 4. (The ln A-versus-r^2 plots in the figure were chosen for ease of presentation only; the actual fits were of the $A(r)$-type and were transformed afterwards).

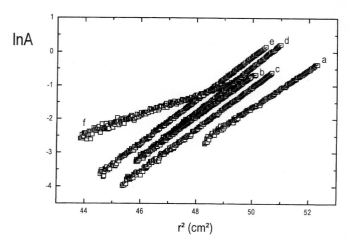

Fig. 4 Sedimentation equilibrium analysis of the Co coordination arrays (cf. Fig. 2): Plots of ln A versus r^2. Solvent: acetone plus 10–50 mM NH$_4$PF$_6$. Sample concentration was approximately 1.5 µM (5 mg/l). Rotor speed: 40,000 rpm

Table 1 Experimental M_{eff}, average $\bar{v}$ and experimental and theoretical M_r values for all compounds of Fig. 2. All figures based on experiment refer to runs at 20 °C using, as a solvent, acetone plus 20 mM NH$_4$PF$_6$

Compound	M_{eff} (Da)	$\bar{v}$ (ml/g)	M_r (exp.)	M_r (calc.)
a	1538	0.624 ± 0.008	3046	3061
b	1702	0.740 ± 0.008	4122	4104
c	1666	0.612 ± 0.010	3238	3189
d	1791	0.594 ± 0.010	3387	3366
e	1993	0.710 ± 0.010	4564	4520
f	520	0.810 ± 0.010	1455	1393

Applying $\bar{v}$ values representing critically weighted averages from different ultracentrifuge methods (see later), the corresponding molar masses were found to be those of the monomeric grids (the deviations from the theoretical value being, with one exception, within the range set by the uncertainty of $\bar{v}$) (Table 1). In some cases, higher compound concentrations led to the appearance of small oligomers or clusters. For compound f, at a 15-fold higher concentration than used in the experiment reported in Fig. 4, this is demonstrated in Fig. 5. Here, a good fit to the data could be obtained by a monomer/trimer model of grid self-association; the trimer content, averaged over the sample volume, was approximately 23%.[1] Virtually the same results were obtained if acetone was replaced with acetonitrile. On the other hand, in propylene carbonate (which is considered as being a good solvent for complexes of

---

[1] A 2% decrease in the sum of the squared residuals, σ, was obtained when a monomer/dimer/tetramer model of self-association was applied, which indicates that this may be the true model of self-association

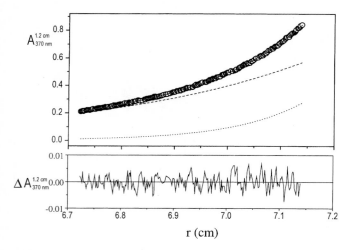

Fig. 5 Sedimentation equilibrium analysis of grid f. *Upper plot*: Experimental absorbance $A_{370\,nm}^{1.2\,cm}$ as a function of the radial position ($r(o)$,) and a least-squares fit to the data assuming the presence of grid monomers and trimers (—). The plot also shows the calculated absorbance contributions of the monomer (- - - -) and the trimer ($\cdots$). *Lower plot*: Local differences $\Delta A(r)$ between experimental and fitted data. Solvent: acetone plus 20 mM NH_4PF_6. Initial compound concentration: 54 μM (75 mg/l), corresponding to $A_{370\,nm}^{1.2\,cm}(r) = 0.4$. Rotor speed: 40,000 rpm

organic molecules and metal ions [17]) low concentrations of the same compound yielded an average molar mass $\bar{M}$ corresponding approximately to the tetramer.

Solute density

The density of the Co coordination arrays or its reciprocal, the partial specific volume, $\bar{v}$, of the compounds (or their apparent specific volume), was found to depend on a variety of parameters. First hints for this behavior were failures to fit experimental absorbance-versus-radius data from sedimentation equilibrium runs by using the $\bar{v}$ values obtained from digital densimetry: deviations of up to more than 20% were found between figures from densimetry and the lowest figures that allowed acceptable fits to the $A(r)$ data. This is demonstrated, for grid f dissolved in acetone, in Table 2. As also shown in the table, $\bar{v}$ values from

digital densimetry measured in acetonitrile solutions allowed very good fits to the $A(r)$ data from acetone solutions. Failures of densimetry to yield figures useful for fitting the $A(r)$ data were observed with other grids as well as other solvents. Checks on the purity of the compounds, including elemental analysis (which agreed with the theoretical values within $\pm 0.4\%$ for each element), showed that these discrepancies were not due to sample impurity. It follows that, at the high compound concentrations required in the density measurements, the Co coordination arrays, at least in acetone, can presumably be in a state (cluster, microcrystal) with a density distinctly different from that at the compound concentrations applied in ultracentrifuge experiments.

The fact that the $\bar{v}$ value obtained in acetonitrile is within the range of the figures determined from sedimentation equilibrium analysis in acetone may indicate that acetonitrile is more efficient in depolymerizing the microcrystals/clusters of the compound. Different attempts were, therefore, made to determine $\bar{v}$ from sedimentation equilibrium measurements. Experiments on the dependency of the effective molar mass of grid f on solvent density, in two different series of mixtures of two solvents, are shown in Fig. 6. Fitting was done with a single exponential; this worked well since the curves were quite flat, but does not prove that the samples were homogeneous. It is obvious from the figure that each series yields a quite accurate value for the density which matches that of the solute, i.e. the reciprocal of the compound's apparent $\bar{v}$. On the other hand, the two $\bar{v}$ values differ from each other by approximately 7%; thus, solute density apparently depends on the nature of the solvent used. To circumvent this problem, we applied the classical Edelstein–Schachman procedure for the simultaneous determination of $\bar{v}$ and M, which uses a pair of sedimentation equilibrium data collected in a solvent's nondeuterated and deuterated form [16]. Due to the small density difference between the two solvents, the results were, however, of relatively low accuracy. For grid f, the resulting $\bar{v}$ value and its experimental error are included in Table 2. The uncertainty in $\bar{v}$ for the other grids studied ranged up to ± 0.06 ml/g.

Table 2 $\bar{v}$ values for the Co coordination arrays based on organic ligand f of Fig. 2, as obtained by different methods (20 °C)

Method	$\bar{v}$ (ml/g)
Digital densimetry in acetone	0.681 $\pm$ 0.019 ($n = 16$)
Digital densimetry in acetonitrile	0.791 $\pm$ 0.006 ($n = 8$)
Determination from fits to $A(r)$ data in acetone	
Assuming a monomeric state of the sample	0.801 $\pm$ 0.006
Assuming the presence of monomers plus oligomers	0.796–0.85
Determination from $M_{eff} = 0$	
In dichlorethane/dichlormethane	0.762 $\pm$ 0.008
In acetonitrile/deuterated tetrachlorethane	0.815 $\pm$ 0.012
Edelstein–Schachman method in acetone/deuterated acetone	0.807 $\pm$ 0.020

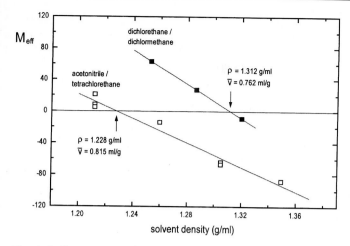

Fig. 6 Sedimentation equilibrium analysis of grid f, in mixtures of dichlorethane/dichlormethane (■) or acetonitrile/deuterated tetrachlorethane (□). The figure shows the dependency of the evaluated effective mass on solvent density, assuming homogeneity of the sample. The solvent density leading to zero effective mass is equal to the reciprocal of the compound's apparent $\bar{v}$; it corresponds to 12:88 (v/v) and 48:52 (v/v) mixtures, respectively. The loading concentrations of grid f were approximately 50 mg/l. Rotor speed: 40,000 rpm

Discussion

The solution properties of the Co coordination arrays studied in this paper were found to be very complex with respect to both main aspects considered: solubility and density. Despite the experience gained, we consider the details of the process of solubilization of the compounds as barely predictable, both with respect to the time course and the final state of association. The history of the sample, the solvent, and the type of salt present clearly represent the most serious among the influencing factors. These problems are quite obstructive, since a uniform state of association, or even a monomeric state, seems to be necessary, for example, when trying to assemble regular surface layers of the compounds on a substrate [18]; thus, control of the compound's state of association apparently has to be an obligatory part of such experiments.

The dependency of the compounds' densities, or their apparent specific volumes, on the solvent and on the state of the sample previously described is not really surprising. Similar findings were reported from macromolecular chemistry [19] and biochemistry (where apparent specific volumes of proteins in the presence of high concentrations of certain additives such as salts, buffer components, etc., can exceed the values determined in their absence by more than 10% [20, 21]). However, the findings also mean that, under most experimental conditions, reliable and, at the same time, accurate $\bar{v}$ data cannot easily be obtained, which distinctly limits the application of analytical ultracen-

trifugation to determine the compounds' state of association. At present, the usefulness of $\bar{v}$ values determined by the standard technique, digital densimetry, does not seem to be granted, because of the high concentrations required for precise experiments. Results of limited reliability are also to be expected from two methods for determining $\bar{v}$ by ultracentrifuge measurements: determination of the solvent density leading to zero effective molar mass of the solute, and application of the Edelstein–Schachman method [16] to the data collected in solvents of different density. At present, the only method we found applicable without principal objections is the original Edelstein–Schachman method, i.e., combining sedimentation equilibrium data collected in the nondeuterated and the deuterated form of the same solvent for the simultaneous determination of $\bar{v}$ and M.[2] However, this method also has distinct disadvantages:

1. Due to the very small density differences, even for the most advantageous solvent pairs (acetone, toluene), the uncertainty in the results is relatively high (in our experience more than 0.015 ml/g even if a number of individual data can be averaged).
2. It requires homogeneity of the solute.

It may be possible, however, to apply the method to a series of related compounds most of which are heterogeneously associated, using a "calculus of differences" [8, 22], provided that at least one member of the series is monomeric or homogeneously associated. At present, weighted average values of the figures for $\bar{v}$ from the different ultracentrifuge methods (see Materials and methods) seem to be the most useful ones. Weighting will have to consider not only the experimental error of the respective figure but also the polarity of the solvents used (e.g., in Table 2 we could reject, besides the figures from digital densimetry, the one obtained in dichlorethane/dichlormethane because of the much lower polarity of that solvent mixture compared to acetone; we would give less weight to the second figure obtained from curve-fitting, due to its low accuracy, ending up with $\bar{v} = 0.810 \pm 0.010$ ml/g). Of course, collecting the different data is a laborious task. On the other hand, the success of the procedure is surprisingly good, as indicated by the close agreement between the experimental and theoretical M_r values (Table 1). We conclude that, despite the problems described, analytical ultracentrifugation is still the best technique among those applicable to the study of the association behav-

---

[2] It has been reported that another supramolecular compound has different densities in nondeuterated and deuterated toluene [23]. The density differences amount, however, only to approximately 0.7%, which is much smaller than the uncertainty in our present $\bar{v}$ values and thus does not significantly hinder application of the method

iour of supramolecular compounds. Besides sedimentation equilibrium analysis, two other ultracentrifuge techniques, sedimentation velocity analysis or the modern version of the "approach to equilibrium" [9], could be applied.

Another important problem revealed by the present study is the instability of the Co coordination arrays in dilute solutions. The conversions occurring may, at least in some of the compounds, be quite drastic. With compounds d and e, we have even observed a distinct decrease in $\bar{v}$ with time by applying the Edelstein–Schachman method (which accounts for the differences between this paper and Ref. [9]). Elucidating the nature of the conversions and designing more stable varieties of the compounds seem to be a worthwhile task.

Acknowledgements This study was supported in part by the Bayerisches Staatsministerium für Wissenschaft, Forschung und Kunst (Bayerischer Habilitationsförderpreis for USS) and the Deutsche Forschungsgemeinschaft (Schu 1229/2-1 and SFB 266, B15).

References

1. Lehn J-M (1995) Supramolecular chemistry: concepts and perspectives. VCH, Weinheim
2. Kotera M, Lehn J-M, Vigneron J-P (1994) J Chem Soc Chem Comm 1994:197
3. Lehn J-M, Rigault A, Siegel J, Harrowfield J, Chevrier B, Moras D (1987) Proc Natl Acad Sci USA 84:2565
4. Baxter PNW, Lehn J-M, DeCian A, Fischer J (1993) Angew Chem 105:92
5. Hanan GS, Schubert US, Volkmer D, Riviere E, Lehn J-M, Kyritsakas N, Fischer J (1997) Can J Chem 75:169
6. Hanan GS, Volkmer D, Schubert US, Lehn J-M, Baum G, Fenske D (1997) Angew Chem 109:1929
7. Schubert US, Weidl CH, Lehn J-M (1999) Design Monom Polym 2:1
8. Schubert D, van den Broek JA, Sell B, Durchschlag H, Mächtle W, Schubert US, Lehn J-M (1997) Prog Colloid Polym Sci 107:166
9. Schubert D, Tziatzios C, Schuck P, Schubert US (1999) Chem Eur J 5:1377
10. Schubert US, Eschbaumer C, Hochwimmer G (1998) Tetrahedron Lett 39:8653
11. Schubert US, Lehn J-M, Hassmann J, Hahn CY, Hallschmidt N, Müller P (1998) In: Patil AO, Schulz DN, Novak BM (eds) Functional polymers. American Chemical Society Symposium Series, vol 704. American Chemical Society, Washington, D.C., p 248
12. Salditt T, An Q, Plech A, Eschbaumer C, Schubert US (1998) Chem Commun 1998:2731
13. Schuck P (1994) Prog Colloid Polym Sci 94:1
14. Schuck P, Legrum B, Passow H, Schubert D (1995) Eur J Biochem 230:806
15. Kratky O, Leopold H, Stabinger H (1973) Methods Enzymol 27:98
16. Edelstein SJ, Schachman HK (1967) J Biol Chem 242:306
17. Nelson RF, Adams RN (1967) J Electroanal Chem 13:184
18. Semenov A, Spatz JP, Möller M, Lehn J-M, Sell B, Schubert D, Weidl CH, Schubert US (1999) Angew Chem 111: (in press)
19. Klärner PEO, Ende HA (1975) In: Brandrup J, Immergut EH (eds) Polymer handbook, 2nd edn. Wiley, New York, IV:61
20. Durchschlag H (1986) In: Hinz H-J (ed) Thermodynamic data for biochemistry and biotechnology. Springer, Berlin Heidelberg New York, p 45
21. Shima S, Tziatzios C, Schubert D, Fugata H, Takahashi K, Ermler U, Thauer R (1998) Eur J Biochem 258:85
22. Durchschlag H, Zipper P (1994) Prog Colloid Polym Sci 94:20
23. Schilling K (1999) PhD thesis. Potsdam

Progr Colloid Polym Sci (1999) 113 : 121–128
© Springer-Verlag 1999

A. Böhm
S. Kielhorn-Bayer
P. Rossmanith

Working with multidetection in the analytical ultracentrifuge: the benefits of the combination of a refractive index detector and an absorption detector for the analysis of colloidal systems

A. Böhm · S. Kielhorn-Bayer
P. Rossmanith (✉)
BASF Aktiengesellschaft
Polymer Research Laboratory
D-67056 Ludwigshafen, Germany
e-mail: peter.rossmanith@basf-ag.de
Tel.: +49-621-6093985
Fax: +49-621-6092181

Abstract The XL-I is an analytical ultracentrifuge equipped with two detection systems: an absorption detector and an interference optical system. For the analysis of colloidal systems the absorption detector is normally denoted as a special detector, since, if Mie scattering is not taken into account, most polymers are transparent to UV and visible light. The interference optics is a detector for measuring the refractive index (RI) difference and is denoted in this context as a universal detector. Most systems in colloidal science of synthetic colloids can only be analyzed with the refractive index detector. If a colloidal (polymeric) system contains components detectable with an absorption detector, the combination of the two detectors of the XL-I will provide additional information about the sample. This paper will show that the specific combination of the de-tectors described above can not only give a deep insight into the colloidal structure of the sample, but can also help to understand the mechanism of the polymerization reaction applied to generate the dispersion under investigation. This can be achieved by thorough sedimentation velocity analysis in both an aqueous and an organic medium. Spectra before and after the runs yield information about the general whereabouts and the spatial distribution of the absorbing component. Density gradient analysis with both detectors can determine the chemical differences between particles containing or not containing the absorbing component.

Key words Analytical ultracentrifu-gation · Interference optics · Absorbance optics · Particle size distribution · Density gradient

Introduction

The Beckman Optima XL-I is an analytical ultracentri-fuge (AUC) equipped with two independent detection systems: an interference optical system and an absorp-tion detector [1, 2]. The interference optical system is a detector sensitive to changes in the refractive index (RI) of the sample. Since normally there is a difference between the refractive index of the sample and that of the solvent, this detection method can be applied very generally in investigations on colloidal systems and the corresponding detector therefore can be denoted as universal [3]. The second detector is an absorption detector with a wavelength range from 180 to 800 nm. This detector is sufficient when dealing with proteins in the analysis of biochemical systems (see, for example, [4]), but in synthetic polymer and dispersion analysis, most systems made from standard monomers are transparent for UV and visible light, and the applicability of the absorption detector is limited to special cases [5].

If there are absorbing components present, the combination of the absorption and RI detectors can supply information not only on the structure of the particles [6] or macromolecules [7–10] under investigation. It will be shown in this article that it is also possible to obtain insights into the mechanism of the polymerization reaction leading to colored particles.

The analysis of colored systems in an AUC with a Schlieren optical system (the Beckman Model E or the XL with the Schlieren system [11, 12]) can be performed in different ways. In the conventional experiment, the Schlieren line provides information on the RI distribution. This is done using monochromatic light flashes. Since this particular optical system works with a flash lamp, it is also possible to illuminate the sample with white light. If photographs are taken with a color film, the distribution of the color throughout the sample cell can be seen directly on the image, with the Schlieren curve superimposed. So both types of information are combined within one picture.

Colorant-containing nanoscopic polymer lattices, so called NanoColorants [13], exhibit a unique property profile, combining the coloristic advantages of classical dyestuffs (brilliance, color strength, stability of shade) with those of organic pigments (migration, light and weather fastness). They are therefore suitable for a number of commercial applications, ranging from the coloration of ink-jet inks, printing inks, plastics, and coatings to the mass coloration of paper and textiles.

Sample preparation

A monomer solution (comprising methyl methacrylate or styrene, co-surfactant, cross-linker, and dyestuff) was added to an aqueous solution of a surfactant (e.g. sodium dodecyl sulfate) and, if desired, a water soluble comonomer like acrylic acid. The stirred mixture was homogenized with ultrasound to prepare a miniemulsion. This aqueous monomer miniemulsion was charged into a polymerization vessel and heated to 80 °C. Subsequently, a free-radical polymerization initiator was added. The resulting mixture was polymerized at 80 °C for 3.5 h. The resultant aqueous polymer dispersions (20% solids by weight) were diluted by a water/ K30 mixture to appropriate concentrations for the AUC experiments.

AUC analysis of the samples

Solubilization of the dye

According to the NanoColorant concept, the dye should be completely incorporated in the latex particles. Therefore, before dealing with the analysis of the particles themselves, it is reasonable to look whether

or not the dye is really located inside the particles. The easiest way of doing this is by looking at the absorption spectrum before and after a sedimentation velocity run (s-run). The two spectra in Fig. 1 can be taken as a proof of concept. Before the s-run (solid line), two types of signals can be found in the resulting spectrum. There are peaks at 450 nm and between 500 and 600 nm, which can be attributed to the absorption of the dye. At wavelengths below 300 nm, transmission is essentially zero. Above 300 nm, the absorption signal looks like a decaying exponential function. This is a typical signal from Mie scattering: Mie scattering is a unique property of particles; the light is not absorbed by the sample but scattered away from the detector. Before the s-run there are particles and dye at the radial position of detection. The spectrum after the run (dashed line) shows no absorption over the complete spectral range. So neither particles nor dye can be detected at this position. Since the dye has a low molecular weight, it must be completely incorporated in the particles, otherwise at least some absorption would be detectable.

Particle size analysis

An analysis of the particle size distribution (psd) of all particles can be done either by a sedimentation velocity run and detection with the Schlieren or the interference optics, or with a particle sizer via turbidity measurement [14]. For colored systems the use of the turbidity detector is difficult owing to the complex RI of the particles. So the more reliable values result from a detection via the RI, e.g. via interference or Schlieren optics.

The original data of an s-run detected with the interference optics of the XL-I at different times are

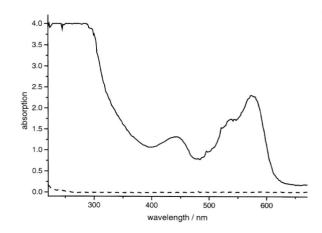

Fig. 1 Absorption spectrum of a colored sample before (*solid line*) and after (*dashed line*) an s-run. The spectra were taken at a radial position of 6.9 cm

plotted in Fig. 2. Two moving boundaries can be extracted from the plot. From these data the particle size distribution was calculated according to Stoke's law[1] [14]. The differential (solid line) and the integral (dashed line) particle size distribution can be seen in Fig. 3. The distribution is bimodal with a narrow line at about 87 nm diameter and a broad line with a shoulder at about 50 nm. From the shape of this distribution, some conclusions as to the mechanism of the polymerization reaction can be drawn. The NanoColorants are made via miniemulsion polymerization, thus implying that every emulsion droplet will independently polymerize to a latex particle. Therefore the particle size distribution of the droplets has to be the same as the particle size distribution of the latex. Since an emulsion always has a unimodal distribution, the outcome of this polymerization reaction is inconsistent with the proposed miniemulsion polymerization mechanism.

The bimodal distribution of the latex particles raised the question of whether or not all particles are colored. This was investigated by an s-run using the absorption detector. A wavelength for the detection was chosen such that there was sufficient absorption from the dye, but negligible contribution from Mie scattering. Under these conditions the particle size distribution showed only one peak at about 85 nm and a shoulder at 50 nm (Fig. 4). In this plot the peak at 85 nm is much broader compared with Fig. 3, where the interference optics was used for detection. This apparent contradiction is due to differences in the time necessary to collect the data

corresponding to one scan. With the absorption detector, one complete scan takes about 1 min. If the boundary is moving fast, it has moved substantially while the data are collected. Since detection starts at the bottom of the cell, and scans down to the meniscus, the resulting band is broader than the real one. This problem does not occur with the interference optics. With this system an image from the CCD camera is analyzed. The time for collecting an image at the CCD camera is exactly fixed.

The particle size distribution of all particles, as collected with an RI detector, is bimodal. Analysis with an absorption detector, where only the colored particles are visible, however, results in a unimodal distribution. This implies that not all particles are colored. This very simple conclusion has severe consequences for the mechanism of the polymerization. The miniemulsion at the beginning of the reaction was homogenous both in droplet size and in composition. On the other hand, the particle size distribution of all particles and of the coloured particles alone are different, thus suggesting inhomogenities in both particle size and composition of the resulting polymer dispersion. So one has to assume that at least part of those particles have been formed via a classical emulsion polymerization with monomer diffusion through the water phase. Further on, the uncolored particles could have only been formed via a homogeneous nucleation process, because the dye itself is completely insoluble in water.

Particle density analysis

From the particle size analysis we know that there are particles with different amounts of dye incorporated. Therefore the question arises of whether this is the only difference in the chemistry of the particle. The density gradient experiment is a very powerful tool for the analysis of chemical inhomogeneities [15]. For the sample under investigation, a gradient was chosen with a mixture of 15 wt% metrizamide and 85 wt% D_2O as gradient building solution. Metrizamide is a very dense nonionic sugar [15], so that no charge is introduced into the system. The concentration of the latex was 0.5 g/l in the sugar solution. To establish the gradient, the centrifuge ran for 96 h at 25 000 rpm. The interference pattern and the fringe displacement at equilibrium are plotted in Fig. 5. The interference pattern shows a very steep gradient with two turbid spots in the cell. The representation of the concentration distribution within the cell as a fringe displacement exhibits tiny peaks at the turbid positions in the cell (Fig. 5). Near the bottom of the cell the gradient is too steep to be properly detected by the interference optics.

In order to determine the density distribution in the cell, the signal without the two tiny peaks, which is the

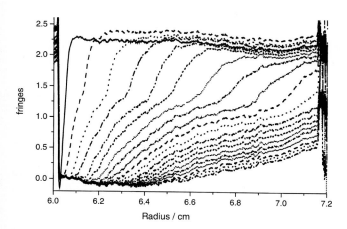

Fig. 2 Sedimentation velocity run detected with the interference optics of the Beckman XL-I. Concentration distribution of the sample in the measuring cell at different times. The time difference between two displayed scans was 2 min. The loading concentration was 1 g/l. The centrifuge ran at 6000 rpm

---

[1] $\rho_{particle} = 1.205$ g/cm^3, $\rho_{disperion\ medium} = 0.997$ g/cm^3, $\eta_{dispersion\ medium} = 0.00891$ g/(s*cm). The density of the particles was calculated via $\bar{v} = \sum_i m_i \bar{v}_i$

Fig. 3 Integral (*dashed line*) and differential (*solid line*) particle size distribution

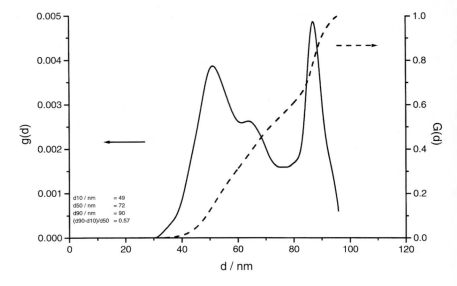

Fig. 4 Integral (*dashed line*) and differential (*solid line*) particle size distribution of the colored particles detected with the absorption detector at 575 nm. The sample under investigation and the conditions of the measurement were the same as for the measurement with the interference optics

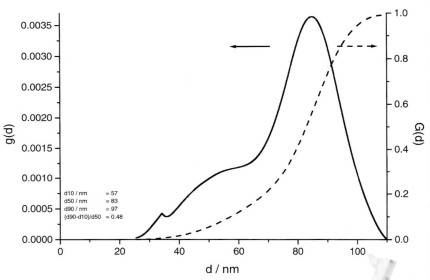

contribution from the sample, is analyzed [3]. The region where the gradient is too steep for a proper detection was cut out, and the signal was extrapolated to the bottom of the cell. Though this procedure seems to be rather error-prone, previous measurements with samples of well-known density showed that the density was calculated correctly by this method (P. Rossmanith, unpublished). Subtraction of the so-determined gradient from the original data yields the density distribution plotted in Fig. 6. Even if an error produced by the extrapolation described above is taken into account, the experiment shows clearly that there are particles with two different densities present in the sample. Since the density is an indicator for the chemical composition, there must be particles with different compositions in the sample.

From the s-run it is already known that the sample consists of particles with dye and others with less or even without dye. With the help of the absorption optics it should again be possible to distinguish between colored and uncolored particles. Therefore, in addition to the interference scan, an absorption scan at a wavelength near the absorption maximum of the dye was made. The results are shown in Fig. 6 as a dashed line. Particles with incorporated dye have a different density compared with particles without the dye. According to Fig. 6 it is also clear that there are particles containing no dye. The density distribution of the colored particles is broader than the one of the uncolored particles. This cannot only be seen from the broader peak of the absorption run, but also from the rising edge of the RI measurement. The steepness of the curve at the low density side, where the colored particles are located, is much smaller than that on the high density side, to where the uncolored particles have moved.

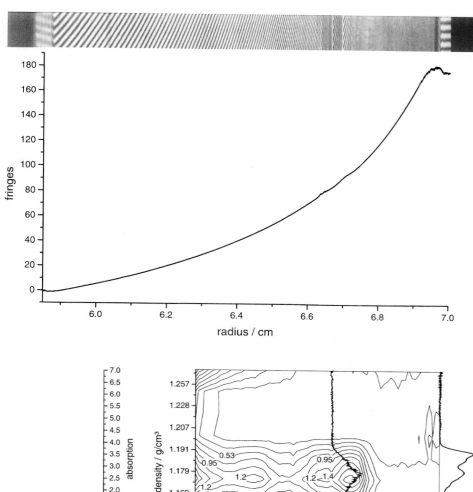

Fig. 5 Interference pattern and raw data of the density gradient run at equilibrium

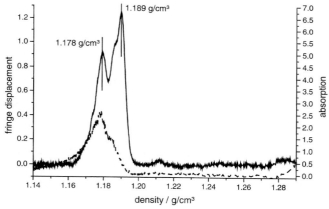

Fig. 6 Density distribution of all the particles (*solid line*) as achieved from the analysis of the data of Fig. 5 and of the colored particles as detected with absorption at 550 nm (*dashed line*)

Fig. 7 Contour plot of the absorption in the cell. Each line represents one absorption level. The absorption of some important lines are indicated in the plot. The radial distribution of the absorption at 550 nm is plotted as a separate line, to show that the maximum of the contour plot is slightly shifted relative to the maximum of the real distribution. This is due to the lack of the density in the collected data points. In order to show the density distribution of the whole dispersion, the result from the refractive index measurement is shown on the right-hand side of the plot

Working with just one wavelength of the absorption optics, there is always the danger of obtaining artifacts. To overcome this problem, spectra were taken at several positions in the cell [16]. Figure 7 shows a contour plot of the absorption distribution in the cell together with the absorption scan at 550 nm and the distribution from the RI scan. The plot shows the spectrum of the dye at the position of the less dense peak from the distribution of the RI scan. At the position of the more dense peak the features of the spectrum are mostly washed out, and there is only a rise of the absorption at low wavelengths that arises from Mie scattering. This plot shows that,

with the absorption detector of the XL-I, chemical information on samples with absorbing components can be retrieved. In this case, the effective distribution of the dye in a density gradient experiment was obtained.

The density gradient experiments produced two results: (1) there are particles with different densities and (2) the density of the colored particles is lower than that of the uncolored ones. The difference in density of these two types of particles is too large for being solely related to the different contributions of the dye. There has to be an additional difference in composition, e.g. a different amount of incorporated co-surfactant. These results can help to reinterpret the particle size distribution:

— Since both particle types have different densities, the psd should be recalculated for both types individually. Fortunately, the difference in densities is small (0.011 g/cm^3), which also makes the shift in particle sizes small (< 3 nm) and a recalculation avoidable.
— The psd of the colored particles (Fig. 4) showed a shoulder at small diameters where the psd of all particles (Fig. 3) had a second maximum. Since it is now clear that there are uncolored particles, this maximum of the overall psd is made up by two different types of particles: colored and uncolored ones. The deconvolution of this peak would require more experiments, without delivering additional information for the understanding of the formation of the NanoColorants.

Micro gel analysis

Up to now we have learned that:

— The dye is incorporated in latex particles.
— There are particles with dye and particles without.
— Colored and uncolored particles have different diameters.
— Colored and uncolored particles have different chemical composition.

However, no information could be gathered so far on how the dye is bound inside the particles. Two different possibilities are imaginable: the dye can either be chemically bound to the polymer chains, or it can be just physically fixed inside the particle. Dissolution of the particles in a good organic solvent should make these two possibilities distinguishable. Unfortunately, the dispersions are made with a certain amount of cross-linker. So the particles can only swell, but not dissolve completely in the solvent [17, 18]. For the systems under investigation, THF is a good solvent and has therefore been used for these experiments. However, if the dye is held inside the particles only by physical fixation, it should still have the ability to migrate into the solvent phase; otherwise it should sediment together with the particles.

Sample 1

Figure 8 shows the sedimentation coefficient distribution of a NanoColorant sample diluted in THF. The mean

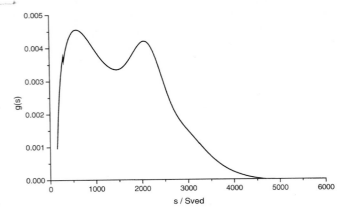

Fig. 8 Sedimentation coefficient distribution of the gel particles. The data were taken from a run at 3000 rpm with the interference optics as detector

sedimentation coefficient of several thousand Sved indicates that the sedimenting species are microgel particles [18]. The distribution has two peaks and a shoulder accounting for fast sedimenting species. The shape of the distribution of the microgel particles corresponds well to the bimodal particle size distribution discussed above. Therefore both types of particles do contain some crosslinker. As both species are sedimenting relatively fast, the amount of crosslinker and the degree of swelling is similar for all particles.

When the same experiment is carried out with the absorption detector working near the absorption maximum, no change in the absorption can be found for the whole run. Absorption spectra taken before and after the run show only small differences (Fig. 9). The solid line (spectrum before sedimentation) indicates a slightly higher absorption at short wavelengths than the dashed

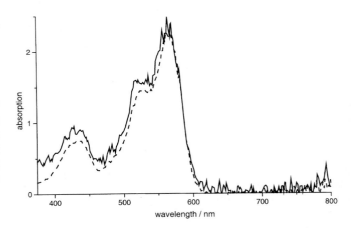

Fig. 9 Absorption of the sample before and after an s-run of the sample diluted in THF. Since there is almost no change in the spectrum, the dye is mobile enough to migrate out of the swollen particles

line (spectrum after the run). These small differences are due to contributions from Mie scattering of the microgel particles in the system.

The combination of the two detectors gave the information that in case of sample 1 the dye was physically held inside the microgel particles. When dissolving the dispersion in an organic solvent, the dye becomes mobile enough to migrate out of the swollen network from the particle into the liquid phase. No dye remained inside the particles.

Sample 2

The experiments described in the last paragraph were also carried out with another sample (sample 2). The sedimentation coefficient distribution, as detected with the RI detector (Fig. 10; wide spaced dots represent the differential distribution, long dashes the integral), is bimodal. This distribution again reflects the bimodal psd of the latex in water. The same experiment with the absorption optics at a wavelength near the absorption maximum of the dye also yields a bimodal sedimentation coefficient distribution, as shown in Fig. 10 (small spaced dots represent the differential distribution, short dashes the integral). The swelling in an organic solvent did not give the dye enough mobility to migrate out of the particle. So the dye in this sample must have been chemically bound to the polymer chains of the latex.

A closer look at Fig. 10 reveals some difference in the distributions. The sedimentation coefficients derived from the peak maximum and the shoulder of the fast peak are slightly shifted to higher s-values when comparing the results from the analysis with the absorption detection with those from RI detection. As discussed before for the differences in the psd measured with these two detectors, this shift is an artifact due to the time needed for detection with the absorption optics. The difference in the broadness of the bands is just an apparent one. More important than this artifact is the difference in the relative intensities of the peaks representing the faster and the slower moving microgel particles. This difference can easily be seen by comparing the two integral distributions. The amount of slow moving colored particles is about 10% (Fig. 10; height of the first step of the short dashed curve, representing the integral distribution of the experiment with the absorption detector). If all particles are taken into account, i.e. the integral distribution from the run detected with the RI detector, a lot more (30%) slow moving particles are detected (Fig. 10; long dashed curve). Therefore there have to be uncolored slow-moving microgel particles. So the combination of the two detectors could prove any differences in the amount of dye chemically bound to different particles. So the results of the s-run in THF underline the results from the same experiment in water.

Conclusions

This paper showed that colloidal samples containing a component absorbing in a wavelength range between 180 and 800 nm can be very efficiently analyzed with the two detectors of the XL-I. As an example for colored systems, a novel class of nanoscopic polymeric colorants, the so-called NanoColorants, were investigated. Beside a lot of information gathered by the use of a single detection unit sensitive to the RI of the sample, the combination of two detectors as described above delivered detailed information about the distribution of the dye in the dispersion. We were able to show that the dye is completely located within the particles. However, not all particles are equally colored. The less dense particles do contain dye, whereas the more dense ones do not. The particle size distributions of the colored and the noncolored particles are different.

In order not to overload this paper, the results from investigations with the Schlieren optics were not depicted. The experiments with this optical system delivered similar results when colored Schlieren photographs were taken, which include the information stored in the color. The Schlieren optics of course fails when dealing with systems showing only UV absorption, but no color on color prints.

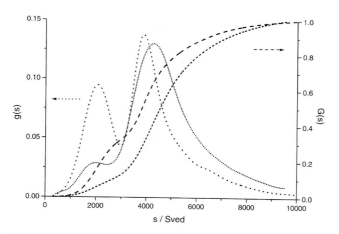

Fig. 10 Differential and integral sedimentation coefficient distribution of sample 2. The *wide spaced line symbols* represent the distribution as detected with the refractive index detector. The *narrow spaced symbols* belong to the distribution from the absorption detector at 553 nm

References

1. Ortlepp B, Panke D (1991) Prog Colloid Polym Sci 86:57–61
2. Mächtle W (1991) Prog Colloid Polym Sci 86:111–118
3. Rossmanith P, Mächtle W (1997) Prog Colloid Polym Sci 107:159–165
4. Winzor DJ, Jacobsen MP, Wills PR (1998) Biochemistry 37:2226–2233
5. Karibyants N, Dautzenberg H, Cölfen H (1997) Macromolecules 30:7803–7809
6. Cölfen H, Pauk T, Antonietti M (1997) Prog Colloid Polym Sci 107:136–147
7. Furst A (1997) Eur Biophys J 25:307–310
8. Voelker P (1995) Prog Colloid Polym Sci 99:162–166
9. Cölfen H, Harding SE, Vårum KM, Winzor DJ (1996) Carbohydr Polym 30:45–53
10. Ortlepp B, Panke D (1992) Makromol Chem Macromol Symp 61:176–184
11. Mächtle W (1999) Prog Colloid Polym Sci 113:1–9
12. Clewlow AC, Errington N, Rowe AJ (1997) Eur Biophys J 25:311–317
13. Iden R, Böhm A, Erk P, Hackmann C, Mronga N, Schmid R, Schumacher P (1998) Presented at the High Performance Pigments Conference, Barcelona, Spain
14. Mächtle W (1992) In: Harding SE, Rowe AJ, Horton JC (eds) Analytical ultracentrifugation in biochemistry and polymer science. Royal Society of Chemistry, Cambridge, pp 147–175
15. Rickwood D (1992) Metrizamide. A gradient medium for centrifugation studies. Nyegraad, Oslo, Norway
16. Schuck P (1994) Prog Colloid Polym Sci 94:1–13
17. Müller HG, Schmidt A, Kranz D (1991) Prog Colloid Polym Sci 86:70–75
18. Mächtle W, Ley G, Streib J (1995) Prog Colloid Polym Sci 99:144–153

Progr Colloid Polym Sci (1999) 113 : 129–134
© Springer-Verlag 1999

N.A. Chebotareva
B.I. Kurganov
A.A. Burlakova

Sedimentation velocity analysis of oligomeric enzymes in hydrated reversed micelles of surfactants in organic solvents

N.A. Chebotareva (✉) · B.I. Kurganov
A.A. Burlakova
A.N. Bach Institute of Biochemistry
Russian Academy of Sciences
Leninsky prospekt 33
Moscow 117071, Russia
e-mail: inbio@glas.apc.org
Tel.: +7-095-9525641
Fax: +7-095-9542732

Abstract The oligomeric state and formation of supramolecular structures of glycogen phosphorylase b from rabbit skeletal muscles have been studied in the system of hydrated reversed micelles of sodium bis-2-ethylhexyl sulfosuccinate (aerosol OT, AOT) in octane. Sedimentation analysis shows that the oligomeric state of the enzyme is controlled by the degree of hydration of the micelles ($[H_2O]/[AOT] = w_0$). The monomeric (in the range of w_0 from 10 to 16), dimeric ($10 < w_0 < 30$), trimeric ($30 < w_0 < 38$), tetrameric ($23 < w_0 < 42$), hexameric ($41 < w_0 < 50$), or octameric forms ($48 < w_0 < 53$) of the enzyme were observed depending on the degree of hydration. Sedimentation behaviour of uridine phosphorylase from Escherichia coli K-12 in the micellar system was studied in the range of w_0 from 8.4 to 23.9. The monomeric (at $w_0 = 8.4$), dimeric (at $w_0 = 12.9$), trimeric (at $w_0 = 16.1$), tetrameric (at $w_0 = 18.6$) and hexameric (at $w_0 = 23.9$) enzyme forms were registered. The results obtained show that the hydrated reversed micelles are a powerful tool for the study of not only dissociated forms of oligomeric enzymes but also supramolecular structures. These latter structures mimic the ordered supramolecular complexes of the enzymes whose formation is favoured by the crowded molecular conditions encountered in vivo.

Key words Sedimentation · Reversed micelles · Phosphorylase b · Uridine phosphorylase

Introduction

Systems of hydrated reversed micelles of surfactants in organic solvents provide an effective tool for the study of the structure and function of enzymes [1–8]. The hydrated reversed micelles of sodium bis-2-ethylhexyl sulfosuccinate (aerosol OT, AOT) in octane are particularly useful in this regard. These micelles are characterized by a relatively narrow distribution of their dimensions, the dimensions of the inner cavity of a micelle being strictly governed by the degree of hydration of the micelles, namely the ratio of the concentration of H_2O to AOT, w_0 [9–10]. Hydrated reversed micelles are able to include enzymes whilst retaining their catalytic activity. In the case of oligomeric enzymes variation of the ratio of the concentration of H_2O to AOT allows dissociation of the enzymes under mild conditions to be induced and the enzymatic characteristics of the individual oligomeric forms to be studied (Table 1) [1–7].

In this paper we report the behaviour of glycogen phosphorylase b (EC 2.4.1.1) from rabbit skeletal muscle in such hydrated reversed micelles. Phosphorylase b consists of two identical subunits with molecular masses of 97.4 kDa each. Dimers of phosphorylase b are very stable at neutral values of pH and simple dilution of the enzyme solution does not result in the dissociation of dimers into monomers. The allosteric activator AMP

Table 1 Oligomeric state of enzymes in the sodium bis-2-ethylhexyl sulfosuccinate (*AOT*) water/octane micellar system

Enzyme	Quaternary structure of the native enzyme	Oligomeric state of the enzyme in the micellar system	References
Lactate dehydrogenase from pig skeletal muscle	Tetramer (144 kDa)	Monomer, dimer, tetramer, octamer	3
Glyceraldehyde-3-phosphate dehydrogenase from rabbit skeletal muscle	Tetramer (140 kDa)	Monomer, dimer, tetramer	1, 3
Alkaline phosphatase from calf intestinal mucosa	Dimer (146 kDa)	Monomer, dimer	3
γ-Glytamyltransferase	Heterodimer (75 kDa)	Light (21 kDa) and heavy (54 kDa) subunits, dimer	6
Penicillin acylase from *Escherichia coli*	Heterodimer (89 kDa)	Light (26 kDa) and heavy (63 kDa) subunits, dimer	4
G_{M1}-galactosidase from human kidney	Octamer (650 kDa)	Monomer, tetramer, octamer	5
Ketoglutarate dehydrogenase from pigeon breast muscle	dimer (150 kDa)	Monomer, dimer	3
Uridine phosphorylase from *Escherichia coli* K-12	Hexamer (165 kDa)	Monomer, dimer, trimer, tetramer, hexamer	This paper
Glycogen phosphorylase *b* from rabbit skeletal muscle	Dimer (195 kDa)	Monomer, dimer, trimer, tetramer, hexamer, octamer	This paper

induces the association of the enzyme with the formation of tetramers [11–16].

We have also studied uridine phosphorylase (EC 2.4.2.3) from *Escherichia coli* K-12 which catalyses phosphorolysis of uridine with formation of ribose 1-phosphate and uracil [17, 18]. This enzyme consists of six identical subunits with molecular masses of 27.5 kDa [19–21]. According to electron microscopy data [22], six subunits are disposed in the apexes of a triangular antiprism, and so the structure of the molecule is two-layered. The diameter of the molecule along the third-order symmetry axis is 9 nm, whereas the height of the two-layer structure is 6 nm. X-ray data support the view that the hexamer is formed from three dimers, two dimers in each trimer being connected by a noncrystallographic second-order symmetry axis [23]. Three dimers in a hexamer are connected by the third-order symmetry axis.

Materials and methods

AOT was purchased from Merck and Serva (Germany) and used without additional purification. According to the IR spectroscopic data, the preparations from Merck and Serva contained 2.5 and 0.85 mol of H_2O per mole of AOT, respectively. When calculating the degree of hydration of reversed micelles of AOT in octane, we took into account these values for the water content. *n*-Octane was purified by distillation over P_2O_5.

Phosphorylase *b* was isolated from rabbit skeletal muscles according to the method described by Fisher and Krebs [24] replacing cysteine by mercaptoethanol. In order to remove AMP, the enzyme solution was passed through a column with a mixture of activated coal Norit A and cellulose powder in the ratio 1:1. A useful criterion of purity of the enzyme was the ratio of the optical absorbances A_{260}/A_{280} which should not exceed 0.55.

Aqueous solutions of phosphorylase *b* with concentrations ranging from 5×10^{-6} to 10^{-4} M were prepared in 0.05 M glycyl glycine buffer (pH 6.8). In order to obtain the micellar system, a definite volume (14–96 μl) of the water enzyme solution was added to 1 ml 0.1 M solution of AOT in octane. The mixture obtained was vigorously shaken until the solution became transparent.

Uridine phosphorylase was prepared from *E. coli* K-12 (the superproducer of the enzyme) by the method described in Ref. [22] using 1 mM 2-mercaptoethanol for protection of the sulfydryl groups of the enzyme from oxidation. According to the results of electrophoresis in polyacrylamide gel [25], the enzyme preparation was homogeneous. Protein concentrations were determined either spectrophotometrically using an absorbance index of 0.67 cm^{-1} for a 0.1% solution at 280 nm [17] or by the Bradford assay [26].

The oligomeric state of phosphorylase *b* and uridine phosphorylase in the system of the reversed hydrated micelles of AOT in octane was studied by the sedimentation velocity method in a Spinco, model E (Beckman) analytical ultracentrifuge using the absorption scanning system, an An-G six-hole rotor and double-sector cells. The rotor speed was varied from 20 000 to 30 000 rpm. Scanning was carried out at 280 and 290 nm for the protein-containing micelles and at 405 nm for the empty micelles stained by 2,4-dinitrophenol. Prior to ultracentrifugation the samples were incubated for 1–2 h at 20 °C.

Results and discussion

Sedimentation of phosphorylase *b* entrapped in reversed micelles of AOT in octane was studied over a range of the degree of hydration, w_0, from 10 to 53. When changing the content of water in the system (in definite intervals of w_0) we observed the simultaneous presence of several oligomeric forms of the enzyme. As an example, Fig. 1 shows sedimentation of phosphorylase *b* entrapped in reversed micelles at ratios of the concentration of H_2O to AOT of 30 and 40. Apart from the boundary which corresponds to the empty micelles and is represented by the lowest value of the sedimentation coefficient, the sedimentation patterns contain two boundaries corresponding to the protein-containing micelles. At other values of w_0 the sedimentation patterns also included, as a rule, several boundaries of the protein-containing micelles. The values of the sedimentation coefficients for the empty and protein-containing reversed micelles at various values of w_0 are presented in Fig. 2. The errors in the measured sedimentation coefficients do not exceed 1 S. The sedimentation coefficients for the empty micelles

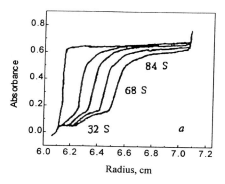

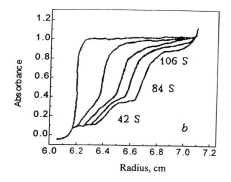

Fig. 1a, b Sedimentation of muscle phosphorylase *b* entrapped in the reversed hydrated micelles of sodium bis-2-ethylhexyl sulfosuccinate (aerosol OT, *AOT*) in octane at 20 °C. Sedimentation was from left to right. **a** The degree of the micelle hydration was 30; sedimentation times were 1, 8, 11, 14.5, and 18 min. **b** The degree of micelle hydration was 40; sedimentation times were 2, 8, 13, 15, and 19 min. The rotor speed was 30 000 rpm. Scanning was carried out at 280 nm. Near the boundaries of the sedimentation patterns the corresponding sedimentation coefficients for the empty and protein-containing micelles are given. The boundaries with the least sedimentation coefficient correspond to the empty micelles

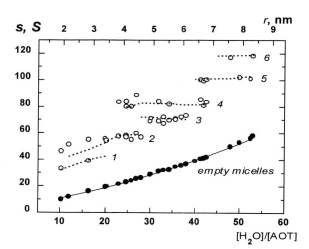

Fig. 2 The dependence of the sedimentation coefficient (*s*) for the reversed hydrated micelles of AOT in octane on the degree of micelle hydration. The *points* are the experimental data obtained for the micelles containing phosphorylase *b*. The *dotted curves* are calculated using Eqs. (2) and (4) for monomer (curve *1*; $M_r = 97\,400$), dimer (*2*; $M_r = 194\,800$), trimer (*3*; $M_r = 292\,200$), tetramer (*4*; $M_r = 389\,600$), hexamer (*5*; $M_r = 584\,400$), and octamer (*6*; $M_r = 779\,200$). The *solid curve* is the dependence of the sedimentation coefficient on w_0 for the empty micelles. The *upper axis* is the radius of the inner cavity of the micelle, *r*, which is related to w_0 by the equation $r\,/\mathrm{nm} = 4 + 0.15\,w_0$ [3, 9]

(s_0) increase with increasing w_0 (Fig. 2) and the experimental data are well represented by the empirical formula (at a constant AOT concentration of 0.1 M):

$$s_0 = 4.5 + 0.55w_0 + 0.0081w_0^2 \ . \tag{1}$$

In order to facilitate the assignment of the sedimentation coefficients to certain oligomeric forms of phosphorylase *b*, we calculated the theoretical values of the sedimentation coefficient of the protein-containing micelles using equations proposed in earlier works [3, 6, 27, 28]. If the dimensions of the inner cavity of the micelle exceed those of the protein molecule, entrapment of the protein occurs without change in the hydrodynamic radius of the micelle. In accordance with the "fixed-sized" model [27, 28], the theoretical value of the sedimentation coefficient of the protein-containing micelle (*s*) may be calculated in this case by the formula [27, 28]

$$s = s_0 \left[1 + \frac{M_r}{M_0}\frac{1}{(1 - \bar{v}_0\rho)} \right] \ , \tag{2}$$

where M_r is the molecular mass of the oligomeric form of the protein, ρ is the solvent density; s_0, M_0, and $\bar{v}_0$ are the sedimentation coefficient, molar mass, and specific partial volume of the empty micelle, respectively. This equation is based on the assumptions that the diffusion coefficient of the empty micelle is equal to that of the protein-containing micelle and that the molecular mass of the protein-containing micelle (M_{pc}) is related to parameters of M_0 and M_r by the expression

$$M_{pc} = M_0 + M_r \ . \tag{3}$$

Equation (2) was deduced from the Svedberg equations written for the protein-containing and empty micelles [28].

When the radius of the inner cavity of the micelle is less than that of the protein globule, in accordance with the "induced-fit" model [3, 27], the solubilized protein creates its own micelle equipped with a hydrated surfactant monolayer with a degree of hydration identical to that of the original empty micelles. The following expression may be used for calculation of the theoretical value of the sedimentation coefficient of the protein-containing micelle [3, 27]:

$$s = s_{opt} \left[1 + \frac{M_r - n_{opt}\left(w_{opt} - w_0\right)m}{M_{opt}}\frac{1}{\left(1 - \bar{v}_{opt}\rho\right)} \right] \ , \tag{4}$$

where w_{opt} is the optimum value of the degree of micelle hydration at which the radius of the inner cavity of the micelle is equal to that of the protein globule; s_{opt}, M_{opt}, n_{opt}, and $\bar{v}_{opt}$ are the values of the sedimentation coefficient, molecular mass, aggregation number (number of AOT molecules in a single micelle), and specific partial volume of the empty reversed micelle at the optimum degree of hydration, respectively, and m is the molecular mass of water. When deriving Eq. (4), it is assumed that the aggregation number for the protein-containing micelle is identical to that calculated for the empty micelle at $w_0 = w_{opt}$. In order to calculate the radius of the protein molecule and the value of w_{opt}, the phosphorylase b molecule was approximated by a solid sphere. The empirical equations [29]

$$r = 0.07(M_r)^{1/3} \text{ nm} \quad \text{and} \quad w_{opt} = 0.47(M_r)^{1/3} - 2.67$$

$$(5)$$

were used then for calculation of parameters r and w_{opt}. The parameters entering Eq. (4) are given in Table 2 for different oligomeric forms of phosphorylase b.

For the dimeric form of phosphorylase b the value of w_{opt} is 24.7; therefore, in the region of w_0 values where $w_0 > 24.7$ the theoretical values of the sedimentation coefficient of the protein-containing micelles were calculated using Eq. (2). In the region where $w_0 < 24.7$ (i.e., where the diameter of the dimeric form of the enzyme exceeds the diameter of the inner cavity of the micelle) we used Eq. (4). The results of our calculations are presented in Fig. 2 (curve 2). As can be seen from Fig. 2, the dimeric form of phosphorylase b is observed in the range of w_0 values from 10 to 30.

It should be noted that the radius of the phosphorylase b molecule which was obtained on the basis of approximation of the dimer by a sphere ($r = 4.1$ nm) corresponds to the most compact arrangement of the dimeric form in the micelle. If we take into account the Stokes radius of the dimeric form determined by the gel-filtration method (4.9 nm [31]), we should accept that the optimum degree of micelle hydration will be essentially greater, namely about 30. When choosing w_{opt} and calculating the theoretical values of the

sedimentation coefficient, we used both variants of the value of the radius of the dimeric form. Comparison of the experimental and theoretical values of the sedimentation coefficient showed that the approximation of dimer by a sphere with a radius of 4.1 nm provides the best description of the experimental data; however, it should be noted that in the region of w_0 values from 10 to 16 the experimental points are placed above the theoretical curve. Equation (4) is probably not applicable for the description of the sedimentation behaviour of the protein-containing micelles in the region $w_0 < w_{opt}$ when w_0 differs substantially from w_{opt}.

The sedimentation coefficient of the monomeric form of phosphorylase b at $w_0 < 19$ was calculated using Eq. (4). The monomeric form was observed in the range of w_0 from 10 to 16 (Fig. 2, curve 1); however, the fraction in the monomeric form is small and apart from this form the micellar system contains dimers and aggregates with a sedimentation coefficient of about 90 S. The appearance of aggregates is probably due to the enhanced ability of the monomeric form of phosphorylase b to aggregate as was shown for the monomeric form of apoenzyme [32, 33]. According to the data obtained by Barford and Johnson [34], 54% of the surface of the isolated subunit has nonpolar character. Dissociation of dimer into subunits is probably accompanied by the exposure of the hydrophobic regions on the surface of the protein globule which are able to interact with the micellar shell, resulting in the aggregation of the enzyme-containing micelle.

As can be seen from Fig. 2, at rather high values of w_0 the supramolecular forms of phosphorylase b appear in the system. According to our calculations, the sedimentation coefficients of 68–72 S (curve 3), 80–84 S (curve 4), 100–102 S (curve 5), and 118 S (curve 6) correspond to the trimeric, tetrameric, hexameric, and octameric forms of phosphorylase b. The trimeric form was observed in the range of w_0 values from 30 to 38, the tetrameric form in the range from 23 to 42, the hexameric form in the range from 41 to 50, and the octameric form in the range from 48 to 53.

Table 2 Parameters of hydrated reversed micelles

Phosphorylase b	M_r	r (nm)	w_{opt}	s_{opt} (S)	M_{opt} [a]	n_{opt} [b]	$(1 - \bar{v}_{opt}\rho)$ [c]
Monomer	97 400	3.22	19.0	17.8	204 000	260	0.3537
Dimer	194 800	4.06	24.7	23.0	356 000	400	0.3465
Trimer	292 200	4.64	28.5	26.7	486 000	508	0.3440
Tetramer	389 600	5.11	31.7	30.0	627 000	617	0.3408
Hexamer	584 400	5.85	36.6	35.5	881 000	798	0.3379
Octamer	779 200	6.44	40.6	40.2	1 133 000	964	0.3350

[a] $M_{opt} = (19 + 2.1 \, w_{opt})^3$ [27]

[b] $n_{opt} = M_{opt}/(M_{AOT} + m w_{opt})$; $M_{AOT} = 445$, $m = 18$

[c] $\bar{v}_{opt}$ was calculated in accordance with data presented in Refs. [10, 30]

Thus, the variation of the degree of micelle hydration allows us to modulate purposefully the oligomeric state of muscle phosphorylase b, which in aqueous solution exists in a stable dimeric form. Using the micellar system with rather low values of the degree of hydration allows us to register the monomeric form of phosphorylase b. When $w_0 > 24.7$ the tetrameric form of the enzyme appears. In aqueous solution the tetrameric form is formed only in the presence of AMP. Moreover, at $w_0 > 41$ phosphorylase b entrapped in the reversed micelles is able to form supramolecular structures of more complex composition (hexamers and octamers).

The sedimentation behaviour of uridine phosphorylase from *E. coli* K-12 in hydrated reversed micelles of AOT in octane was also studied in the range of w_0 values from 8 to 23.9. At $w_0 = 8.4$ the sedimentation coefficient of the protein-containing micelles was found to be 22 S. When $w_0 = 12.9$, we observed the boundary corresponding to the empty micelles and two boundaries of the protein-containing micelles ($s = 23$ and 34.6 S) corresponding, presumably, to different oligomeric forms of uridine phosphorylase. At $w_0 = 16.1$ and 18.6 the sedimentation patterns also included several boundaries of the protein-containing micelles. Experimental values of the sedimentation coefficient for the protein-containing micelles as a function of w_0 are presented in Fig. 3. In order to address the question of which oligomeric forms correspond to the sedimentation boundaries, we calculated the sedimentation coefficients to be expected for definite oligomeric enzyme forms. For the micelles with hydration degrees of 8.4, 12.9, and 16.1 the radii of the inner cavity of the micelle were found to be 1.65, 2.33, and 2.81 nm, respectively. The dimensions of the inner cavity of the micelles allow

the monomeric, dimeric, and trimeric forms of uridine phosphorylase, to be incorporated. Curves 1–4 in Fig. 3 are calculated according to Eq. (2) assuming that the incorporation of monomer (1), dimer (2), trimer (3), and tetramer (4), respectively, does not result in a change in the size of the micelle. Comparison of the experimental and calculated values of the sedimentation coefficients shows that at degrees of hydration of 8.4 and 12.9 the micelles are able to incorporate the monomeric form. At $w_0 = 12.9$, apart from the monomeric form, the dimeric form is incorporated in micelles. At $w_0 = 16.1$ the incorporation of the monomeric, dimeric, and trimeric forms takes place. When $w_0 = 18.6$, apart from the trimeric form, the tetrameric form is incorporated in micelles.

At $w_0 > 20.9$ the micellar system becomes opaque. The appearance of turbidity is probably due to nonspecific aggregation of the enzyme in micelles. To preclude such aggregation, we centrifuged the micellar system in the presence of 42 mM guanidine hydrochloride. The sedimentation coefficient of the protein-containing micelles at $w_0 = 23.9$ was found to be 49.3 S. The theoretical value of the sedimentation coefficient calculated from Eq. (2) for the hexameric enzyme form (50 S) coincides with the experimental value of the sedimentation coefficient; therefore, we assign the protein-containing micelles at this value of w_0 to the hexameric form of the enzyme.

Thus, using hydrated reversed micelles of AOT in octane facilitates the dissociation of uridine phosphorylase with the formation of the monomeric, dimeric, trimeric, and tetrameric forms of the enzyme. Under certain conditions the hexameric form of the enzyme in the micellar system may be registered. Special experiments showed that all the oligomeric forms of uridine phosphorylase incorporated in hydrated reversed micelles are catalytically active [35].

The results obtained confirm that hydrated reversed micelles can be used as microreactors for the design of supramolecular complexes of proteins [3]. The crowded solution environment in a micellar core reminds us of the nanoreactor concept where inorganic colloids of defined sizes are synthesized in the cores of block copolymer micelles [36, 37]. The environment within cells is often vastly different from the very dilute solutions studied in laboratories. It is becoming increasingly apparent that the crowded molecular conditions encountered in vivo favour the formation of macromolecular assemblies [38]. The supramolecular structures registered by us in the micellar systems may serve as a model of ordered supramolecular complexes of the enzymes in the cell.

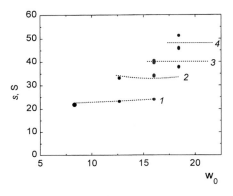

Fig. 3 The dependence of the sedimentation coefficient (s) of hydrated reversed micelles containing entrapped uridine phosphorylase on w_0. The *points* are the experimental data. The *dotted lines 1–4* are calculated according to Eq. (2) for monomer (*1*; $M_r = 27\,500$), dimer (*2*; $M_r = 55\,000$), trimer (*3*; $M_r = 82\,500$), and tetramer of uridine phosphorylase (*4*; $M_r = 110\,000$)

Acknowledgements This work was supported by grant no. 99-04-48639 from the Russian Foundation for Basic Research.

134

References

1. Klyachko NL, Ugolnikova AV, Ivanov MV, Levashov AV (1995) Biokhimiya 60:1048–1054 (Russian)
2. Khmelnitsky YuL, Kabanov AV, Klyachko NL, Levashov AV, Martinek K (1989) In: Pileni MP (ed) Structure and reactivity in reversed micelles. Elsevier, Amsterdam, pp 230–260
3. Kabanov AV, Klyachko NL, Nametkin SN, Merker SK, Zaroza AV, Bunik VI, Ivanov MV, Levashov AV (1991) Prot Eng 4: 1009–1017
4. Kabakov VE, Merker Sh, Kollova EV, Pshezhetskii AV, Shvyadas VK, Martinek K, Klyacho NL, Levashov AV (1992) Bioorg Khim 18: 1073–1080 (Russian)
5. Pshezhetsky AV, Levashov AV, Wiederschain GYa (1992) Biochim Biophys Acta 1122: 154–160
6. Kabanov AV, Nametkin SN, Evtushenko GN, Chernov NN, Klyachko NL, Levashov AV, Martinek K (1989) Biochim Biophys Acta 996: 147–152
7. Martinek K, Klyacho NL, Kabanov AV, Khmelnitsky YuL, Levashov AV (1989) Biochim Biophys Acta 981: 161–172
8. Klyachko NL, Levashov AV, Kabanov AV, Khmelnitsky YuL, Martinek K (1991) In : Gratzel M, Kalyanasundaram K (eds) Kinetics and catalysis in microheterogeneous systems. Dekker, New York, pp 135–181
9. Eicke H-F, Rehak J (1976) Helv Chim Acta 59: 2883–2891
10. Zulauf M, Eicke H-F (1979) J Phys Chem 83: 480–486
11. Graves DJ, Wang JH (1972) In: Boyer PD (ed) The enzymes. Academic Press, New York, pp 435–482
12. Buc MH, Buc H (1968) In: Kvamme E, Pihl A (eds) Regulation of enzyme activity and allosteric interactions. Academic Press, London, pp 109–130
13. Silonova GV, Kurganov BI (1970) Mol Biol 4: 445–458 (Russian)
14. Chebotareva NA, Kurganov BI, Lubarev AE, Davydov DR (1991) Biochimie 73: 1339–1343
15. Kurganov BI, Mitskevich LG, Fedurkina NV, Chebotareva NA (1996) Biokhimiya 61: 912–918 (Russian)
16. Dombradi V (1981) Int J Biochem 13: 125–139
17. Leer JC, Hammer-Jespersen K, Schwartz M (1977) Eur J Biochem 75: 217–224
18. Vita A, Amici A, Cacciamani T, Lanciotti M, Magni G (1986) J Biochem 18: 431–436
19. Cook WJ, Kosalka GW, Hall WW, Narayana SVL, Ealick SE (1987) J Biol Chem 262: 2852–2853
20. Tsuprun VL, Tagunova IL, Linkova EV, Mironov AS (1991) Biokhimiya 56: 930–934 (Russian)
21. Walton L, Richards CA, Elwell LP (1989) Nucleic Acids Res 17: 6741
22. Mikhailov AM, Smirnova EA, Tsuprun VL, Tagunova IV, Vainshtein BK, Linkova EV, Komissarov AA, Siprashvili ZZ, Mironov AS (1992) Biochem Int 26: 607–615
23. Morgunova EYu, Mikhailov AM, Popov AN, Blagova EV, Smirnova EA, Vainshtein BK, Mao Ch, Armstrong ShR, Ealick SE, Komissarov AA, Linkova EV, Burlakova AA, Mironov AS, Debabov VG (1995) FEBS Lett 367: 183–187
24. Fischer EH, Krebs EG (1958) J Biol Chem 231: 65–71
25. Laemmli UK (1970) Nature 277: 680–685
26. Bradford MM (1976) Anal Biochem 2: 248–254
27. Levashov AV, Khmelnitsky YuL, Klyachko NL, Chernyak VYa, Martinek K (1982) J Colloid Interface Sci 88: 444–457
28. Levashov AV, Khmelnitsky YuL, Klyachko NL, Chernyak VYa, Martinek K (1981) Anal Biochem 118: 42–46
29. Klyachko NL, Pshezhetskii AV, Kabanov AV, Vakula SV, Martinek K, Levashov AV (1990) Biol Membr 7: 467–472 (Russian)
30. Robinson BH, Steytler DC, Tack RD (1979) J Chem Soc Faraday Trans I 75: 481–496
31. De Vincenzi DL, Hedrick JL (1970) Biochemistry 9: 2048–2058
32. Gunar VI, Sugrobova NP, Chebotareva NA, Poznanskaya AA, Kurganov BI (1990) In: Fukui T, Kagamiyama H, Soda K, Wada H (eds) Enzymes dependent on pyridoxal phosphate and other carbonil compounds as cofactors. Pergamon, Oxford, pp 417–420
33. Chebotareva NA, Sugrobova NP, Bulanova LN, Poznanskaya AA, Kurganov BI, Gunar VI (1995) Biokhimiya 60: 2030–2039 (Russian)
34. Barford D, Johnson LN (1992) Prot Sci 1: 472–493
35. Burlakova AA, Kurganov BI, Chebotareva NA, Debabov VG (1996) Biol Membr 13: 504–511
36. Antonietti M, Forster S, Oestreich S (1997) Macromol Symp 121: 75–88
37. Spatz JP, Mossmer S, Moller M (1996) Chem Eur J 2: 1552–1555
38. Minton AP (1997) Curr Opin Biotechnol 8: 65–69

Progr Colloid Polym Sci (1999) 113:135–141
© Springer-Verlag 1999

K.-J. Tiefenbach
H. Durchschlag
R. Jaenicke

Spectroscopic and hydrodynamic investigations of nonionic and zwitterionic detergents

K.-J. Tiefenbach · H. Durchschlag (⊠)
R. Jaenicke
Institute of Biophysics
and Physical Biochemistry
University of Regensburg
Universitätsstrasse 31
D-93040 Regensburg, Germany
e-mail: helmut.durchschlag@biologie.uni-regensburg.de
Tel.: +49-941-943 3041
Fax: +49-941-943 2813

Abstract Selected nonionic and zwitterionic detergents have been investigated by sedimentation velocity and equilibrium runs in the analytical ultracentrifuge in a systematic manner. UV-visible absorption spectroscopy and determinations of the partial specific volumes were used as complementary techniques. Pilot tests comprised the following detergents: Nonidet P40, Triton X-100 and X-114, Tween 20 and 80, C_7-, C_8-, C_9-, and C_{10}-glucosides, C_8-thioglucoside, and Zwittergents 3-08, 3-10, 3-12, 3-14, and 3-16. Application of analytical ultracentrifugation to detergents requires a series of adaptations and improvements of the conventional techniques, in addition to the consideration of possible pitfalls and precautions required for certain detergents. In this context, the necessity to differentiate between the signals of monomeric and micellar detergents, on the one hand, and (macro)solutes of different size, on the other, has to be mentioned. Monitoring the sedimentation profiles of weakly absorbing detergents becomes feasible by labeling the detergent micelles using the fluorescent dye N-phenyl-1-naphthylamine. Owing to the chemical nature of the detergents under analysis and differences in size and shape of the corresponding micelles, the results reveal significant differences in the behavior of the detergents, depending on the experimental conditions. The detergents exhibit a broad mass distribution, caused by heterogeneity and impurities of commercial samples.

Key words Nonionic and zwitterionic detergents · Micelles · Aggregation number · Analytical ultracentrifugation · Absorption spectroscopy

Abbreviations *AF* amplification factor · *AUC* analytical ultracentrifugation · *CMC* critical micelle concentration · *CMT* critical micelle temperature · *HSSE* high-speed sedimentation equilibrium · *NaP* sodium phosphate buffer · *NPN* N-phenyl-1-naphthylamine · *SDS* sodium dodecyl sulfate

Introduction

Ionic and nonionic detergents are widely used in the (bio)sciences and in (bio)technological processes, to solve problems such as solubilization, unfolding, and characterization of simple and conjugated proteins (e.g., membrane proteins) (cf. [1–9]). The investigation of monomeric and micellar detergents by means of physicochemical techniques yields a variety of molecular characteristics. As follows from a critical inspection of the relevant literature, however, most investigations are restricted to characteristic numbers such as critical micelle concentration (CMC), critical micelle temperature (CMT), aggregation number, cloud point, etc.

By contrast, studies on structural properties and other more complex features of detergents are rather scarce. For a deeper understanding of the interaction of detergents with biomolecules and for the utilization of detergents on a more rational basis, the comparative analysis of various types and classes of detergents is required. Therefore, we started a detailed physicochemical investigation of cationic, anionic, zwitterionic, and nonionic detergents under various experimental conditions. Results may be used as starting points for the appropriate choice of a suitable detergent for a given problem.

Experiments on anionic sodium dodecyl sulfate (SDS) and protein-SDS complexes have been reported [10–16]. The present study is concerned with an ultracentrifugal analysis of nonionic and zwitterionic detergents, together with supplementary spectroscopic and densimetric measurements concerning their absorption and buoyancy behavior.

In the past, for the mass estimation and structural analysis of detergent micelles, primarily techniques such as solution scattering (light, small-angle X-ray, and neutron scattering) were applied (e.g. [17–21]). Though manifold investigations have been performed on certain micelles, the precise size and structure of the various types and classes of micelles and their behavior under varying environmental conditions are not yet fully understood. We chose analytical ultracentrifugation (AUC) as an efficient screening method to estimate the size of detergents and to investigate their mass distributions. Applying AUC, however, necessitated the adaptation of existing evaluation procedures to these types of labile and heterogeneous compounds.

Materials and methods

Polyoxyethylene alkylphenols (Nonidet P40, Triton X-100, Triton X-114), polyoxyethylene monoacyl sorbitans (Tween 20, Tween 80), and alkylthioglucoside (C_8TGS) were obtained from Boehringer (Mannheim), alkylglucosides (C_7GS, C_8GS, C_9GS, C_{10}GS) and dimethylalkylammoniopropanesulfonates (Zwittergents 3-08, 3-10, 3-12, 3-14, 3-16) from Calbiochem (Bad Soden), and N-phenyl-1-naphthylamine (NPN) from Sigma (Munich); all other reagents were of analytical grade.

Detergents were dissolved in bidistilled water or aqueous solutions containing 50 mM sodium phosphate pH 7.0 (NaP, $I \approx 0.1$ mol/l) and the fluorescent dye NPN for labeling (primarily micellar) detergents [22].

Sedimentation velocity and equilibrium experiments were performed in a Beckman model E analytical ultracentrifuge equipped with a high-sensitivity photoelectric scanner and multiplexer system and a 10-inch recorder; runs were performed in a six-hole rotor (AnG), using 12 mm double-sector cells (charcoal-filled epon) and sapphire windows. Densities and viscosities of solvents and detergent solutions were determined by means of a Paar digital density meter (DMA 02) and an Ostwald viscometer (5 ml, flow time for water about 320 s), respectively. Absorption spectra were recorded with a Perkin-Elmer Lambda 5 spectrophotometer. Investigations were performed at 25 °C.

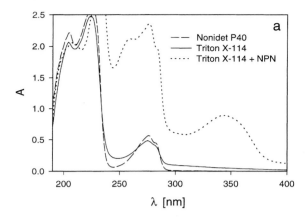

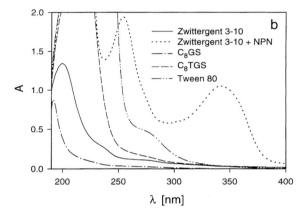

Fig. 1 Absorption spectra of aqueous solutions of nonionic and zwitterionic detergents in the absence and in the presence of 0.1 mM NPN: **a** detergents ($c = 0.05\%$) containing aromatic residues; **b** detergents ($c = 1\%$) devoid of aromatic residues

AUC experiments made use of sedimentation velocity and equilibrium runs, allowing the characterization of detergent micelles in terms of sedimentation coefficients and molar mass weight averages of the macromolecular species [8, 13, 14, 23–27]. Apparent sedimentation coefficients were obtained from ln r versus t plots; they were converted to standard conditions ($s_{20,w}$), making use of corrections for solvent viscosity and density. Molar masses (M) were determined by high-speed sedimentation equilibrium (HSSE), using the meniscus depletion technique [24].

Partial specific volumes of micellar detergents were calculated on the basis of the volume increments of the constituents according to [28, 29]. Experimental volumes were obtained from observed solution densities (cf. [30]).

Special problems with micellar detergents

The investigation of micellar systems by means of AUC is affected by various problems and pitfalls (cf. [13, 14]). In the context of the analysis of detergent micelles, the following aspects are of importance:

1. Frequently, detergents contain impurities and are heterogenous (presence of a variety of homologs).

Table 1 Comparison of calculated and experimental partial specific volumes of selected nonionic and zwitterionic detergents

Detergent[a]	M[b] (g/mol)	$\bar{v}_{calc}$[c] (cm³/g)	$\bar{v}_{exp}$[d] (cm³/g)
Nonionic detergents			
Polyoxyethylene alkylphenols			
tC$_2\phi$E$_{11}$ (Nonidet P40)[e]	606.8	0.876	0.916 ± 0.003[f]
tC$_8\phi$E$_{7.5}$ (Triton X-114)	536.7	0.924	0.925 ± 0.004[f]
tC$_8\phi$E$_{10.0}$ (Triton X-100)	646.9	0.913 [28, 29]	0.905 ± 0.001[f]
			0.908–0.918 [20, 32, 33]
Polyoxyethylene monoacyl sorbitans			
C$_{12}$SorbE$_{20}$ (Tween 20)	1227.5	0.869 [28, 29]	0.876 ± 0.002[f], 0.869 [33]
C$_{18:1}$SorbE$_{20}$ (Tween 80)	1309.7	0.889 [29]	0.896 ± 0.002[f], 0.896 [33]
Alkylglucosides			
C$_7$GS	278.3	0.826	–
C$_8$GS	292.4	0.842 [28, 29]	0.859–0.867 [26, 34]
C$_9$GS	306.4	0.856	
C$_{10}$GS	320.4	0.869	–
Alkylthioglucoside			
C$_8$TGS	308.4	0.830	0.833 ± 0.005[f]
Zwitterionic detergents			
Dimethylalkylammoniopropanesulfonates			
DC$_8$APS (Zwittergent 3-08)	279.4	0.918	–
DC$_{10}$APS (Zwittergent 3-10)	307.5	0.939 [29]	0.920[g] [32]
DC$_{12}$APS (Zwittergent 3-12)	335.6	0.956 [29]	0.957[g] [32]
DC$_{14}$APS (Zwittergent 3-14)	363.6	0.971	–
DC$_{16}$APS (Zwittergent 3-16)	391.7	0.984 [29]	0.986[g] [32]

[a] Trade/trivial names are added in parentheses
[b] The molar mass is given for the monomeric unit
[c] Calculations were performed according to [28, 29] and refer to the micellar state and 25 °C; some values were taken from the given references
[d] Experimental values refer to the micellar state and 20–25 °C; some values were taken from the literature
[e] Different trade names and chemical structures are given by manufacturers: Nonidet P40, Nonidet NP40, NP40, etc.; tC$_2\phi$E$_{11}$, tC$_8\phi$E$_9$, etc. Consequently, observed results differ considerably. Varying composition may also be responsible for the discrepancies between calculated and experimental partial specific volumes or molar masses, on the one hand, and similarities in the properties between Nonidet P40 and Triton X-100, on the other
[f] This study ($\bar{v}_{exp}^0$)
[g] Values measured at 30 °C

This holds especially for commercially available samples.

2. Appropriate experimental conditions regarding concentration, temperature, ionic strength, etc., have to be adopted (consideration of CMC, CMT, and cloud point), to guarantee micelle formation, on the one hand, and to avoid disturbances caused by crystallization of detergents or formation of colloidal inhomogeneities, on the other.

3. Detergent micelles exhibit sensitive equilibria, highly influenced by environmental conditions such as concentration of solutes and cosolutes, ionic strength, temperature, and pressure. Both size and shape of micelles may change as a consequence of the conditions applied.

4. The absorbance of most detergents in the accessible wavelength range is rather weak. This impedes the detection and precise analysis of detergent micelles in the absence of appropriate labels.

5. Labeling the detergent micelles by certain dyes may enable monitoring the micelles. In the context of ultracentrifugation studies, 1,6-diphenyl-1,3,5-hexatriene (DPH) has already been used as an efficient label [31]. We have applied NPN advantageously for spectroscopic and ultracentrifuge studies [13, 14]. Both labels have a limited solubility in water and partition primarily into an apolar environment such as the micellar interior [22, 31]. Obvious deficiencies of labeling procedures, however, are the possible influence of such dyes on micelle formation (CMC, mass, shape) and a pressure dependence of the partition coefficients.

6. The occurrence of heterogeneity prevents the analysis of structural details, unless experiments at different rotor speeds are performed to achieve full separation of all macrosolutes present in solution.

7. Mass determination by HSSE experiments requires an appropriate choice of the baseline. This turns out to be a crucial point, because different baseline settings cause different mass estimates. If ambiguous results are obtained, the baseline may be used as a free parameter within narrow limits.

Table 2 Sedimentation coefficients of selected nonionic detergents in water, monitored in the absence of NPN. AUC conditions: 40 000 rpm, 25 °C, λ = 230 or 275 nm, amplification factor (AF) = 2 or 4

Detergent	Ultracentrifugation			
	c (%)	λ (nm)	AF	$s_{20,w}^c$ (S)
Nonidet P40[a]	0.02	275	2	1.59 ± 0.16
Triton X-114	0.02	275	2	1.88
Triton X-100	0.02	275	2	1.35 ± 0.10[b]
Tween 20	1.0	275	2	1.66 ± 0.11
Tween 80	1.0	275	2	2.82
C_8GS	0.5	230	2, 4	1.28 ± 0.06[c]
C_9GS	0.5	230	2, 4	3.03 ± 0.08
$C_{10}GS$	0.5	230	2, 4	5.21 ± 0.79
C_8TGS	1.0	275	2	2.87

[a] Cf. annotation e in Table 1
[b] A value of 1.32 S is reported in [35]
[c] A value of 1.3 S is reported in [34]

Results and discussion

The present investigation is concerned with spectroscopic and ultracentrifugal properties of nonionic and zwitterionic detergent micelles under various experimental conditions, in particular with the elucidation of their molar masses.

Absorption spectroscopy

Under the common experimental conditions, the majority of detergents in the concentration range of their CMCs exhibit only weak absorption in the UV range > 250 nm (Fig. 1), i.e. at wavelengths usually accessible to precise AUC experiments. However, in the presence of certain dyes such as NPN, detergent micelles bind and/or include this label, thereby shifting spectra to longer wavelengths (about 346 nm in the case of NPN) and enhancing the absorbance considerably [13, 14].

Determination of partial specific volumes

As is well known, knowledge of the absolute values of $\bar{v}$ is a necessary prerequisite for mass estimations by HSSE. Since the $\bar{v}$ values of many detergents are in the vicinity of 1 cm³/g, serious problems arise in solvents of a density of about 1 g/cm³. This holds especially for aqueous solutions, where the buoyancy term $(1 - \bar{v}\rho)$ approaches zero. Since precise experimental determinations of partial specific volumes $\bar{v}$ by densimetry are frequently not feasible (e.g. because of scarce amounts of material available, insufficient purity, adsorption), appropriate calculation approaches have to be adopted.

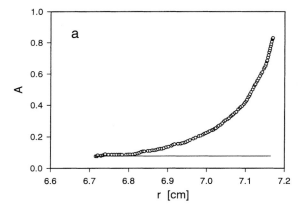

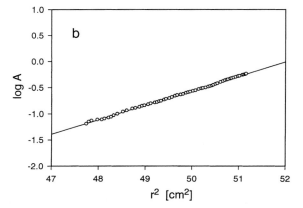

Fig. 2 HSSE of unlabeled Triton X-100 micelles in **a** A vs. r and **b** log A vs. r^2 plots. Recording of micelles (c = 0.05%, NaP) was performed at 26 000 rpm and 245 nm. Observed molar mass: 51.3 ± 1.9 kg/mol

Fortunately, the prediction of partial specific volumes of micellar detergents on the basis of the increments of the constituents is accurate to ±2% [29]; this corresponds to the usual range of experimental accuracy of volume measurements.

A comparison of the partial specific volumes of nonionic and zwitterionic detergents proves satisfactory coincidence between calculated and experimental values (Table 1). This justifies the application of the calculative procedure for estimating partial specific volumes of the detergents used in this study. An inspection of the values predicted for different detergents shows that the partial specific volumes vary with the nature of the detergent class and the chain length.

Analytical ultracentrifugation

Sedimentation velocity

The sedimentation coefficients of nonionic detergent micelles under analysis vary between 1 and 5 S,

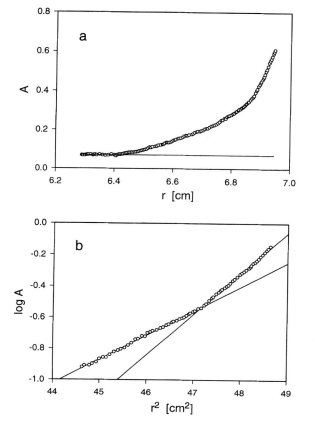

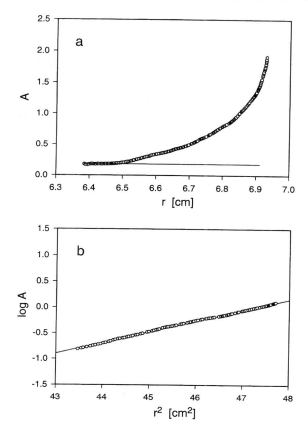

Fig. 3 HSSE of C$_9$GS micelles, labeled with 0.1 mM NPN, in **a** A vs. r and **b** log A vs. r^2 plots. Recording of micelles (c = 0.5%, NaP) was performed at 16 000 rpm and 250 nm. Observed molar masses: 44.3 ± 1.7 and 75.2 ± 1.5 kg/mol in the meniscus and bottom regions of the cell, respectively

Fig. 4 HSSE of Zwittergent 3-12 micelles, labeled with 0.1 mM NPN, in **a** A vs. r and **b** log A vs. r^2 plots. Recording of micelles (c = 1.0%, NaP) was performed at 40 000 rpm and 346 nm. Observed molar mass: 33.7 ± 1.6 kg/mol

depending on the detergents used (Table 2). Differences in the values are due to differences in chemical structure (detergent class, chain length) and size and shape of the micelles. For example, for the Tween and alkylglucoside series, a clear correlation between increase in s and chain length is indicated. So far, the analysis of zwitterionic detergents failed, presumably owing to perturbations by NPN.

Sedimentation equilibrium

A great variety of HSSE experiments were performed with nonionic and zwitterionic detergent micelles under a variety of ultracentrifugal conditions (rotor speed, wavelength and amplification of registration, solvent, labeling). Typical data are illustrated in Figs. 2–4. Results are summarized in Table 3.

The estimation of the particle size of both nonionic and zwitterionic detergents shows that the micelles analysed exhibit a variety of masses, considerably influenced by the rotor speed used (Table 3). Conse-

quently, the masses found at the highest speed show the lowest masses, and vice versa. This behavior is in marked contrast to previous findings for micellar SDS [14].

In the present study, both the chemical heterogeneity of the samples investigated and the size and shape changes of the micelles as a consequence of centrifugation conditions (e.g., influence of pressure on the monomer/micelle balance and shape of micelles) may be responsible for the observed mass changes. This behavior may be demonstrated clearly for the detergents in buffer (Table 3).

Sometimes we observed a relatively pronounced break in the log A versus r^2 plots and apparent molar masses in the bottom region of the centrifuge cell, differing by a factor of about two from those in the meniscus region (cf. Fig. 3). This phenomenon, of course, increases the heterogeneity, caused by the presence of various homologs, further and makes the interpretation of the results much more difficult. A possible interpretation might be a concentration-dependent association of certain micelles.

Table 3 Molar masses M of selected detergents in water or buffer, obtained by HSSE under various conditions of centrifugation and registration in the absence or in the presence of the fluorescent dye NPN

Detergent	c (%)	NPN (mM)	Solvent	Ultracentrifugation					Literature
				Rotor speed (rpm $\times 10^{-3}$)	λ (nm)	AF	M_{mic}^a (kg/mol)	n^b	n^b
Nonionic detergents									
Nonidet P40[c]	0.05	–	H_2O	22	245, 290	1, 4	74 ± 5	122	–
				26	245, 280, 290	1, 4	39 ± 5	65	
				40	245, 280, 290	1, 4	32 ± 6	52	
		–	NaP	16	245, 280, 290	1, 4	84 ± 10	139	–
				22	245, 280, 290	1, 4	58 ± 12	95	
				26	245, 280, 290	1, 4	43 ± 6	70	
				40	245, 280, 290	1, 4	33 ± 8	53	
Triton X-114	0.05	–	H_2O	26	245, 280, 290	1, 4	91 ± 23	170	–
				40	245, 280, 290	1, 4	104 ± 20	193	
Triton X-100	0.05	–	H_2O	26	245, 280, 290	1, 4	57 ± 7	88	100–155 [18, 20, 35–37]
				40	245, 280, 290	1, 4	52 ± 3	80	
		–	NaP	16	245, 290	1, 4	66 ± 12	102	100–155 [18, 20, 35–37]
				22	245, 280, 290	1, 4	50 ± 11	78	
				26	245, 280, 290	1, 4	50 ± 6	77	
				40	245, 280, 290	1, 4	36 ± 3	56	
Tween 20	1.0	–	H_2O	26	230, 275, 290	1, 4	36 ± 11	29	–
				40	230	1	29	24	
Tween 80	1.0	–	H_2O	26	275, 290	1, 4	87 ± 8	66	38–58 [3, 4]
				40	230	1	61	46	
C_7GS	1.0	0.1	H_2O	26	250, 295, 346	1, 4	27 ± 9	97	–
				40	250, 295, 346	1	5 ± 1	18	
C_8GS	0.5	0.1	H_2O	10	250, 295, 346	1, 4	55 ± 9	189	27–85 [34, 36, 38]
				16	250, 295, 346	1, 4	39 ± 4	133	
				26	250, 295, 346	1, 4	32 ± 3	111	
				40	250, 295, 346	1, 4	28 ± 2	95	
		0.1	NaP	10	250, 295, 346	1, 4	84 ± 25	287	27–85 [34, 36, 38]
				16	250, 295, 346	1, 4	35 ± 8	119	
				26	250, 295, 346	1, 4	30 ± 2	101	
				40	250, 295, 346	1, 4	26 ± 4	88	
C_9GS	0.5	0.1	NaP	10	250, 295, 346	1, 4	169 ± 14	550	–
				16	250, 295, 346	1, 4	90 ± 26	294	
				26	250, 295, 346	1, 4	64 ± 16	210	
				40	250, 295, 346	1, 4	31 ± 7	102	
$C_{10}GS$	0.5	0.1	NaP	16	250, 295, 346	1, 4	58 ± 28	180	–
				26	250, 295, 346	1, 4	33 ± 13	103	
				40	250, 295, 346	1, 4	38 ± 6	119	
C_8TGS	1.0	–	H_2O	26	230	1	67	217	–
Zwitterionic detergents									
Zwittergent 3-08	1.0	0.1	NaP	26	250	4	13 ± 3	47	–
				40	250, 295, 346	1, 4	8 ± 3	29	
Zwittergent 3-10	1.0	0.1	NaP	26	295, 346	1, 4	46 ± 2	149	42 [17]
				40	250, 295, 346	1, 4	35 ± 2	113	
Zwittergent 3-12	1.0	0.1	NaP	26	295, 346	1	66 ± 19	195	54 [17]
				40	250, 295, 346	1, 4	37 ± 8	111	
Zwittergent 3-14	1.0	0.1	NaP	26	295, 346	1	120 ± 48	330	83 [36]
				40	250, 295, 346	1, 4	65 ± 14	180	
Zwittergent 3-16	0.2	0.1	NaP	26	250, 295, 346	1, 4	167 ± 64	426	156 [17]
				40	250, 295, 346	1, 4	76 ± 13	195	

[a] The mentioned mean values for M_{mic} are obtained from averaging the results obtained at different AUC conditions (λ, AF). In the case of nonlinear curves in log A vs. r^2 plots (cf. Fig. 3), the given mean values are characterized by large standard deviations

[b] Aggregation number, calculated on the basis of the masses of the detergent monomers given in Table 1

[c] Cf. annotation e in Table 1

Comparing the experimental results listed in Table 3 with some values found in the relevant literature reveals that the masses found in this study include in most cases the values reported by other authors. The diversity of observed and reported molar masses and aggregation numbers may be due to heterogeneity of the samples. In this context, however, it has to be mentioned that different techniques yield different mass averages. A special advantage of AUC is the possibility to separate heterogenous particles, with the consequence that different rotor speeds result in different mass estimates.

Conclusions

Irrespective of a variety of problems involved in the investigation of detergent micelles, their analysis in the analytical ultracentrifuge turns out to be successful, provided some precautions are taken into account. Particularly, the investigation under various experimental conditions, especially at variable rotor speeds, is required.

The results on a selected set of nonionic and zwitterionic detergents clearly unveil the heterogeneous nature of commercial preparations and the occurrence of micelle-micelle interactions. Obviously, the separation process in the centrifuge cell, which may be improved by applying varying centrifugal forces, allows the clear discrimination between particles of different size. The values reported in the literature are in fair agreement with our results.

References

1. Helenius A, Simons K (1975) Biochim Biophys Acta 415:29–79
2. Israelachvili JN, Mitchell DJ, Ninham BW (1976) J Chem Soc Faraday Trans II 72:1525–1568
3. Tanford C, Reynolds JA (1976) Biochim Biophys Acta 457:133–170
4. Helenius A, McCaslin DR, Fries E, Tanford C (1979) Methods Enzymol 56:734–749
5. Tanford C (1980) The hydrophobic effect: formation of micelles and biological membranes, 2nd edn. Wiley, New York
6. Luisi PL, Magid LJ (1986) CRC Crit Rev Biochem 20:409–474
7. Israelachvili JN (1991) Intermolecular and surface forces: with applications to colloidal and biological systems, 2nd edn. Academic Press, London
8. Schubert D, Schuck P (1991) Prog Colloid Polym Sci 86:12–22
9. Jones MN, Chapman D (1995) Micelles, monolayers, and biomembranes. Wiley-Liss, New York
10. Durchschlag H, Christl P, Jaenicke R (1991) Prog Colloid Polym Sci 86:41–56
11. Durchschlag H, Binder S, Christl P, Jaenicke R (1994) Jorn Com Esp Deterg 25:407–422 and Anexo 26–27
12. Durchschlag H, Weber R, Jaenicke R (1996) In: Proceedings of the 4th World Surfactants Congress, vol 1. AEPSAT, Barcelona, pp 519–534
13. Durchschlag H, Tiefenbach K-J, Jaenicke R (1997) Jorn Com Esp Deterg 27:185–196 and Anexo 35–36
14. Tiefenbach K-J, Durchschlag H, Jaenicke R (1997) Prog Colloid Polym Sci 107:102–114
15. Durchschlag H, Jaenicke R (1997) Chim Oggi – Chem Today 15, No 9/10, pp 15–24
16. Durchschlag H, Kuchenmüller B, Tiefenbach K-J, Jaenicke R (1998) Jorn Com Esp Deterg 28:353–365
17. Herrmann KW (1966) J Colloid Interface Sci 22:352–359
18. Robson RJ, Dennis EA (1977) J Phys Chem 81:1075–1078
19. Kratky O, Müller K (1982) In: Glatter O, Kratky O (eds) Small angle X-ray scattering. Academic Press, London, pp 499–510
20. Stubičar N, Matejaš J, Zipper P, Wilfing R (1989) In: Mittal KL (ed) Surfactants in solution, vol 7. Plenum Press, New York, pp 181–195
21. Vass S (1991) Struct Chem 2:(167)375–(189)397
22. Brito RMM, Vaz WLC (1986) Anal Biochem 152:250–255
23. Chervenka CH (1973) A manual of methods for the analytical ultracentrifuge. Spinco Division of Beckman Instruments, Palo Alto
24. Yphantis DA (1964) Biochemistry 3:297–317
25. Reynolds JA, Tanford C (1976) Proc Natl Acad Sci USA 73:4467–4470
26. Reynolds JA, McCaslin DR (1985) Methods Enzymol 117:41–53
27. Roxby RW (1992) In: Harding SE, Rowe AJ, Horton JC (eds) Analytical ultracentrifugation in biochemistry and polymer science. Royal Society of Chemistry, Cambridge, pp 609–618
28. Durchschlag H, Zipper P (1994) Prog Colloid Polym Sci 94:20–39
29. Durchschlag H, Zipper P (1995) Jorn Com Esp Deterg 26:275–292
30. Durchschlag H (1986) In: Hinz H-J (ed) Thermodynamic data for biochemistry and biotechnology. Springer, Berlin Heidelberg New York, pp 45–128
31. Schubert D, Tziatzios C, van den Broek JA, Schuck P, Germeroth L, Michel H (1994) Prog Colloid Polym Sci 94:14–19
32. Benjamin L (1966) J Phys Chem 70:3790–3797
33. Steele JCH, Tanford C, Reynolds JA (1978) Methods Enzymol 48:11–23
34. Kameyama K, Takagi T (1990) J Colloid Interface Sci 137:1–10
35. Yedgar S, Barenholz Y, Cooper VG (1974) Biochim Biophys Acta 363:98–111
36. Hjelmeland LM, Chrambach A (1984) Methods Enzymol 104:305–318
37. Paradies HH (1989) In: Mittal KL (ed) Surfactants in solution, vol 7. Plenum Press, New York, pp 159–180
38. VanAken T, Foxall-VanAken S, Castleman S, Ferguson-Miller S (1986) Methods Enzymol 125:27–35

Progr Colloid Polym Sci (1999) 113 : 142–149
© Springer-Verlag 1999

A. Hammond
P.M. Budd
C. Price

Microemulsion polymerization
of butyl acrylate and methyl methacrylate

A. Hammond · P.M. Budd (✉) · C. Price
Department of Chemistry
University of Manchester
Manchester M13 9PL, UK
Tel.: + 44-161-2754711
Fax: +44-161-2754598
e-mail: peter.budd@man.ac.uk

Abstract Microemulsions of n-butyl acrylate (BA) were prepared with the following surfactant/cosurfactant systems: ethoxylated nonylphenol (EONP30)/1-pentanol, EONP30/acrylic acid, sodium dodecyl sulphate/1-pentanol and cetyltrimethylammonium bromide (CTAB)/1-pentanol. Microemulsions of BA and methyl methacrylate (MMA) were also prepared using CTAB without cosurfactant. Polymerizations were carried out by both batch and semi continuous methods with the lipophilic initiators 2,2′-azobis(isobutyronitrile) and benzoyl peroxide and with the hydrophilic initiator ammonium persulphate. Core/shell polymer particles were obtained using crosslinked BA particles as

seeds for a second stage polymerization of MMA. Phase diagrams were investigated for BA microemulsions. Particle radii for microemulsions and products, determined by light scattering, sedimentation velocity and transmission electron microscopy, were in the range 8–50 nm. For BA microemulsions, an increase in particle size was observed on increasing the ratio of the number of moles of monomer to the number of moles of surfactant and cosurfactant, as predicted by a simple model.

Key words Microemulsion polymerization · Particle size · Light scattering · Sedimentation velocity · Electron microscopy

Introduction

Certain mixtures of oil, water and surfactants spontaneously form transparent or translucent systems, which are often referred to as "microemulsions" or may be regarded as "swollen micelles" [1–3]. Many surfactants require the presence of a cosurfactant to achieve the ultralow interfacial tensions that are required for microemulsion formation. In recent years there has been considerable interest in carrying out polymerizations in microemulsions, in order to obtain polymer dispersions with small particle sizes (radius < 50 nm) and narrow particle size distributions [4–10].

The present contribution describes a systematic study of the microemulsion polymerization of n-butyl acrylate (BA). The use of three surfactants was investigated:

1. The nonionic surfactant ethoxylated nonylphenol (EONP30), which has the formula $C_9H_{19}C_6H_4O(CH_2CH_2O)_nH$, where n has an average value of 4.5.
2. The anionic surfactant sodium dodecyl sulphate (SDS), $C_{12}H_{25}OSO_3Na$.
3. The cationic surfactant cetyltrimethylammonium bromide (CTAB), $C_{16}H_{33}(CH_3)_3NBr$.

1-Pentanol was employed as a cosurfactant with each of these surfactants. In addition, EONP30 was investigated with acrylic acid as a cosurfactant, and CTAB was used without any cosurfactant. Three different free radical initiators were employed in these studies:

1. The lipophilic initiator 2,2′-azobis(isobutyronitrile) (AIBN), $(CH_3)_2C(CN)N{=}NC(CH_3)_2CN$.
2. The lipophilic initiator benzoyl peroxide (BPO), $C_6H_4C(O)OOC(O)C_6H_4$.

3. The hydrophilic initiator ammonium persulphate (APS), $NH_4O_3SOOSO_3NH_4$.

Initially, phase diagrams for BA/surfactant/cosurfactant/water systems were investigated, in order to establish the composition range within which a stable microemulsion polymerization was feasible. Further studies were carried out with varying amounts of surfactant and cosurfactant at the maximum monomer concentration that could reliably be polymerized as a microemulsion. Microemulsion composition is expressed in this work in terms of $C_W = m_{mon}/m_{surf}$, where m_{mon} and m_{surf} are the masses of monomer and surfactant, respectively, and $C_N = n_{mon}/(n_{surf} + n_{cosurf})$, where n_{mon}, n_{surf} and n_{cosurf} are the numbers of moles of monomer, surfactant and cosurfactant, respectively. Products were characterized by dynamic light scattering (DLS), static light scattering (SLS), sedimentation velocity (SV) and transmission electron microscopy (TEM).

The microemulsion polymerization of methyl methacrylate (MMA) was also studied, with CTAB alone as the surfactant. In addition, the formation of core/shell poly(BA)/poly(MMA) (PBA/PMMA) dispersions was investigated.

Experimental

Materials

Monomers, BA (Aldrich) and MMA (Aldrich), were distilled and stored in the dark until used. Surfactants, EONP30 (Kasei Chemicals, 99.5%), SDS (Aldrich, 98%) and CTAB (BDH, 99%), were used as received. Cosurfactants, 1-pentanol (Aldrich, 99%) and acrylic acid (Fluka, 99%) were distilled before use. Initiators, AIBN, BPO and APS, were used as received. The crosslinking agent, divinylbenzene (Aldrich, 99.5%), was used as received. Water was doubly distilled, boiled and bubbled with oxygen-free nitrogen.

Determination of phase diagrams

Experiments were conducted at 60 °C in a glass vessel which was fitted with a stirrer and purged with oxygen-free nitrogen before use. A microemulsion was characterized by the formation of a single transparent phase. Phase boundaries for unpolymerized BA microemulsions were determined by titration of various mixtures with water, surfactant or cosurfactant. Phase boundaries for polymerized systems were determined by polymerizing various formulations with AIBN as initiator.

Batch polymerization

Batch polymerizations were carried out in a five-neck glass vessel fitted with a stirrer, a thermometer, an inlet through which oxygen-free nitrogen could be passed and a solid CO_2/acetone cold finger. The vessel was suspended in a thermostatted oil bath. A typical polymerization is described below.

BA monomer (9.5 cm³) and lipophilic initiator AIBN (0.1 g) were stirred (200–250 rpm) at 30 °C to form the oil phase.

A solution of surfactant (e.g., EONP30, 15 g), cosurfactant (e.g., 1-pentanol, 17 cm³) and degassed water (43 cm³) was prepared separately. The oil and aqueous phases were dispersed with high-speed stirring (2500 rpm) and then stirred (200–250 rpm) for 10 h at room temperature. With continuous slow stirring, the reaction mixture was heated to 60–70 °C and maintained at that temperature for 48 h.

In every case, an aliquot of the polymerized microemulsion system was kept for characterization by DLS, SV and TEM, and stored under nitrogen. In some cases, polymer was isolated by dissolving the crude product in tetrahydrofuran (100 cm³) and reprecipitating in methanol (500 cm³). The resulting white powder was oven-dried for 48 h at 50 °C and the procedure repeated. Conversions of 80–85% were obtained, based on the mass of isolated polymer.

The batch polymerization in microemulsion of BA was investigated for the surfactant/cosurfactant systems EONP30/1-pentanol, EONP30/acrylic acid, SDS/1-pentanol and CTAB/1-pentanol for various microemulsion compositions ($C_W = 0.806$, $C_N = 0.493$; $C_W = 0.781$, $C_N = 0.437$; $C_W = 0.578$, $C_N = 0.408$; $C_W = 0.645$, $C_N = 0.394$; $C_W = 0.459$, $C_N = 0.340$; $C_W = 0.559$, $C_N = 0.313$) and with CTAB without cosurfactant at the same values of C_W. The batch polymerization of MMA in microemulsion with CTAB was investigated at compositions corresponding to $C_W = 1.0$, 0.67 and 0.5. Amounts of initiator were also varied (7×10^{-4}–1×10^{-2} mol) but no significant dependence of particle size on initiator concentration was observed in this range.

Semicontinuous polymerization

Semicontinuous polymerizations were carried out in apparatus similar to that previously described. A typical polymerization is described below.

The components for the recipe were divided into four parts:

1. A bulk phase consisting of water (about two-thirds total water), monomer (about one-third total monomer) and emulsifier mixture (about two-thirds total emulsifier).
2. Remaining monomer.
3. Emulsifier feed comprising the remaining surfactant and cosurfactant in water.
4. Initiator in aqueous solution.

The reaction vessel was purged with nitrogen, the bulk phase added and the temperature raised to 60 °C. The reaction was started by allowing 0.1 cm³ initiator solution into the stirred vessel. The monomer, emulsifier and initiator feeds were added dropwise with addition rates of 0.1 cm³ min⁻¹ for monomer and emulsifier, and 0.01 cm³ min⁻¹ for initiator. After all reagents had been added, stirring was continued for 24 h at 65 °C.

Semicontinuous polymerizations in microemulsion of BA were carried out for the surfactant/cosurfactant systems EONP30/1-pentanol, EONP30/acrylic acid, SDS/1-pentanol and CTAB/1-pentanol at the same overall microemulsion compositions as used in the batch polymerizations.

Core/shell polymerization

Core/shell polymers were prepared by a two-stage process. A modified version of the batch polymerization procedure gave a crosslinked PBA microemulsion seed, which was used subsequently for polymerization of MMA. Initial experiments established levels of emulsifier mixture sufficiently low that a second crop of PMMA particles was not obtained, as observed by TEM.

For preparation of the seed, divinylbenzene crosslinker (5 wt% relative to BA) was included in the formulation. Seeds were prepared with an EONP30/1-pentanol microemulsion (composition: 2.00 g EONP30, 7.00 g 1-pentanol, 4.25 g BA, 0.21 g divinylbenzene, 0.50 g initiator, 27.65 g water) and an EONP30/acrylic acid microemulsion (composition: 1.50 g EONP30, 4.25 g acrylic acid, 3.13 g BA, 0.16 g divinylbenzene, 0.11 g initiator, 20.95 g water). Polymerizations were carried out both with AIBN and with BPO as initiator. Water, initiator and emulsifier were placed in the reaction vessel, which was purged with dry nitrogen. About a quarter of the monomer mixture (BA/divinylbenzene) was added, the temperature raised to 60 °C and the remaining monomer mixture then added dropwise. After all the monomer had been added, the temperature was maintained at 60 °C for a further 17 h and then allowed to cool. The seed was stored under nitrogen for no more than 24 h before the second-stage polymerization was commenced.

For the second-stage polymerization, the seed was placed in the reaction vessel along with initiator and water, such that the final solids content would be about 10%. For the EONP30/1-pentanol seed, the composition for the second-stage polymerization was 20.40 cm^3 seed, 2.13 g MMA, 0.20 g initiator and 20.00 g water. For the EONP30/acrylic acid seed, the composition for the second-stage polymerization was 14.90 cm^3 seed, 1.57 g MMA, 0.11 g initiator and 14.95 g water. In both cases, the weight ratio BA:MMA was 1:1. In a reaction similar to the preparation of the seed, the vessel was heated to 60 °C and MMA was added dropwise, with an addition rate of 0.1 cm^3 min^{-1}.

Light scattering

DLS and SLS experiments were carried out using an Otsuka DLS-700 light scattering photometer equipped with a 5 mW He-Ne laser (wavelength 632.8 nm) and a digital correlator. Microemulsions for analysis were diluted with filtered (Millipore 0.22 μm) water and the diluted samples were also filtered.

For DLS, initial experiments demonstrated little dependence of apparent hydrodynamic radius on concentration in the range 0.5–5 × 10^{-6} g cm^{-3}, and further experiments were undertaken at a concentration of 1 × 10^{-6} g cm^{-3}. Measurements were made at an angle of 90°. Data were analysed by the method of cumulants to obtain a diffusion coefficient, D, and the hydrodynamic radius, r_h, was evaluated using the Stokes–Einstein relationship

$$r_h = \frac{kT}{6\pi\eta D} , \qquad (1)$$

where k is the Boltzmann constant, T is the absolute temperature and η is the viscosity of the medium. This procedure gives a z-average D and consequently a reciprocal z-average reciprocal r_h. The estimated uncertainty in r_h from DLS is ±10%.

For SLS, measurements were made as a function of angle and concentration. Data were analysed by means of a Zimm plot to determine the weight-average molar mass, M_W, and the z-average root-mean-square radius of gyration, $\langle r_g^2 \rangle_z^{1/2}$. Estimated uncertainties in the results are ±10%. Refractive index increments (dn/dc) were determined using an Abbe 60 refractometer and are given in Table 1.

Analytical ultracentrifugation

SV experiments were carried out using a Beckman Model E analytical ultracentrifuge equipped with Schlieren optics and operating at a rotor speed of 40 000 rpm. Microemulsion latices were diluted to a volume fraction, ϕ, of 0.01, transferred to a cell with a single-sector aluminium centrepiece, and centrifuged for several hours. The sedimentation coefficient, s, was evaluated from the rate of movement of the peak in the Schlieren image. Values of the hydrodynamic radius, r_h, were obtained using [1]

Table 1 Refractive index increments at 25 °C for latices of poly (butyl acrylate) (*PBA*) and poly (methyl methacrylate) (*PMMA*) with surfactants ethoxylated nonylphenol (*EONP3O*), sodium dodecyl sulphate (*SDS*) and cetyltrimethylammonium bromide (*CTAB*)

Polymer	Surfactant	Cosurfactant	dn/dc (cm^3 g^{-1})
PBA	EONP30	1-Pentanol	0.128
PBA	EONP30	Acrylic acid	0.124
PBA	SDS	1-Pentanol	0.130
PBA	CTAB	1-Pentanol	0.128
PBA	CTAB	–	0.129
PMMA	CTAB	–	0.125

$$r_h = \left[\frac{9\eta s}{2(\rho_D - \rho_M)} \times f(\phi) \right]^{1/2} , \qquad (2)$$

where $(\rho_D - \rho_M)$ is the difference in density between the dispersed phase and the medium (estimated as 1.7×10^{-4} g cm^{-3}) and $f(\phi)$ was calculated using the Batchelor equation [11]

$$f(\phi) = \frac{1}{1 - 6.55\phi} . \qquad (3)$$

The validity of this relationship is questionable for very precise work [12], but it is adequate for the present purposes. This procedure is expected to give an average r_h smaller than that obtained from DLS. The estimated uncertainty in r_h from SV is ±15%.

Transmission electron microscopy

TEM was carried out using a JEOL 100CX electron microscope operating at 80 kV. Microemulsion latices were examined using a cryoscopic technique for uncrosslinked samples and a conventional technique for crosslinked samples. Samples were mounted on a carbon substrate supported by a copper grid. In some cases, samples were carbon-shadowed using a Nanotech M250 coating unit. Reciprocal z-average reciprocal radius values, r, were evaluated, for comparison with DLS results. The estimated uncertainty in r from TEM is ± 10%.

Results and discussion

Phase diagrams

Polymerization of BA microemulsions at monomer concentrations greater than about 10% w/w gave rise to turbidity or phase separation. Ternary (surfactant/cosurfactant/water) phase diagrams are shown in Fig. 1 for the maximum BA concentrations that could be polymerized without incurring instability (10.4% w/w with EONP30/1-pentanol, 10.5% w/w with EONP30/acrylic acid, 9.9% w/w with SDS/1-pentanol and 9.3% w/w with CTAB/1-pentanol). The phase diagrams indicate the composition range (% w/w) within which a microemulsion polymerization can be achieved for these systems. The different emulsifier systems show broadly similar behaviour,

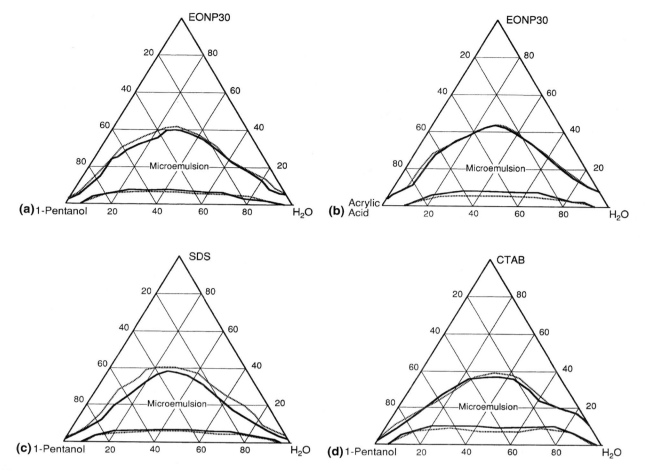

Fig. 1 Ternary phase diagrams for **a** 10.4% w/w butyl acrylate (*BA*) with ethoxylated nonylphenol (*EONP30*), 1-pentanol and water, **b** 10.5% w/w BA with EONP30, acrylic acid and water, **c** 9.9% w/w BA with sodium dodecyl sulphate (*SDS*), 1-pentanol and water and **d** 9.3% w/w BA with cetyltrimethylammonium bromide (*CTAB*), 1-pentanol and water. *Dotted lines* refer to initial microemulsions and *solid lines* to polymerized systems

except that with CTAB (Fig. 1d) microemulsions can be achieved without cosurfactant over a relatively wide composition range. CTAB was used both with and without cosurfactant in the present work.

Microemulsion polymerization of BA

The dependence of r_h, determined by DLS, on the composition parameter C_N for EONP30/1-pentanol microemulsions prior to polymerization by batch and semicontinuous method is shown in Fig. 2a. A marked increase in r_h was observed on increasing the total amount of monomer relative to surfactant and cosurfactant. Results for products obtained from polymerizations initiated by AIBN, BPO and APS are shown in Fig. 2b–d. The different initiators gave broadly similar

results, with arguably slightly larger particle sizes for products initiated with the water-soluble initiator APS. Batch polymerizations gave similar results to semicontinuous polymerizations. Radii determined by SV and TEM are seen to agree reasonably with those from DLS.

Particle radii determined for EONP30/acrylic acid microemulsions are shown in Fig. 3a, and for the products of polymerization with AIBN in Fig. 3b. Other initiators gave very similar results. Use of acrylic acid as a cosurfactant is seen to give results similar, in terms of particle radii, to those obtained with 1-pentanol (Fig. 2), even though there is likely to be a high proportion of the acrylic acid in the aqueous phase. Acrylic acid is itself polymerizable and may become incorporated into the polymer particles or appear as poly(acrylic acid) in the aqueous medium. A series of experiments was undertaken to investigate the extent of acrylic acid polymerization, with $C_W = 0.46$ and C_N values in the range 0.16–0.34. The percentage conversion of acrylic acid to polymer was estimated from infrared spectroscopy data as 40% at $C_N = 0.16$, increasing to 55% at $C_N = 0.34$. Particle size distributions from DLS for products gave a small peak below 4 nm, which was attributed to free poly(acrylic acid) chains. The proportion of the initial

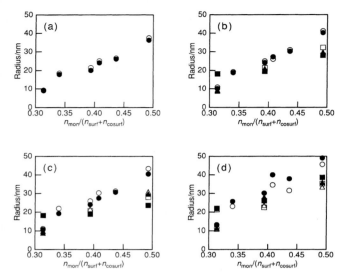

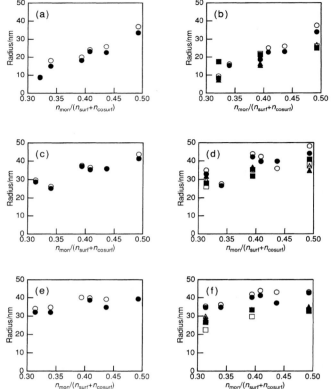

Fig. 2a–d BA/EONP30/1-pentanol microemulsions: dependence on composition of particle radius determined by dynamic light scattering (*DLS*) (○, ●), sedimentation velocity (*SV*) (□, ■) and transmission electron microscopy (*TEM*) (△, ▲). **a** Unpolymerized, **b** polymerized with 2,2'-azobis(isobutyronitrile) (*AIBN*) initiator, **c** polymerized with benzoyl peroxide initiator and **d** polymerized with ammonium persulphate initiator. Polymerizations were carried out by both batch (○, □, △) and semicontinuous (●, ■, ▲) methods

Fig. 3 Dependence on composition of particle radius determined by DLS (○, ●), SV (□, ■) and TEM (△, ▲) for **a** BA/EONP30/acrylic acid microemulsions, **b** BA/EONP30/acrylic acid microemulsions polymerized with AIBN initiator, **c** BA/SDS/1-pentanol microemulsions, **d** BA/SDS/1-pentanol microemulsions polymerized with AIBN initiator, **e** BA/CTAB/1-pentanol microemulsions and **f** BA/CTAB/1-pentanol microemulsions polymerized with AIBN initiator. Polymerizations were carried out by both batch (○, □, △) and semicontinuous (●, ■, ▲) methods

acrylic acid existing as free chains in the microemulsion products was estimated to range between 14% at $C_N = 0.16$ and 3% at $C_N = 0.34$.

Particle radii determined for the SDS/1-pentanol microemulsions are shown in Fig. 3c and for products polymerized with AIBN in Fig. 3d. An increase in particle radius with increasing C_N may be seen for the initial microemulsion and the polymerized products, although the dependence is not as pronounced as was observed with the nonionic surfactant EONP30. For the CTAB/1-pentanol system (Fig. 3e, f) and for CTAB without cosurfactant (Fig. 4) there is only a slight dependence of particle radius on composition over the range studied. With all these systems there was little effect of the type of initiator on the particle sizes of the products.

Table 2 gives values of sedimentation coefficient, s, and hydrodynamic radius, r_h, from SV, particle radius, r, from TEM, hydrodynamic radius, r_h, from DLS, z-average radius of gyration, $\langle r_g^2 \rangle_z^{1/2}$, and weight-average molar mass, $\bar{M}_W$, from SLS, for BA microemulsion latices prepared with various surfactant/cosurfactant systems at $C_W = 0.645$, $C_N = 0.394$. Reasonable agreement can be seen between the various methods of particle size analysis. TEM indicated that particle size distributions were narrow.

Antonietti et al. [4] explained the dependence of particle size on composition for styrene microemulsions

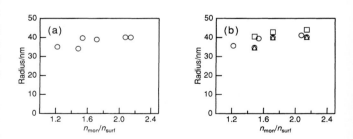

Fig. 4 Dependence on composition of particle radius determined by DLS (○), SV (□) and TEM (△) for BA/CTAB microemulsions: **a** unpolymerized and **b** polymerized by a batch method with AIBN initiator

stabilized by a single surfactant, without cosurfactant, in terms of a simple model of a monomer core surrounded by a surfactant shell. They found

Table 2 Particle characteristics for butyl acrylate latices prepared at $C_W = 0.645$, $C_N = 0.394$ with initiators 2,2′-azobis(isobutyronitrile) (AIBN), benzoyl peroxide (BPO) and ammonium persulphate (APS), obtained by sedimentation velocity (SV), transmission electron microscopy (TEM), dynamic light scattering (DLS) and static light scattering (SLS)

Surfactant	Cosurfactant	Initiator	Method	SV s $(10^{-13}\,s)$	SV r_h (nm)	TEM r (nm)	DLS r_h (nm)	SLS $\langle r_g^2 \rangle_z^{1/2}$ (nm)	SLS $\bar{M}_w$ (g mol^{-1})
EONP30	1-Pentanol	AIBN	Batch	0.18	21	21	25	23	115 000
EONP30	1-Pentanol	AIBN	Semicontinuous	0.15	20	20	24	22	110 000
EONP30	1-Pentanol	BPO	Batch	0.17	21	22	26	24	121 000
EONP30	1-Pentanol	BPO	Semicontinuous	0.14	19	19	24	21	116 000
EONP30	1-Pentanol	APS	Batch	0.20	23	24	26	25	157 000
EONP30	1-Pentanol	APS	Semicontinuous	0.27	26	28	30	28	146 000
EONP30	Acrylic acid	AIBN	Batch	0.19	22	16	21	19	105 000
EONP30	Acrylic acid	AIBN	Semicontinuous	0.18	21	15	18	17	99 000
EONP30	Acrylic acid	BPO	Batch	0.22	24	18	23	22	111 000
EONP30	Acrylic acid	BPO	Semicontinuous	0.18	21	17	20	19	104 000
EONP30	Acrylic acid	APS	Batch	0.20	23	24	28	28	140 000
EONP30	Acrylic acid	APS	Semicontinuous	0.19	22	20	24	23	131 000
SDS	1-Pentanol	AIBN	Batch	0.49	35	37	44	37	116 000
SDS	1-Pentanol	AIBN	Semicontinuous	0.40	32	36	42	36	125 000
SDS	1-Pentanol	BPO	Batch	0.39	31	34	38	33	123 000
SDS	1-Pentanol	BPO	Semicontinuous	0.34	29	34	42	35	132 000
SDS	1-Pentanol	APS	Batch	1.12	53	45	50	43	155 000
SDS	1-Pentanol	APS	Semicontinuous	0.36	30	37	41	36	163 000
CTAB	1-Pentanol	AIBN	Batch	0.35	30	34	40	32	101 000
CTAB	1-Pentanol	AIBN	Semicontinuous	0.44	33	33	40	32	101 000
CTAB	1-Pentanol	BPO	Batch	0.39	31	34	41	32	105 000
CTAB	1-Pentanol	BPO	Semicontinuous	0.35	30	34	40	32	107 000
CTAB	1-Pentanol	APS	Batch	0.46	34	38	42	34	132 000
CTAB	1-Pentanol	APS	Semicontinuous	0.46	34	38	41	34	135 000

$$r \approx \frac{b}{1 - (1+S)^{-1/3}} \,, \tag{4}$$

where r is the radius of the particle, b is the shell thickness and $S = m_{surf}/m_{mon} = 1/C_W$. Wu [5] suggested that the surfactant should partly penetrate the core, so the shell thickness b is less than the length of the surfactant molecule. He obtained the relationship

$$C_W = a\left(\frac{N_A \rho}{3 M_{surf}}\right) r + \text{constant} \,, \tag{5}$$

where a is the average surface area per surfactant molecule, N_A is the Avogadro number, ρ is the particle density, M_{surf} is the molar mass of the surfactant and the constant includes the shell thickness b.

Here we develop a simple model to take account of the presence of cosurfactant. Assuming that all the monomer is contained within microemulsion particles, the volume of a particle core is

$$\frac{4}{3}\pi r_{core}^3 = \frac{n_{mon} M_{mon}}{N_p \rho_{core}} \,, \tag{6}$$

where r_{core} and ρ_{core} are the radius and density, respectively, of the core, n_{mon} and M_{mon} are the number of moles and molar mass, respectively, of monomer, and

N_p is the number of particles. The surface area of the core is

$$4\pi r_{core}^2 = \frac{a_{surf} n_{surf} f_{surf} + a_{cosurf} n_{cosurf} f_{cosurf}}{N_p} \,, \tag{7}$$

where a_{surf} and a_{cosurf} represent the average cross-sectional areas of the surfactant and cosurfactant, respectively, at the core surface, n_{surf} and n_{cosurf} are the numbers of moles of surfactant and cosurfactant, respectively, and f_{surf} and f_{cosurf} are the fractions of total surfactant and cosurfactant, respectively, that are contained within microemulsion particles. Combination of Eqs. (6) and (7) gives

$$r_{core} = \frac{3 M_{mon}}{\rho_{core}} \frac{n_{mon}}{(a_{surf} n_{surf} f_{surf} + a_{cosurf} n_{cosurf} f_{cosurf})} \,, \tag{8}$$

The overall particle radius, r, exceeds r_{core} by the shell thickness b. If it is assumed that all the surfactant and cosurfactant are contained within microemulsion particles ($f_{surf} = f_{cosurf} = 1$), and that the average area per molecule is similar for surfactant and cosurfactant ($a_{surf} = a_{cosurf} = a$).

$$r = r_{core} + b = \frac{3 M_{mon}}{\rho_{core} a} \frac{n_{mon}}{n_{surf} + n_{cosurf}} + b \,, \tag{9}$$

which suggests that in the simplest case r depends linearly on C_N. The data in Figs. 2 and 3 agree approximately with this linear dependence. A fuller analysis of the results would require additional information about partitioning of the monomer, surfactant and cosurfactant between the aqueous phase and microemulsion particles.

The microemulsion polymerizations studied here are ideal in the sense that the sizes of the final products closely reflect those of the initial microemulsions. It is tempting to think of each droplet essentially as a minute reactor. However, exchange of monomer, surfactant and cosurfactant via the aqueous phase is possible. In some other microemulsion systems, growth has been observed on polymerization. For example, Kuo et al. [9] reported growth during the photoinitiated polymerization in microemulsion of styrene with dibenzylketone initiator. Ming et al. [10] observed an increase in particle size in microemulsion polymerizations of styrene and butyl methacrylate, but found a constant size throughout polymerization for MMA, methyl acrylate and BA.

Microemulsion polymerization of MMA

Particle radii for MMA microemulsions, stabilized by CTAB without cosurfactant, and for products polymerized with AIBN are shown in Fig. 5. Other initiators gave similar results. Small particle sizes (r_h in the range 15–20 nm) were obtained regardless of composition over the range studied. Similar results have been obtained previously for microemulsion polymerizations of MMA

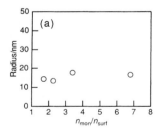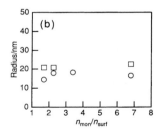

Fig. 5 Dependence on composition of particle radius determined by DLS (○) and SV (□) for methyl methacrylate/CTAB microemulsions: **a** unpolymerized and **b** polymerized by a batch method with AIBN initiator

[4]. MMA is a significantly more hydrophilic monomer than BA and polymerization within the aqueous medium is a possibility in this case.

Core/shell latices of BA and MMA

Particle radii for crosslinked PBA seeds and PBA/PMMA core/shell latices, determined by DLS and TEM, are given in Table 3. An increase in particle radius on second-stage polymerization was observed, indicating that composite particles were produced. Core/shell volume ratios, which should approximately equal the weight ratio of monomers used (1:1), are indicated in Table 3. For the EONP30/acrylic acid system initiated by BPO the core/shell volume ratio is rather high, possibly because of incorporation of polyacrylic acid into the core (see earlier).

For these latices, the particle size distributions were narrow, with ratios of weight-average to number-average radius less than 1.03, as determined by TEM. Since MMA is relatively hydrophilic, and the PBA seed was crosslinked, the particles are expected to comprise a core of crosslinked PBA and a shell of PMMA.

Conclusions

Phase diagrams were determined for BA with EONP30/1-pentanol, EONP30/acrylic acid, SDS/1-pentanol and CTAB/1-pentanol. Microemulsions of BA had hydrodynamic radii in the range 8–50 nm, the size increasing with the increase in the amount of monomer relative to the total amount of surfactant and cosurfactant. The magnitude of the dependence of hydrodynamic size on composition varied according to the surfactant used, dimishing in the sequence EONP30 > SDS > CTAB. Microemulsions of MMA with CTAB had hydrodynamic radii in the range 15–20 nm, the size being essentially independent of composition over the range studied.

Polymerization in microemulsion, for the systems studied, gave products with particle sizes comparable to the initial microemulsions. No dependence of particle size on initiator concentration was observed and the type of initiator had little effect. Similar results were obtained from both batch and semicontinuous polymerizations.

Table 3 Particle characteristics of PBA seeds and PBA/PMMA core/shell latices prepared with EONP30 surfactant

Cosurfactant	Initiator	Seed		Core/shell product		Core/shell volume ratio
		TEM r (nm)	DLS r_h (nm)	TEM r (nm)	DLS r_h (nm)	
1-Pentanol	AIBN	24	25	33	35	0.6:1
1-Pentanol	BPO	25	26	34	35	0.7:1
Acrylic acid	AIBN	19	20	25	26	0.8:1
Acrylic acid	BPO	22	23	25	26	2.2:1

Composite core/shell particles with narrow particle size distributions were obtained by preparation of a cross-linked PBA seed followed by a second-stage polymerization of MMA.

Reasonable agreement was achieved between different methods of characterization: light scattering, analytical ultracentrifugation and electron microscopy.

References

1. Budd PM, Pinfield RK, Price C (1997) Prog Colloid Polym Sci 107:189
2. Hoar TP, Schulman JH (1943) Nature 152:102
3. Schulman JH, Stoeckenius W, Prince LM (1959) J Phys Chem 63:1679
4. Antonietti M, Bremser W, Müschenborn D, Rosenauer C, Schupp B, Schmidt M (1991) Macromolecules 24:6636
5. Wu C (1994) Macromolecules 27:298; 27:7099
6. Antonietti M, Basten R, Lohmann S (1995) Macromol Chem Phys 196:441
7. Roy S, Devi S (1997) Polymer 38:3325
8. Capek I, Juranicova V (1998) Eur Polym J 34:783
9. Kuo PL, Turro NJ, Tseng CM, El-Aasser MS, Vanderhoff JW (1987) Macromolecules 20:1216
10. Ming W, Jones FN, Fu S (1998) Macromol Chem Phys 199:1075
11. Batchelor GK (1972) J Fluid Mech 52:245
12. Rowe AJ (1992) In: Harding SE, Rowe AJ, Horton JC (eds) Analytical ultracentrifugation in biochemistry and polymer science. Royal Society of Chemistry Cambridge, p 394

Progr Colloid Polym Sci (1999) 113:150–157
© Springer-Verlag 1999

SYNTHETIC POLYMERS, COLLOIDS AND SUPRAMOLECULAR SYSTEMS

G.M. Pavlov
E.V. Korneeva
R. Roy
N.A. Michailova
P.C. Ortega
M.A. Perez

Sedimentation, translational diffusion, and viscosity of lactosylated polyamidoamine dendrimers

This paper is dedicated to the memory of Professor Victor N. Tsvetkov.

G.M. Pavlov (✉) · N.A. Michailova
Institute of Physics, University
Ulianovskaya str. 1, 198904
St.Petersburg, Russia
e-mail: gpolym@onti.niif.spb.su
Tel.: +7-812-4284365
Fax: +7-812-4287240

E.V. Korneeva
Institute of Macromolecular Compounds
RAS, Bolshoi pr. 31, 199004
St.Petersburg, Russia

R. Roy · P.C. Ortega · M.A. Perez
Department of Chemistry
University of Ottawa, Ottawa
Ontario K1N 6N5, Canada

Abstract The hydrodynamic characteristics of six generations of lactosylated polyamidoamine dendrimers (LacPAMAM) in 0.165% NaCl have been investigated and their molecular characteristics determined. Experimental values varied over the following ranges: sedimentation velocity coefficient $0.65 < s < 6.2$, translational diffusion coefficient ($\times 10^7$ cm^2 s^{-1}) $19.1 > D > 4.9$, which corresponds to a change in molecular weight ($\times 10^3$) in the range $2.5 < M_{SD} < 93$ and which is in good agreements with LacPAMAM chemical structures. The intrinsic viscosity of LacPAMAMs practically did not change and the average value was (4.25 ± 0.45) cm^3/g. The following scaling relationships for hydrodynamic values were thus established:

$$s = 4.84 \times 10^{-16} M^{0.63 \pm 0.015} \ ,$$

$$D = 3.56 \times 10^{-5} M^{-(0.37 \pm 0.015)} \ ,$$

$$[\eta] = 2.59 \times M^{0.05 \pm 0.05} \ .$$

The hydrodynamic invariant is $A_0 = (2.61 \pm 0.07)10^{-10}$. In the case of the dendrimers the value of the intrinsic viscosity became insensitive to molecular weight changes. In contrast, the sedimentation velocity coefficient became more sensitive to M in comparison to that of linear molecules. The hydrodynamic values of LacPAMAM molecules were compared to the values obtained for lactosylated dendrimers based on a poly(propylene imine) core.

Key words Carbohydrate dendrimers · Hydrodynamic properties · Polyamidoamine dendrimers

Introduction

In light of the potential biomedical applications of carbohydrate-containing macromolecules and glycodendrimers in particular [1–3], the biophysical characterization of these novel biopolymers has become a research area of considerable interest. The motivation toward the study of glycodendrimers has emerged out of the necessity to better understand the roles played by carbohydrates in living systems at the molecular level. Glycodendrimers have recently broadened the range of hydrophilic biopolymers that might serve in the preparation of synthetic vaccines and in the modification of polymers that can be used as carriers of biologically active groups [4, 5].

The peculiar appeal manifested toward the modification of regularly branched molecules – dendrimers – is based on the presumption that they may be more effective carriers for biologically active groups in comparison to linear polymers. Dendrimers are also promising molecules that may find wide applications in electronic devices, catalysis, etc., [6–8] depending on their specific structures. This is mainly because of the large number of reactive end groups that are assumed to be equally accessible.

Moreover, owing to their exceptional structures, dendrimers are interesting objects for molecular investigation. It may be expected that the establishment of quantitative correlation (regular branched structural properties) will lead to a better understanding of irregularly branched molecules.

The difficulties arising in dendrimer studies are directly related to fundamental problems already existing in research on branched macromolecules in polymer science. The first theoretical considerations of regularly branched macromolecules were by Flory [9], Kuhn and Kuhn [10], and Tsvetkov [11]. These studies were triggered to some extent by other investigations on naturally occurring branched polymers, such as glycogen, starch, and lignin, which principally exhibit short-chain branches and are randomly branched [12].

Some interest in dendrimer studies may also be due to their intrinsic artistic beauty and charm representing ideal molecular models. This might also explain a few of the poetic terms introduced to describe dendrimers: arborols [8], starburst molecules [13], little prince baobabs [14], sugar ball [15] or a less poetical name – cauliflower [16].

At present, many papers are available on the synthesis of dendrimers, their chemical modifications, and their computer simulation [17, 18]. Dendrimers based on polyamidoamine (PAMAM) [19] and poly(propylene imine) (DAB) [20] are the structures investigated in greatest detail, presumably because they are now commercially available. However, it should be noted that there are a few physicochemical studies in which molecular weights and other molecular characteristics of these molecules have been determined using absolute methods. Furthermore, some fundamental problems, in particular the distribution of surface groups on dendrimers and their type of homology, are not yet fully understood.

In fact, the concept behind the construction, modification, and use of these molecules is based on the following assumptions: first, it is generally taken as granted that reactive groups have equal accessibility in all stages of growth and modification and, second, the density of the dendritic core is usually lower than that of the periphery, i.e. the molecule is relatively hollow. These concepts were confirmed by the first analytical calculations [16].

However, subsequent computer simulations [14, 21, 22], calculations using molecular dynamic methods [23, 24], and calculations based on the equilibrium self-consistent field [25] led to opposite conclusions: namely, a dendrimer's density is a maximum in the center and decreases toward the periphery, and surface groups on dendrimers are distributed throughout the entire volume of the molecule and are not located in the external layer.

It should be noted that all theoretical calculations have been carried out on uncharged homodendrimers with flexible and relatively long spacers between consecutive branching points. Moreover, it is assumed that thermodynamic chain rigidity does not change on passing to homodendrimers of higher generations. It is also clear that the situation becomes more complicated when the chemical nature of the surface groups is different from that of the core.

Experimental work in which the size and molecular weight of dendrimers have been systematically studied is not very numerous [26–30]. Data on the investigation of hybrid dendrimers are virtually absent. In the present work, we wish to describe the hydrodynamic study of hybrid carbohydrate-based dendrimers made of lactoside residues covalently attached to the surface group of PAMAM dendrimers using combined methods of molecular hydrodynamics [31–33]. These studies include simultaneous investigations of sedimentation velocity, translational diffusion, and viscous flow of dilute solutions. Hence, it is possible to study quantitatively molecules and molecular systems with different structures and molecular weight over almost the entire molecular weight range. It must also be emphasized that hydrodynamic characteristics are obtained in independent experiments and certain correlations between them should be obeyed on the basis of fundamental relationships. This serves as an additional check of the results.

Experimental

Samples and solvents

Lactosylated dendrimers are a new type of neoglycoconjugates of considerable interest in the study of multivalent carbohydrate–protein interactions [1, 34]. Using commercially available starburst PAMAM dendrimers containing amine residues for surface functionality, it was possible to readily synthesize a wide range of sugar-based "glycodendrimers" using unprotected p-isothiocyanatophenyl glycosides in aqueous media. The synthesis of lactosylated starburst PAMAM (LacPAMAM) dendrimers up to generation 5 with an unprotected lactoside derivative [35] followed the strategy described for mannosylated and sialylated dendrimers [36, 37]. Treatment of amine 2 with excess thiophosgene in 80% aqueous ethanol according to ref. [38] resulted in its conversion to isothiocyanate 3 with 69% yield after crystallization from water. Its treatment with ethylenediamine or dendritic PAMAM provided access to the dimer and G0–G5 glycodendrimers in yields of more than 80%. The dimer was purified by size-exclusion chromatography on a Biogel P2 column after lyophilization of the reaction mixture, while the glycodendrimers were purified by removal of residual reactants by dialysis against water. The products gave consistent high-field 1H NMR and 13C NMR as well as MALDI (Matrix-Assisted Laser Desorption Ionization) time-of-flight spectra [35].

Thus, LacPAMAM dendrimers up to generation 5 possessing symmetrical structures were prepared (Fig. 1).

Hydrodynamic investigations were carried out in 0.165% NaCl with the following characteristics at 25 °C: density $\rho_0 = 0.9982$ g/cm^3, viscosity $\eta_0 = 0.893$ cP. The solvent with this composition has been previously used in the investigation of lactose dendrimers based on DAB [29, 30].

Methods

Sedimentation velocity

Sedimentation velocity was investigated on an analytical MOM 3180 ultracentrifuge (Budapest) in cells with artificial

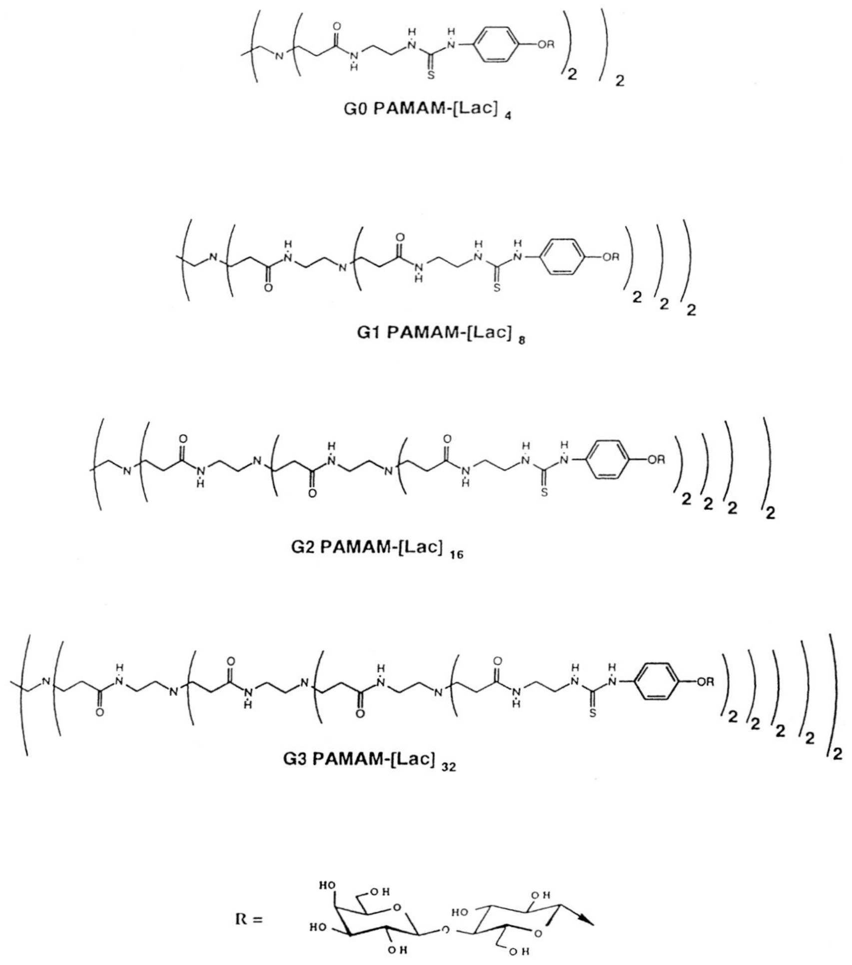

Fig. 1 Structural formulae of thiolactosyl modified polyamido-amine dendrimers (lactosylated dendrimers, *LacPAMAM*) for generations 0, 1, 2, and 3

G0 PAMAM-[Lac] $_4$

G1 PAMAM-[Lac] $_8$

G2 PAMAM-[Lac] $_{16}$

G3 PAMAM-[Lac] $_{32}$

R =

boundary formation of solution at a concentration of $0.4 < c \times 10^2$ g/cm^3 < 1 with solvent. At this concentration dendrimer solutions have limiting dilution ($c[\eta] < 0.04$), and sedimentation coefficients, s, calculated from the displacement of the sedimentation boundary X ($s = \Delta \ln X / \omega^2 \Delta t$; Fig. 2) can be considered to be independent of concentration.

Translational diffusion

Translational diffusion was studied on a polarizing diffusometer [32, 39] in a glass cell. The diffusion boundary was formed by introducing the solution at a concentration of $0.14 < c \times 10^2$ g/ cm^3 < 0.22 under the solvent ($c[\eta] < 0.02$). The diffusion boundary dispersion, σ^2, was calculated from the area and the maximum ordinate of the interference curves according to the equation [32]

$$H = (Q/a) \times \Phi(a/2^{3/2}\sigma) \ ,$$

where H is the maximum ordinate, Q is the area under the interference curve, a is the twinning value of Iceland spars, σ is the standard deviation, and $\Phi(a/2^{3/2}\sigma)$ is a probability integral. The diffusion coefficients, D, were calculated from the time dependence of the diffusion boundary dispersion (Fig. 3)

$$\sigma^2 = \sigma_0^2 + 2Dt$$

where σ_0^2 is the zero dispersion characterizing the quality of the boundary formation, and t is the diffusion time.

The refractive index increment $\Delta n/\Delta c$ (Table 1) was determined from the area spanned by the interference curve, and its average value $(\Delta n/\Delta c)$ was (0.148 ± 0.004) cm^3/g at $\lambda = 550$ nm. The optical system used for recording the solution–solvent boundary in the both diffusion and sedimentation experiments was a Lebedev polarizing interferometer [32, 39]. Experiments were carried out at 26 °C, and diffusion coefficients were corrected to 25 °C according to a standard procedure [31, 32]. The values of the diffusion coefficient obtained at these conditions were assumed to be values extrapolated to zero concentration. Table 1 gives sedimentation and translational diffusion coefficients and their mean-square errors calculated using a linear correlation coefficient, r, [40].

Intrinsic viscosity and buoyancy factor

Intrinsic viscosities, $[\eta]$, were determined from a Huggin's plot [41] (Fig. 4). Relative viscosities, η_r, were determined from the ratio of the flow time of the solution, τ, to that of the solvent, τ_0, ($\tau_0 = 77.9$ s) in the range $1.15 \leq \eta_r \leq 1.45$. The flow time was measured in an Oswald viscometer.

The specific partial volume, v, was calculated from the buoyancy factor, $(1 - v\rho_0)$,

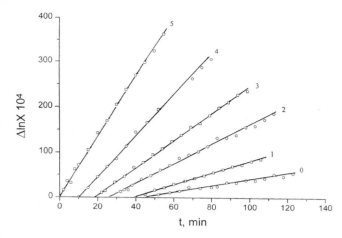

Fig. 2 Dependences of $\Delta \ln x$ on sedimentation time Δt, where x is the position of the sedimentation peak maximum for solutions of LacPAMAM dendrimers 0–5 dispersed in 0.165% NaCl solutions. Rotor speed: 40 000 rpm. For clarity, each curve is displaced along the Δt axis by 10 min relative to the previous curve

$$(1 - \upsilon \rho_0) = \Delta \rho / \Delta c \equiv \Delta m / mw \ ,$$

where $\Delta \rho / \Delta c$ is the density increment, and $\Delta m = m - m_0$, where m and m_0 are the masses of the same volume of solution and solvent, respectively. w is the weight concentration of the solution, and its value was determined by pycnometry from solution measurements in pure water. Table 1 lists individual values of υ. As follows from Fig. 5, all pycnometric measurements for all dendrimer generations fall on a single curve; therefore, for further interpretation the average value was used: $(1 - \upsilon \rho_0) = 0.336 \pm 0.011$.

Conductivity

The conductivity of dendrimer solutions in pure water without salt was studied with a Tacussel type CD78 conductimeter. The measurements were performed with a CM02/55/G platinum platined cell at a frequency of 250 Hz in concentration range $1 < c \times 10^5 \text{ g/cm}^3 < 40$ at 25 °C. The conductivity exceeding that of water has a linear concentration dependence (Fig. 6), which makes it possible to evaluate the summary mobility of the dendrimers. The data obtained confirm the conclusion that lactosylated dendrimers containing tertiary amine groups inside the core of native dendrimers bear a positive charge under the above-mentioned conditions. The conductivity molar coefficient (or summary mobility) increases with generation number and is a maximum for generations 4 and 5. This fact implies that the molecules of higher generations were more charged than those of lower generations. To suppress electrostatic interaction between dendrimers, a low-molecular-weight salt, NaCl, was introduced into the solution.

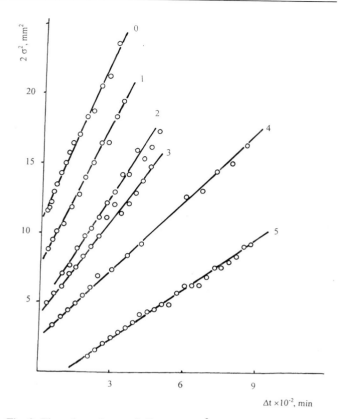

Fig. 3 Time dependence of dispersion (σ^2, second central moment) of the x spectrum of macromolecules in the region of diffusion boundaries in 0.165% NaCl solutions of LacPAMAM dendrimers. Each curve is displaced along the $2\sigma^2$ axis by 3 mm^2 relative to the previous curve

Discussion and results

Molecular weight and scaling relationship

Hydrodynamic investigations of dendrimer solutions with suppressed electrostatic interactions make it possible to determine the molecular weights and the dimensions of the molecules. Molecular weights, M, were calculated according to Svedberg's equation [42] from experimental values of s, D, and the average value of the specific partial volume, $\upsilon = (0.66 \pm 0.012) \text{ cm}^3/\text{g}$:

Table 1 Hydrodynamic characteristics of lactosylated polyamidoamine (*PAMAM*) dendrimers

N	$[\eta]$ cm^3/g	s Sv	Δs Sv	r_s	D cm^2/s	ΔD cm^2/s	r_D	$\Delta n / \Delta c$ cm^3/g	υ cm^3/g
4	4.1	0.65	0.03	0.9835	19.1	0.5	0.9954	0.145	0.644
8	3.2	1.25	0.02	0.9985	14.0	0.4	0.9967	0.151	0.649
16	5.1	2.0	0.04	0.9972	10.7	0.4	0.9922	0.140	0.672
32	4.4	2.85	0.03	0.9994	9.2	0.2	0.9958	0.147	0.684
64	4.1	4.2	0.06	0.9989	6.6	0.2	0.9958	0.153	0.662
128	4.6	6.2	0.08	0.9990	4.9	0.1	0.9966	0.150	0.651

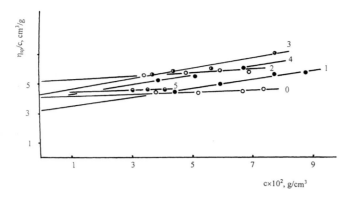

Fig. 4 Plots of η_{sp}/c against c for LacPAMAM solutions in 0.165% NaCl at 25 °C. *Figures* on the curves are the numbers of LacPAMAM generations

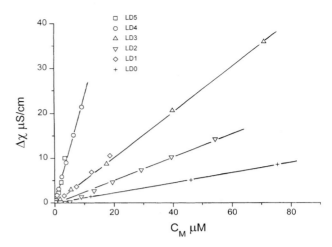

Fig. 6 Variation of the conductivity $\Delta\chi$ of aqueous LacPAMAM dendrimer solutions (without salt) as a function of LacPAMAM concentration (mol/dm^3), where the molarity is related to the molecular weight of the dendrimer

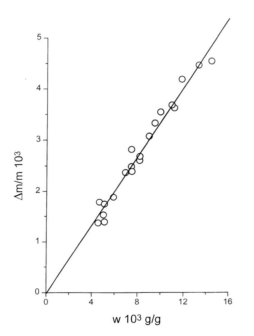

Fig. 5 Dependence of $\Delta m/m$ on weight concentration, w, for all LacPAMAM generations in pure water

$$M_{SD} = [RT/(1 - \upsilon\rho_0)]s/D \ .$$

These values are given in Table 2. The same table lists mean-square errors in molecular weights determined from the corresponding errors in the determination of s, D, and $(1 - \upsilon\rho_0)$. The comparison of M_{SD} with theoretical values, M_{theor}, shows that they are in satisfactory agreement. The values of M_{theor} were calculated on the assumption that the dendrimer's propagation was quantitative, just as the attachment of lactose units. A certain difference is observed between M_{SD} and M_{theor} for generations 2 and 3.

Correlations that can be established between hydrodynamic values and M_{SD} with the assumption of their homologous character lead to the following scaling relations:

$$s = 4.84 \times 10^{-16} M^{0.63\pm0.015}, \ r = 0.9989$$

$$D = 3.56 \times 10^{-5} M^{-(0.37\pm0.015)}, \ r = 0.9969$$

$$[\eta] = 2.59 \times M^{0.05\pm0.05}, \ r = 0.3927 \ .$$

There are the so-called Mark–Kuhn–Houwink–Sakurada plots (Fig. 7). Within experimental error scaling indexes correlate with each other but differ slightly from the theoretical value for a rigid sphere ($b_S = 0.667$, $b_D = 0.333$, $b_\eta = 0$). Their numerical values reflect the fact that for regularly short-branched molecules hydrodynamic values are arranged in the following sequence with respect to their sensitivity to changes in molecular weight: $s > D > [\eta] \sim M^0$. In contrast, in the case of linear molecules, the hydrodynamic values have the following sequence: $[\eta] > D > s$.

An important conclusion from these experimental results is the establishment of the fact that in the case of dendrimer investigations the intrinsic viscosity cannot be considered as a quantitative characteristic. It is only a qualitative characteristic indicating that the density (compactness) of a dendrimer molecule is greater than that of a linear molecule having the same molecular weight. This is analogous to the situation occurring in the investigation of globular proteins when the value of the intrinsic viscosity does not make it possible to distinguish between proteins with different molecular weights [31, 33].

If the dendrimer molecules were hard spherical particles with a specific partial volume υ, they would

Table 2 Molecular weight, hydrodynamic invariant A_0, and hydrodynamic radii of lactosylated PAMAM dendrimers

N	M_{SD}	ΔM_{SD}	M_{theor}	M_{SD}/M_{theor}	$A_0 \times 10^{10}$	$R_D \times 10^8$ cm	$R_{s\eta} \times 10^8$ cm	$R_{s\eta/D} \equiv R_\eta \times 10^8$ cm
4	2 500	200	2 420	1.03	2.68	12.8	11.3	11.8
8	6 600	300	5 230	1.26	2.50	17.5	13.8	15.0
16	13 700	700	10 840	1.26	2.85	22.8	22.1	22.3
32	22 800	900	22 120	1.03	2.76	26.6	24.5	25.2
64	47 000	2000	44 640	1.05	2.46	37.0	28.7	31.2
128	93 000	4000	89 690	1.04	2.39	49.9	36.9	40.8

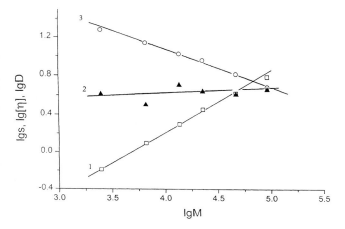

Fig. 7 Double-logarithmic plot of sedimentation coefficient (*1*), intrinsic viscosity (*2*), and translational diffusion coefficient (*3*) versus molecular weights M_{SD}, evaluated from the Svedberg equation for LacPAMAM dendrimers 0–5

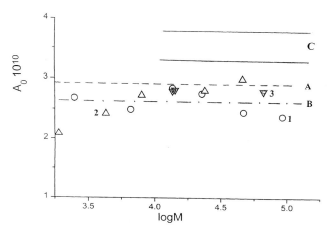

Fig. 8 Comparison the hydrodynamic invariant values for Lac-PAMAM dendrimers (*1*), for lactosylated dendrimers based on poly(propyleneimine) (*DAB*) (*2*), and for proteins (*3*) with the theoretical value for a rigid impermeable sphere (*A*) and with the value for linear polymers (*C*). *Line B* corresponds to the mean value of A_0 for lactosylated dendrimers

have a lower expected intrinsic viscosity $[\eta] = 2.5\upsilon = 1.7$–$1.6$ cm^3/g. By comparing the experimental values of $[\eta]_{exp}$ (Table 1) with the above estimate of $[\eta]_\upsilon$, we may estimate the volume fraction ($[\eta]_\upsilon/[\eta]_{exp}$) of dendrimer substance in the model sphere. This estimation procedure gives values of the order of 40%, which are ten times higher compared to analogous estimates for linear macromolecules [40]. This result is indicative of a greater density of substance in the volume occupied by a true dendrimer molecule (dendrimers of high generations).

It follows from the fundamental equations [31–33]

$$[\eta] = \Phi' \langle R^2 \rangle^{3/2}/M$$

$$s = (M/f)(1 - \upsilon\rho_0)N_A$$
$$= M(1 - \upsilon\rho_0)/\langle R^2 \rangle^{1/2} P' \eta_0 N_A$$

that in dendrimer studies, the sedimentation velocity coefficient is greater than those of their linear analogues (where f is the translational frictional coefficient, $\langle R^2 \rangle$ is the mean-square radius of gyration, N_A is Avogadro's number, and P' and Φ' are Flory hydrodynamic parameters). Moreover, s becomes more sensitive to changes in molecular weight. This is reflected in the

results given in Table 1 and also in the results obtained earlier [29, 30] on the investigation of lactosylated dendrimers based on a DAB core.

Hydrodynamic invariant

The hydrodynamic invariant A_0 [32, 43] was calculated from the experimental values of s, D, $[\eta]$, and $(1 - \upsilon\rho_0)$.

$$A_0 = (R[D]^2[s][\eta])^{1/3} ,$$

where $[D] \equiv D\eta_0/T$, $[s] \equiv s\eta_0/(1 - \upsilon\rho_0)$, and R is the universal gas constant. These values are listed in Table 2. The values of A_0 for different generations differ only slightly, and their average value is $(2.61 \pm 0.07) \times 10^{-10}$. It is easy to compare the values of A_0 for LacPAMAM dendrimers, LacDAB dendrimers [29, 30], and also for globular proteins [31, 33]. These comparisons (Fig. 8) showed us that

1. The values are practically unchanged as a function of M.
2. The values are similar for all molecular systems under consideration.

3. The experimental values are lower that the theoretical value $A_0 = 2.914 \times 10^{-10}$ for rigid impermeable spheres.

A tentative explanation for such a low value of A_0 for globular proteins was made in terms of the hydrodynamics of a porous sphere [44]. (The first attempt to apply the Debye–Bueche model of porous spheres [45] for the hydrodynamic parameter calculations of macromolecules was made by Tsvetkov and Klenin [46]). Although the calculations [44] predict values of A_0 below the value of an impermeable sphere, the observed magnitude of the effect is much larger than that which can be expected theoretically; therefore, the puzzlingly low values of hydrodynamic invariant for proteins and dendrimers must also be regarded as not resolved. Theoretical and experimental values of A_0 are compared for different kinds of lactosylated dendrimers and proteins in Fig. 8.

Hydrodynamic dimensions

Equivalent hydrodynamic dimensions (radii) can be determined on the basis of either the data on translational friction or viscometry by using relations for impermeable spheres [31]:

$$[\eta] = 10\pi N_A R_\eta^3 / 3M \tag{1}$$

$$D = kT / 6\pi\eta_0 R_D . \tag{2}$$

Moreover, it is possible to evaluate R from a direct comparison of the values s and $[\eta]$ [42]. Since

$$s = [(1 - \upsilon\rho_0)/6\pi N_A \eta_0]M/R \tag{3}$$

then

$$s[\eta] = [5(1 - \upsilon\rho_0)9\eta_0]R_{s\eta}^2 .$$

It follows from Eqs. (1)–(3) that R_η, R_D, and $R_{s\eta}$ are related to each other by the equation $R_\eta^3 = R_{s\eta}^2 \times R_D$; hence, only two values of R are independent in this method of M determination.

The change in R_η for dendrimer molecules depending on generation number is virtually completely determined by the change in M because $R_\eta \sim (M[\eta])^{1/3}$ and $[\eta]$ change only slightly on passing from one generation to another. Therefore, if in calculating R_η the value of M_{theor} is used, the comparison of R_η and M_{theor} does not provide original information.

Comparison of the characteristics of lactosylated dendrimers based on a DAB core [29, 30] and those based on PAMAM are worthy of comment. These comparisons are carried out in a system of coordinates such that $R_D \sim M_{\text{SD}}$ (Fig. 9). All lactosylated dendrimer generations are virtually described by a single dependence. The core effect in this case is probably screened by the lactose environment (shell).

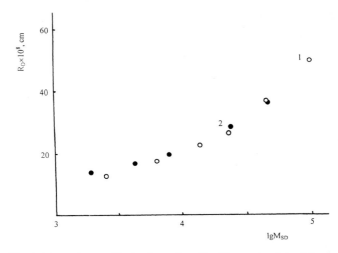

Fig. 9 Dependence of hydrodynamic radii of lactosylated dendrimers LacPAMAM (*1*) and lactosylated dendrimers based on DAB (*2*) on $\log M_{\text{SD}}$

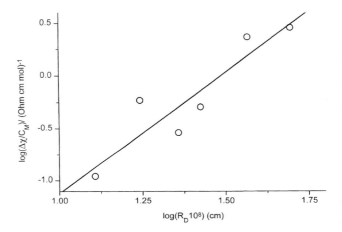

Fig. 10 Double-logarithmic dependence of $(\Delta\chi/C_M)$ on the hydrodynamic radius, R_D, for LacPAMAM in pure water

Now, it is useful to reconsider the conductivity data. Let us examine the dependence of the slope of the $\Delta\chi$ against C_M plot (Fig. 6) as a function of the radius of LacPAMAM molecules' R_D. The slope of $\Delta\chi$ against C_M determines the summary mobility of the dendritic molecules and the counterions around them. It turns out that this value ($\Delta\chi/C_M$) is roughly proportional to R_D^2 (Fig. 10), i.e. to the surface of the molecules. This means that the more important contribution to the conductivity is only from the charges situated near the surface of the dendrimers.

Conclusion

The application of a full set of hydrodynamic methods for studying dendritic molecules allows us to evaluate the molecular weight and dimensions of LacPAMAM dendrimers for generations 0 (tetramer) to 5 (128-mer). It was clearly demonstrated that from comparisons between the intrinsic viscosity values, that are practically insensitive to molecular weight changes, the sedimentation velocity coefficient became more sensitive toward experimental hydrodynamic values (sedimentation velocity coefficient, translational diffuson coefficient, intrinsic viscosity). This corresponds to the behavior of compact spherelike molecules. Experimental molecular weight values were in good agreement with the calculated molecular weight, thus confirming complete substitution of the amine end groups of the initial PAMAM dendrimers by lactose residues. The data obtained are in good correlation with those obtained for lactosylated dendrimers based on a DAB core. This is evidence for the weak influence of the core structures on the hydrodynamic dimensions of lactosylated dendrimers. By their compactness, dendritic lactosylated molecules are similar to globular proteins.

Acknowledgements G.M.P. is grateful to the Max-Planck-Institut and the Organizing Committee of 11th International Symposium on Analytical Ultracentrifugation for their financial support. G.M.P. is extremly grateful to M. Rinaudo for the possibilities to perform the conductivity measurements during his stay at CERMAV (Grenoble) in her laboratory. We are thankful to R. Spindler from Dendritech (Midland, Michigan) for a generous supply of PAMAM. R. R. is also thankful to NSERC for financial support of the work.

References

1. Roy R (1996) Polym News 21:226
2. Roy R (1998) In: Boon G-J (ed) Carbohydrate chemistry. Thomson, London, 241
3. Ashton PR, Boyd SE, Brown CL, Jayaraman N, Nepogodiev SA, Stoddart JF (1996) Chem Eur J 2:1115
4. Klein J, Herzog D (1987) Makromol Chem 188:1217
5. Pavlov G, Ivanova N, Korneeva E, Michailova N, Panarin E (1996) J Carbohydr Chem 15:419
6. Tomalia D, Durst H (1993) Top Curr Chem 165:193
7. Jansen J, de Brabander-van-den Berg E, Meijer E (1994) Science 266:1226
8. Newkome G, Moorefield C, Vogtle F (1996) Dendritic molecules. VCH, Weinheim
9. (a) Flory P (1941) J Am Chem Soc 63:3083; (b) Flory P (1941) J Am Chem Soc 63:3091; (c) Flory P (1941) J Am Chem Soc 63:3096
10. Kuhn W, Kuhn H (1947) Helv Chim Acta 30:1233
11. Tsvetkov V (1951) DAN SSSR 78:1123
12. Burchard W (1983) Adv Polym Sci 48:1
13. Tomalia D, Naylor A, Goddard W (1990) Angew Chem Int Ed Engl 29:138
14. Mansfield M, Klushin L (1993) Macromolecules 26:4262
15. Aoi K, Itoh K, Okada M (1997) Macromolecules 30:8074
16. de Gennes P-G, Hervet H (1983) J Phys (Paris) 44:L351
17. Mathews OA, Shipway AN, Stoddart JF (1998) Prog Polym Sci 23:1
18. Roovers J, Comanita B (1999) Adv Polym Sci 142:179
19. Tomalia D, Baker H, Dewald JR, Hall M, Kallos G, Martin S, Roeck J, Ryder J, Smith P (1985) Polym J 17:117
20. de Brabander-van den Berg EMM, Meijer EW (1993) Angew Chem Int Ed Engl 32:1308
21. Lescanec RL, Muthukumar M (1990) Macromolecules 23:2280
22. Chen ZYu, Cui S-M (1996) Macromolecules 29:7943
23. Murat M, Grest GS (1996) Macromolecules 29:1278
24. La Ferla R (1997) J Chem Phys 106:688
25. Boris D, Rubinstein M (1996) Macromolecules 29:7251
26. Aharoni SM, Crosby Ch R, Walsh EK (1982) Macromolecules 15:1093
27. Mourey TH, Turner SR, Rubinstein M, Frechet JMJ, Hawker CJ, Wooly KL (1992) Macromolecules 25:2401
28. Ihre H, Hult A, Soederlind E (1996) J Am Chem Soc 118:6388
29. Pavlov G, Korneeva E, Nepogodiev S, Jumel K, Harding S (1998) Vysokomol Soedin 40:2056
30. Pavlov G, Korneeva E, Jumel K, Harding S, Meijer E, Peerling H, Stoddart J, Nepogodiev S (1999) Carbohydr Polym 38:195
31. Tanford C (1961) Physical chemistry of macromolecules. Wiley, New York
32. Tsvetkov VN, Eskin VE, Frenkel SYa (1970) Structure of macromolecules in solution. Butterworths, London
33. Cantor CR, Schimmel PR (1980) Biophysical chemistry, part II. Freeman, San Francisco
34. Roy R (1997) Top Curr Chem 187:242
35. Andre S, Cejas Ortega PJ, Alamino Perez M, Roy R, Gabius H-J (1999) Glycobiology (in press)
36. Pagé D, Roy R (1997) Bioconjugate Chem 8:714
37. Zanini D, Roy R (1998) J Org Chem 63:3486
38. McBroom CR, Samanen CH, Goldstein IJ (1972) Methods Enzymol 28:212
39. Tsvetkov VN (1989) Rigid-chain polymers. Consultants Bureau, New York
40. Pavlov G, Panarin E, Korneeva E, Kurochkin C, Baikov V, Ushakova V (1990) Makromol Chem 191:2889
41. Huggins ML (1942) J Am Chem Soc 64:2716
42. Svedberg T, Pedersen KO (1940) The ultracentrifuge. Oxford University Press, Oxford
43. Tsvetkov VN, Lavrenko PN, Bushin SV (1984) J Polym Sci Polym Chem Ed 22:3447
44. McCammon JA, Deutch JM, Bloomfield VA (1975) Biopolymers14:2479
45. Debye P, Bueche AM (1948) J Chem Phys 16:573
46. Tsvetkov VN, Klenin SI (1953) DAN SSSR 88:49

Progr Colloid Polym Sci (1999) 113 : 158–163
© Springer-Verlag 1999

R.I. Bayliss
N. Errington
O. Byron
A. Svensson
A. Rowe

A conformation spectrum analysis of the morphological states of myosin S1 in the presence of effectors

R.I. Bayliss · A. Svensson
Department of Physics and Astronomy
University of Leicester
Leicester LE1 7RH, UK

N. Errington · A. Rowe (✉)
NCMH Business Centre
University of Nottingham
Sutton Bonington LE12 5RD, UK
e-mail: arthur.rowe@nottingham.ac.uk
Tel.: +44-115-9516156
Fax: +44-115-9516157

O. Byron
Division of Infection & Immunity
ILBS, University of Glasgow
Glasgow G12 8QQ, UK

Abstract The subfragment-1 (S1) of the contractile protein myosin is the postulated site of energy transduction in muscle. It is currently considered that a large conformational change in this moeity, which forms 'cross-bridges' between the thick (myosin based) and thin (actin based) filaments, is the mechanical driving force which leads to mutual sliding of the two types of filaments, and hence to contraction.

We have studied the possibility that S1 from skeletal myosin can in free solution (and in the absence of actin) be induced to undergo related changes in conformation, in the presence of a range of effectors whose action mimics stages of the contractile cycle. Analysis of the $g(s^*)$ profiles of S1 under these conditions, displayed after force-fitting of the known monomer mass as conformation spectra (CON-SPECs-[7]), shows that changes of the type associated with the contractile cycle are readily detected. There is full qualitative and extensive quantitative agreement between the magnitude of the changes seen and those predicted on the basis of hydrodynamic bead modeling and high resolution electron microscopy. Results from recently published X-ray crystallography of smooth muscle S1 [8] are also in general agreement with our findings.

Key words Myosin S1 · EM · CON-SPEC · Analogues · Modeling

Introduction

The subfragment-1 (S1) of skeletal myosin is generally considered to be the site of energy transduction in muscle. The S1 units project from the thick filaments, which are composed chiefly of myosin, and make contact with the thin filaments, comprised chiefly of actin. The individual S1 units possess ATPase activity, and are able to couple the loss of chemical potential associated with the hydrolysis of the nucleotide substrate to the mechanical potential required for relative sliding of the two types of filament with respect to each other. This is the classical "sliding filament" theory of muscle contraction [1, 2]. Calculations based upon the known velocity of shortening and the known number of such cross-bridges shows that the "power stroke" typical of the operation of one cross-bridge must be of the approximate order of 10 nm [2, 3].

Crystallographic evidence [4] has indicated the presence of a "lever arm" in S1, of which the central feature is a single α-helix, to which the "light chains" of myosin are attached. By the swiveling of this lever arm about the larger mass of S1, which is bound to the actin filament, relative movement of the filaments could in principle be generated, based upon a large conformational change in the S1 unit [5]. This concept has generally replaced the older "swinging cross-bridge" hypothesis, in which the entire S1 unit was supposed to rotate about the point of its attachment to the rest (or rod portion) of the myosin molecule.

The question arises as to whether or not a conformational change of this magnitude can be detected when purified S1 is exposed in solution to nucleotide. Limited evidence from low-angle X-ray diffraction [6] has suggested that an appreciable conformational change does indeed occur. This method gave values for the radius of gyration under various conditions, which were interpreted by the authors in terms of very simple models. The method is open to the criticism that exposure to high-intensity X-ray beams can produce very significant specimen damage. We have sought to monitor the conformational state of S1 under a range of conditions using hydrodynamic techniques. Bead modeling, based upon the crystallographic structure and thus capable of greater sophistication than the X-ray modeling of Wakabayashi et al. [6], has been employed to predict the level of change expected on the "rotating lever arm" hypothesis, and $g(s^*)$ profiles obtained under the range of conditions simulating physiological states have been transformed into "conformation spectra" [7], or CON-SPECs, to make clear the extent to which significant changes have been detected. This approach has additionally been reinforced by the use of high-resolution electron microscopy, which has been employed in the direct visualization of these morphological changes.

The solvent conditions used in this study follow those employed by others [6, 8] to mimic the various states postulated to exist in the cross-bridge cycle: namely, extreme relaxation with lever arm extended with a nonhydrolyzable ATP analogue, ATPγS bound; end of the power stroke with nucleotide released with no bound nucleotide; bound product prior to the power stroke, with ADP present; and maximum compaction of the S1 prior to the commencement of the power stroke, with VO_4 or AlF_4 present. The effect of the presence of BeF_x, has been a matter of dispute. Our study addresses the issue as to whether BeF_x mimics the ATP (extended) state [5] or the compact, prepower stroke state [9].

Materials and methods

Preparation of myosin S1

Rabbit skeletal myosin was prepared by standard techniques as used in our laboratory [10]. Myosin in synthetic filament form (loc. cit) was then digested using chymotrypsin in the absence of divalent metal ions, for a period of time which control experiments showed (by SDS-PAGE) to be adequate to give a 30–50% yield, without significant degradation of the S1 heavy chain [11]. The action of chymotrypsin was stopped by addition of inhibitor (p-methyl sulfonyl fluoride). The final preparation of S1 contained approximately 1.6 mg ml^{-1}, in a volume of 5–10 ml. Analytical ultracentrifugation showed a single peak in the $g(s^*)$ profile, with a width which could be accurately fitted by a single species of molecular mass of 110 000 Da. This is plausible for a mixed isomer S1 preparation, with an unknown degree of light-chain degradation. A small amount of dimer also appeared to be present, approximately constant at around 15% by mass under all conditions studied. We believe this estimated proportion is likely to be artifactually higher than the actual value in solution, being due in part at least to an elevation of the leading baseline of the $g(s^*)$ profile caused by radial dilution of the plateau region.

Analytical ultracentrifugation

Samples of S1 in the presence of the appropriate solute components were run on a Beckman Optima XL-I analytical ultracentrifuge employing Rayleigh interference optics. Rotor speeds in the range 34 000–45 000 rpm were employed. The data logged into the computer were transformed into $g(s^*)$ against s^* profiles and thence into CON-SPEC(s^*) profiles using standard procedures for the former [12] and a recently derived approach for the latter. In fitting the Gaussian function to the $g(s^*)$ profile, the molecular mass was fixed using a procedure which we have derived [7], to give maximum precision in the estimates for $g(s^*)_{max}$. The CON-SPEC (s^*) function is given by the transformation of the $g(s^*)$ function for a component of known mass out of direct s^* space into $(s^*)_{max}$ error space. This CON-SPEC (s^*) function is a very narrow Gaussian for a single, well-defined component: the presence of multiple, resolved peaks in the total CON-SPEC is indicative of multiple conformational states.

Routine runs with identical components in different cells in the same rotor were performed to confirm that essentially identical spectra were produced. Where results were to be compared between cells in different runs, a common reference standard in a single channel was employed for normalization purposes.

Electron microscopy

Initial experiments are reported in which a new approach to high-resolution electron microscopy has been employed. This approach, called "microcrystallite decoration", is based upon very low-dose evaporation of metal onto a specimen surface, such that only very small microcrystallites of the metal (< 0.7 nm) are deposited. By computer summation of identically oriented images, an "averaged" image is obtained. Test specimens using tobacco mosaic virus as a standard have shown that this method can yield resolutions of the order of 1.1 nm, which is a factor of at least 2 better than conventional metal shadowing [13]. Results from two conditions only are reported here: more detailed results from electron microscopy of myosin S1 under a full range of conditions will be reported elsewhere.

Myosin S1 preparations, with or without added ADP, were sprayed at a protein concentration of 0.15 mg ml^{-1} in 50% glycerol onto freshly cleaved mica surfaces, using high-pressure nitrogen to drive a modified artist's airbrush. The mica fragments bearing droplets were then subjected to rotary evaporation of Pt from a thermal source at an angle of 5°, for a period of time known from control experiments to restrict the growth of Pt crystals to the microcrystallite (< 0.7 nm) range. The replicas thus produced were stripped off onto a water surface, picked up on copper grids, and examined at 66 000× in a JEOL 100 CX transmission electron microscope at an accelerating voltage of 80 keV. The magnification of the microscope was calibrated using a diffraction grating replica.

Images of S1, selected for having a clearly definable orientation, were digitized using a UMAX Powerlok scanner with transparency hood, interfaced to an Apple Macintosh 4400 PowerPC. The image analysis software NIH-IMAGE was employed to orientate and integrate up to 30 multiple images.

Hydrodynamic bead modeling

Predicted sedimentation coefficients of conformers of myosin S1 were computed using the programs AtoB and HYDRO The file containing the coordinates for myosin S1 (code 2MYS in the Brookhaven Protein Data Bank), which contains data for most of the residues in S1, was converted using AtoB to a bead model with a nominal resolution of 10 Å consisting of 272 beads. The angle between the long axes of head and tail segments was then varied from an initial 117° (for S1 without ADP) to 65° (the most extreme angle observed under the electron microscope for S1.ADP) with four intermediate conformations. The coordinates for these bead models were then used as inputs for HYDRO. The partial specific volume ($\bar{v}$) required by HYDRO was calculated from the known amino acid composition of myosin S1. The presence of ADP was assumed to have a minimal effect on $\bar{v}$. In all cases the models were anhydrous; the sedimentation coefficient generated by HYDRO was later adjusted for hydration using the relationship

$$s_h = s_a \left(\frac{\bar{v}}{\bar{v} + \delta v_1^0} \right)^{1/3} , \qquad [1]$$

where s is the sedimentation coefficient and the subscripts h and a refer to the hydrated and anhydrous states, respectively. The level of hydration is given by δ (g water/g protein) and v_1^0 is the specific volume of water.

Results

Electron microscopy of S1 ± ADP

Computer averaged images of S1 obtained using magnetic circular dichroism (MCD) in the absence of added nucleotide (Fig. 1a) show an extended, slightly curved

High resolution (MCD) electron microscopy of S1 ± ADP reveals a large change in the angle of the lever arm

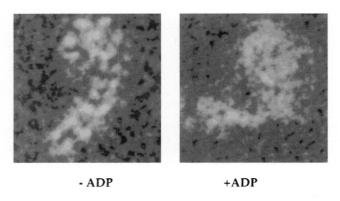

- ADP   +ADP

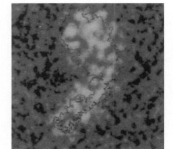

The structure -ADP matches the outline of the S1 crystal structure (little more extended)

Fig. 1 a,b High-resolution magnetic circular dichroism electron microscope images of S1 ± ADP revealing a large change in the angle of the lever arm. **c** The structure-ADP matches the outline of the S1 crystal structure (little more extended)

structure, of a size compatible with the crystal structure [4]. Indeed the outline of the latter could be almost precisely superimposed on the MCD image (not shown). The orientation of the lever arm in solution with respect to the "head" may thus be taken as corresponding closely to that observed in the crystalline state. Statistical analysis of the results (not shown) indicates only a very limited spread of measured head–lever angles about the angle of 117° measured from the crystalline structure.

In the presence of ADP, however, many of the S1 units visualized had a more compact configuration, with the putative lever arm being inclined at a much smaller angle with respect to the head (Fig. 1b); however, statistical analysis (not shown) in this case indicated that only around half the S1 units had wholly adopted this compact configuration. The angle of this more compact configuration measured from the MCD images is 65°. The remainder was distributed about a range of intermediate states. This fact is of importance in interpretation of hydrodynamic data.

CON-SPEC analysis

The results from the CON-SPEC analysis of myosin S1 under various conditions are shown in Fig. 2. The results in the presence of AMP-PNP, a slowly released analogue of ADP, show a marked change in $g(s^*)_{max}$, of +4.9%. This is in agreement with our previous observations [7] that in presence of ADP the sedimentation coefficient increases by around 3%. This implies a degree of compaction of the molecule, and is qualitatively compatible with a rotation of the lever arm of S1 as suggested by electron microscopy. The fact that the effect seen with AMP-PNP is a little larger than that seen with ADP [7] could be due to either the extent of compaction being larger, or to the proportion of the population having a compact form being larger. Given the tight-binding nature of AMP-PNP, the latter explanation is more plausible.

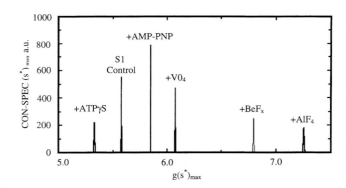

Fig. 2 A conformation spectrum diagram showing the S1 $g(s^*)_{max}$ under various conditions

In the presence of the nonhydrolyzable ATP analogue (ATPγS) the sedimentation coefficient decreases, by approximately 4.4%, indicating that the molecule adopts a more extended shape than the unaffected S1 when it binds to the actin filament. The possibility arises that the change in sedimentation coefficient could also be associated with a change in the bulk "head" portion of the S1 molecule, i.e. it would become more compact than the unaffected S1 molecule.

In the presence of VO$_4$, and especially of BeF$_x$ or AlF$_4$, the sedimentation coefficient increased markedly. This would correspond to the compaction of the S1 unit, by the lever arm adopting an even more acute angle than in the ADP-bound state. This may well be the case for VO$_4$, but the degree of increase seen using BeF$_x$ or AlF$_4$ is of a magnitude which could call into question whether the S1 remains monomeric. As only a single peak is seen in the g(s*) profile, corresponding to a single sedimenting boundary, could it be that in the presence of these polyvalent metal ions, especially AlF$_4$, the protein undergoes reversible dimerization?

Detailed consideration of the g(s*) profiles makes it in fact very unlikely that the main component seen in the presence of BeF$_x$ or AlF$_4$ is a monomer–dimer reaction boundary.

The fit to the g(s*) profiles obtained on the assumption of a fixed, monomer mass, is excellent, even in the presence of these effectors (Fig. 3). Moreover, the

frictional ratios computed for S1 under these conditions (Table 1) remain in the upper range (1.14–1.22) of values generally estimated for globular proteins, and are in that sense compatible with a strongly compacted conformation being present.

Hydrodynamic bead modeling

The results of hydrodynamic bead modeling of the possible states of S1 are shown in column 4 of Table 1. The predicted values for the sedimentation coefficient agree plausibly with the range of experimental values seen (Table 1) for the control S1 and for S1 in the presence of ADP and AMP-PNP, on the assumption that the latter effector produces more uniform compaction of the population than the former. By simple extrapolation one can consider that the s values seen in the presence of ATPγS and of VO$_4$ are reasonable. This is dependent upon the hypothesis that the lever arm in the S1 monomer is extended beyond the 119° angle in the presence of ATPγS, as seen by electron microscopy in control S1. In the presence of VO$_4$ the lever arm adopts a smaller angle than the 64° seen in the ADP state. The simple modeling which we have employed, with rotation in a single plane of the lever arm around a central swivel, is, however, not adequate to model the much higher degree of compaction seen with BeF$_x$ or AlF$_4$ (Table 1). As already noted, however, the values for the computed frictional ratio (Table 1) are consistent with those expected for a slightly noncompact globular protein; hence the assumption that the lever arm has largely folded back onto the main body of the S1, as

Fig. 3a–c A Selection of the analysis data sets. Each data set has been corrected to standard temperature ans solvent conditions. The *fit line* on all graphs shows the total of a fit that was forced into a monomer/dimer system. **a** The curve for myosin S1 (the control). **b** The data for myosin S1 + AlF$_4$. **c** The data for myosin S1 + BeF$_x$

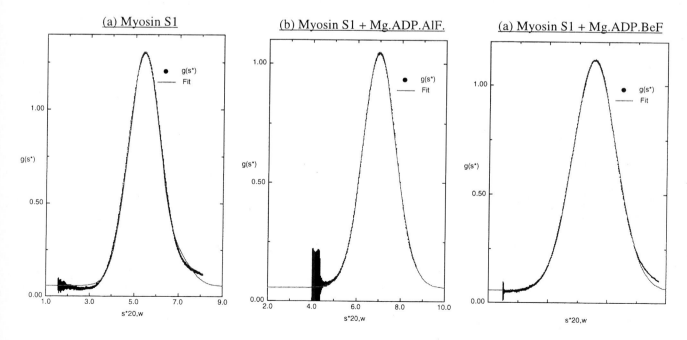

Table 1 The hydrodynamic parameters for skeletal myosin S1 in the presence of effectors. The $g(s^*)_{max}$ value is equivalent to the conventional sedimentation motion coefficient. The second column shows the sign and magnitude of the change induced by the effector, and the third column shows the conventionally calculated frictional ratio. The fourth column shows which each effector is believed to be an analogue

Sample	$g(s^*)_{max}$	$\Delta g(s^*)_{max}(\%)$ measured	$\Delta g(s^*)_{max}(\%)$ computed	f/f_0	Analogue
Control S1	5.577	–	–	1.48	Power-stroke end
+ AMP-PNP	5.848	+4.86	+4.2	1.42	ADP
+ VO$_4$	6.075	+8.93	+8.8	1.36	ADP.P$_i$
+ ATPγS	5.330	−4.44	−1.9	1.55	ATP
+ BeF$_x$	6.795	+21.8	–	1.22	ATP or ADP.P$_i$[a]
+ AlF$_4$	7.254	+30.1	–	1.14	ADP.P$_i$

[a] Differing opinions [5, 9]

suggested by crystallographic work with smooth muscle S1 [9].

Discussion

The current theory of muscle contraction envisages a range of conformational states for the myosin S1 unit depending upon whether it is bound or not bound to the actin filaments, and whether it is bound to adenine nucleotide in tri- or diphosphate form. A summary of this theory, in outline, is shown in Fig. 4. Our results are generally in agreement with this scheme. The release of the product of hydrolysis of ATP, namely ADP, has long been considered to be associated with the "power stroke" [14]. Comparison of $g(s^*)_{max}$ of S1, with no bound nucleotides, (i.e. a condition equivalent to end of the power stroke) with that of S1 in the presence of ADP [7] or of AMP-PNP, which essentially "locks" the conformation at the start of power stroke, shows a change which by hydrodynamic bead modeling and by electron microscopy alike can be interpreted in terms of a rotation of the lever arm by around 55°. Assuming a rigid S1 molecule this translates to a power stroke of approximately 10 nm, which agrees with previous work [2, 3].

By contrast, the effect of ATPγS is to decrease $g(s^*)_{max}$ of S1. This implies that the lever arm is capable of adopting an even more extended form than we have observed in the absence of effectors. From this we could infer that the ATP-bound S1 state might not truly model the end of the power stroke, where ATP binds to S1, initiating hydrolysis and the re-formation of the compact state (Fig. 4). This reduction in the sedimentation coefficient could hypothetically be explained by involving a degree of rearrangement within the "head" portion of the S1 molecule, possibly "preparing" the molecule for the binding to the actin filament.

At the other extreme, is it possible that at the commencement of the power stroke, the S1 unit is even more compact than our basic models, based upon electron microscopy (Fig. 1). This suggests that the result obtained in the presence of VO$_4$ supports this and the results obtained in the presence of BeF$_x$ or AlF$_4$

Fig. 4 Diagram to illustrate the principal stages currently believed to be involved in the generation of sliding motion between thick and filaments in muscle by means of a series of conformation changes in myosin. S1. The release of the product of hydrolysis (ADP) is shown to be associated with a 'power stroke' following which the extended myosin head detaches from the actin filament on binding of ATP whose hydrolysis is associated with the re-attainment by S1 of a compact form

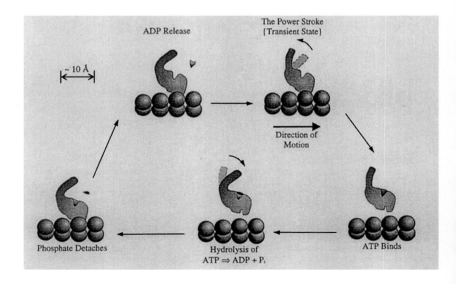

seem to provide even stronger evidence for a compact molecule. As previously noted, we do need to give careful consideration to the possibility that we are seeing a reaction (monomer–dimer) boundary; however, there are good reasons for rejecting this supposition, it being noted that even in the case of the most extreme apparent compaction (with AlF$_4$), the frictional ratio value (1.13) is in the range usual for a typical globular protein, and that a good fit to the $g(s^*)$ profile is obtained with a single Gaussian, due to allowance being made for the presence, noted earlier, of a small amount of the aggregated, dimeric form. It should be possible to test this matter further, but certainly with VO$_4$ we feel confident that the evidence supports the existence of states more compact than the 64°-angle state (Fig. 1) found for S1 in the presence of ADP.

Hydrodynamic bead modeling results (Table 1) support these interpretations. The actual changes seen with AMP–PNP, with ATPγS and with VO$_4$ are predicted quite closely (Table 1), if one assumes that the VO$_4$ state represents the smallest physical angle between lever and head which is physically plausible (40°). However, the changes seen with AlF$_4$ and with BeF$_x$ are larger than can be modeled in simple terms, and suggest that other factors such as a change in solvation – not implausible if two surfaces come together – may need to be considered.

Strong support for the existence of the more highly extended or compacted states (more extended: ATPγS; more compact: AlF$_4$, BeF$_x$) is provided by the recent demonstration that such states can exist; in the presence of the above-mentioned effectors, in the crystalline form for the case of smooth muscle S1 [9]. Whilst one cannot simply equate the two entirely, the two S1 units from skeletal and smooth muscle myosin are very similar in architecture, and these findings do indicate that such large transitions are geometrically possible for a molecule of this type. Our results with S1 in the presence of BeF$_x$ are wholly consistent with this effector causing compaction [9], and not with the suggestion of Holmes [5] that extreme extension results, i.e. this is an analogue for the ATP-bound state. Detailed electron microscopy of skeletal myosin S1 under all the above conditions is currently in progress in our laboratories in an attempt to observe these highly compact and extended states.

Thus, overall, our results demonstrate that skeletal myosin S1 monomers, in free solution, can in the presence of a range of known physiological effectors adopt a range of configurations, which are very similar to those expected on the basis of current theories as to the physical basis of the contraction cycle (Fig. 5). The range of angles observed in the electron microscopy and bead modeling of the analytical ultracentrifuge data suggests a morphological change consistent with a 10 nm "power stroke" as discussed by Dominguez et al. [9]. Our results indicate that the presence of bound actin is not needed for these changes to occur in S1 and that the binding of actin is simply the anchoring needed to produce the relative sliding of the filaments. Although similar changes under a limited range of conditions have been seen by electron microscopy using smooth muscle S1 bound to actin filaments [15, 16], these studies failed to find any changes in skeletal myosin S1 bound to actin. The reason for this is not clear, but the imaging noise level in such a system is inevitably high.

It can be deduced that whilst it is certain that the binding of actin must cause some degree of conformational change in S1, leading to strong activation of the ATPase activity, it does not seem to be tightly coupled to a change in the lever arm angle relative to the main mass of S1, due to the observation of the "swinging" of the lever arm when S1 is in solution, without the presence of actin.

References

1. Huxley AF, Niedergerke R (1954) Nature 173: 973
2. Huxley AF, Simmons RM (1971) Nature 233: 533
3. Finer JT, Simmons RM, Spudich JA (1994) 368: 113
4. Rayment I, Rypiewski WR, Schmidt-Base K, Smith R, Tomchick DR, Benning MW, Winkelmann DA, Wesenberg G, Holden HM (1993) Science 261: 50
5. Holmes KC (1997) Curr Biol 7: R112
6. Wakabayashi K, Tokunaga M, Kohno I, Sugimoto Y, Hamaka T, Tukezawa Y, Wakabayashi T, Amemya Y (1992) Science 258: 443
7. Errington N, Rowe AJ (1999) Biophys Chem (in press)
8. Goodno GC (1982) Methods Enzymol 85: 116
9. Dominguez R, Freyson Y, Trybus KM, Cohen C (1998) Cell 94: 559
10. Persechini AJ, Rowe AJ (1984) J Mol Biol 172: 23
11. Margossian SS, Lowey S (1982) Methods Enzymol 85: 63
12. Stafford WF (1992) Anal Biochem 203: 295
13. Willison JHM, Rowe AJ (1980) In: Glauert A (ed) Practical methods in electron microscopy, vol. 8. North-Holland, Amsterdam
14. Lymm RW, Taylor EW (1971) Biochemistry 10: 4617
15. Jontes JD et al (1995) Nature 378: 751
16. Whittaker M et al (1995) Nature 378: 748

Progr Colloid Polym Sci (1999) 113 : 164–167
© Springer-Verlag 1999

BIOLOGICAL SYSTEMS

M.D. Kirkitadze
K. Jumel
S.E. Harding
D.T.F. Dryden
M. Krych
J.P. Atkinson
P.N. Barlow

Combining ultracentrifugation with fluorescence to follow the unfolding of modules 16–17 of complement receptor type 1

M.D. Kirkitadze (✉) · P.N. Barlow
The Edinburgh Centre for Protein
Technology Department of Chemistry
University of Edinburgh
West Mains Road
Edinburgh EH9 3JJ, UK
e-mail: marina@chem.ed.ac.uk
Tel.: +44-131-6504704
Fax: +44-131-6507155

K. Jumel · S.E. Harding
NCMH Unit, University of Nottingham
School of Biological Sciences
Sutton Bonington
Leicestershire, LE12 5RD, UK

D.T.F. Dryden
Institute of Cell and Molecular Biology
University of Edinburgh
West Mains Road
Edinburgh, EH9 3JR, UK

M. Krych · J.P. Atkinson
Department of Internal Medicine
Division of Rheumatology
Washington University School of Medicine
660 S. Euclid, Box 8045
St. Louis MO, 63110, USA

Abstract Complement receptor type 1 (CR1) is a member of a family of regulators of complement activation with therapeutic potential. A fragment of CR1 comprising modules 16 and 17 was overexpressed as a recombinant nonglycosylated protein in *Pichia pastoris*. Intrinsic fluorescence studies of the unfolding of recombinant CR1~16–17 caused by increasing concentrations of guanidinium chloride revealed that the intermodular junction unfolds first, followed by module 17, with module 16 the last to unfold. Sedimentation velocity studies in the analytical ultracentrifuge revealed a corresponding clear change in conformation from the native macromolecules with an axial ratio of about 5:1 to a much more extended conformation.

Key words Complement protein modules · Intrinsic fluorescence · Sedimentation velocity · Guanidinium chloride · Unfolding

Introduction

The extracellular part of complement receptor type 1 (CR1) is a representative example of mosaic proteins [1]. It consists of 30 modules all belonging to the complement protein (CP) type also called short consensus repeats, and is a member of the regulators of complement activation (RCA) family [2]. CP modules usually consist of 60 residues and have four conserved cysteines disulfides bonded in 1–3 and 2–4 fashion, several conserved glycines, prolines, hydrophobic residues and a virtually invariant tryptophan. A previous study [3] of modules 15–17 of CR1, which is a biologically active fragment that binds C3b of the complement cascade, along with modules 15–16 and 16 of CR1, was designed to establish the extent of the junction between modules by monitoring the unfolding of these fragments using different techniques (differential scanning calorimetry, nuclear magnetic resonance spectroscopy, circular dichroism and fluorescence). The results revealed that as temperature or denaturant concentration was increased, the 16–17 junction appeared to melt first, followed by the 15–16 junction, module 17 itself, and finally by modules 15 and 16. Modules 15 and 16 appear to form intermediate states prior to total denaturation. In this paper we present unfolding studies on a fragment of CR1 comprising modules 16–17. The results presented here complement those obtained previously for 15–17, 15–16 and 16 module fragments of CR1. Taken together they yield insight into the extent of intermodular interactions between modules 15, 16 and 17. This knowledge could be helpful in better understanding the structure–function relations among RCA proteins as an approach towards the design and assessment of more possible inhibitors with potential in a range of clinical settings, such as xenotransplantation [4]. In addition, many RCA proteins are viral targets [5–6].

Materials and methods

CR~16–17

Recombinant CR~16–17 was overexpressed in *Pichia pastoris*. The purification procedure was essentially as described in Ref. [3]. Sodium phosphate buffer (20 mM, pH 6.5) was used in all experiments.

Fluorescence measurements

Intrinsic fluorescence spectra, using the conserved tryptophan residue present in each module, were recorded on a Perkin-Elmer LS-50B spectrofluorimeter in a 100-μl cuvette of 0.3-cm pathlength at 25.0 °C unless otherwise stated. The excitation wavelength was 295 nm and the emission spectra were recorded between 300 and 500 nm. The spectral bandwidth was 10 nm. The protein concentration was 0.3 g/l.

Ultracentrifugation

Sedimentation velocity experiments were performed on an Optima XL-A (Beckman Instruments, Palo Alto) analytical ultracentrifuge equipped with scanning ultraviolet optics set at a wavelength of 280 nm. An optical pathlength cell (12 mm) with a 400-μl sample was used and the experiment was performed at 4.0 °C with a rotor speed of 55 000 rpm. Sedimentation coefficients were evaluated from the absorption concentration distribution data using the Svedberg procedure as described by Philo [7] based on a solution of the Lamm equation, and the DCDT procedure (time derivative evaluation of apparent distribution of apparent sedimentation coefficients) as described by Stafford [8], to yield the sedimentation coefficient, $s_{T,b}$, at a finite concentration, c, where the subscripts T and b correspond to temperature (in this case 4.0 °C) and buffer, respectively. These values were corrected to standard solvent conditions (the density and viscosity of water at 20.0 °C) in the standard way [9]. For these corrections, the solvent density (in native and denaturing solvents) and the partial specific volume (from the amino acid sequence) of the protein were evaluated using the routine Sednterp based on the procedure described in Ref. [10]. $s_{20,w}$ values determined at several loading concentrations, c, (ranging from 0.5 to 4.0 g/l) were then plotted against concentration and the zero concentration (i.e. non-ideality-free) value, $s^0_{20,w}$, was obtained by linear regression.

Results and discussion

Intrinsic fluorescence studies

The denaturation of CR1~16,17 was followed by intrinsic tryptophan fluorescence. It is known that each module has a conserved tryptophan residue and there are no tryptophan residues elsewhere in the sequence. On the basis of homology with experimentally derived structures, the conserved tryptophan side chains can be expected to occupy largely buried positions within the hydrophobic cores of the modules close to the cysteine II-cysteine IV disulphide bond and the junction with the previous module. Excitation at 295 nm resulted in a fluorescence emission spectrum with a maximum at 340 nm. The increase in the intensity of tryptophan fluorescence upon complete denaturation of CR1~16,17 (Fig. 1a) was about tenfold. As previously shown in Ref. [3] upon denaturation of CR1~15–17 in 7 M guanidinium chloride (GdmCl) there was an increase in fluorescence emission intensity at 340 nm of about 14-fold (Fig. 1b), a consequence of solvent exposure of the tryptophan side chain, and a shift in the maximum to 351 nm. The increase in the intensity of tryptophan fluorescence upon denaturation for CR1~15,16 was about tenfold (Fig. 1b), and smaller for the single module CR1~16, about sevenfold (Fig. 1a). The fluorescence intensity changes were reversible. Figure 1b indicates there are three resolved transitions for CR1~15–17 and two resolved transitions for CR1~15,16, which correspond closely to the second and third transitions for CR1~15–17. Thus the first transition on the CR1~15–17 plot corresponds to the unfolding of the 16–17 junction and module 17 itself. This conclusion was also supported by collecting Hetero nuclear Single Quantum Coherence (HSQC) spectra at 2.5 M GdmCl for CR~15–17, where it was shown that almost no crosspeaks were left which corresponded to module 17 [3]. GdmCl – induced denaturation of CR1~16 takes place in the range 2–6 M denaturant and appears to be a gradual process. When the relative intensity is plotted against GdmCl concentration for CR1~16,17, a first transition is observed at 1 M denaturant. This may correlate with the first transition on the plot for CR1~15–17. Unfolding of module 17 takes place at a denaturant concentration slightly higher than that required for denaturation of the 16–17 junction; thus, these two processes cannot be completely resolved. Above 3.5 M GdmCl a large change in relative intensity is observed. By comparison of the CR1~16,17 GdmCl denaturation plot with the results obtained previously for module fragments 15–17, 15,16 (Fig. 1b) and 16 of CR1, it can be concluded that the transition above 3.5 M GdmCl corresponds to the unfolding of module 16. Further increase in the relative intensity of the plot for CR1~16,17 is gradual, which correlates with the previously observed behaviour of module 16. Comparing the results of fluorescence studies between 3 M and 4.5 M GdmCl for CR1~16–17 with those obtained for the CR1~15–17, 15–16 and 16 module fragments, we can make the following conclusions:

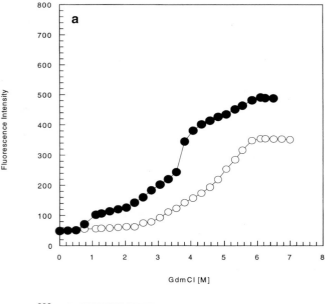

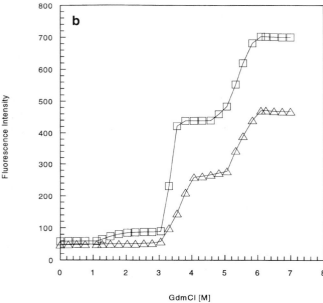

Fig. 1a, b Fluorescence studies on complement receptor type 1 (*CR1*) fragments. **a** Fluorescence intensity at 340 nm as a function of guanidinium chloride (*GdmCl*) concentration for CR1~16,17 (●) and CR1~16 (○). **b** Fluorescence intensity at 340 nm as a function of GdmCl concentration for CR1~15–17 (□) and CR1~15,16 (△)

1. CR1~16 shows the least cooperative folding amongst all modules investigated.
2. Modules 16–17 also have a broad unfolding transition, but module 16 unfolds more sharply as a part of CR1~16–17 than when on its own.
3. Modules 15–16 show a sharper transition during unfolding than 16–17.
4. CR1~15–17 has the sharpest unfolding transition of all.

Ultracentrifugation studies: native buffer at 4.0 °C

Single symmetric boundaries were observed with single symmetric peaks on the $g*(s*)$ plots. An $s^0_{20,w}$ value of (1.48 ± 0.04) S was obtained from linear regression analysis. To interpret the sedimentation coefficient in terms of conformation, we first of all evaluate the frictional ratio f/f_o defined by [11]

$$f/f_o = [M(1 - \bar{v}\rho_{20,w})/N_A 6\pi\eta_{20,w}S^0_{20,w}](4\pi N_A/3M\bar{v})^{1/3} \ ,$$

(1)

where f is the frictional coefficient of the macromolecule, f_o is the frictional coefficient of a spherical particle of the same mass and anhydrous volume, N_A is Avogadro's number and $\rho_{20,w}$ and $\eta_{20,w}$ are the density and viscosity, respectively, of water at 20.0 °C. M, the molecular weight and $\bar{v}$, the partial specific volume, were both evaluated from the amino acid sequence using the programme Sednterp [10] and were found to be 14513 Da and 0.725 ml/g, respectively. Using these values, a value for f/f_o of (1.42 ± 0.05) was thus obtained.

The frictional ratio is an intrinsic function of molecular shape and hydration (solvation). To eliminate the solvation contribution we used the universal Perrin function, P, or "frictional ratio due to shape" [11, 12] defined by

$$P = f/f_o(1 + \delta/\bar{v}\rho_{20,w})^{-1/3} \ .$$

(2)

To get an unequivocal estimate for the shape parameter P one needs δ. Unfortunately this is a notoriously difficult "parameter" to measure with any precision, but what we can do is to choose a range of hydrations, δ, and see the corresponding effect on P and the axial ratio. In this way, if we set δ to 0.3, 0.35 and 0.4 the following estimates for P are obtained: 1.28, 1.26 and 1.23.

One can then use these values of P to get an overall idea of the shape of the protein in terms of the axial ratio a/b of the hydrodynamically equivalent ellipsoid of revolution, with one major semiaxis, a, and two equal minor semiaxes, b, using the simple-to-use routine Ellips1 [11, 13]. Corresponding values for the axial ratio for CR1~16–17 for the three different values of δ were thus 5.5:1, 5.2:1 and 4.6:1. One can say quite reasonably therefore that CR1~16–17 under native conditions is clearly an asymmetric molecule of axial ratio of about 5:1. Since an individual module is known from NMR studies [14] to have an axial ratio of about 3:1 we can say that modules 16–17 are linearly arrayed in an end-to-end fashion. To attempt a more refined analysis using triaxial ellipsoid or bead/bead-shell modelling [15], other hydrodynamic data are required [11, 15] which are not currently available for this protein.

Ultracentrifugation studies: 2 M GdmCl at 4.0 °C

It is also interesting to observe that on dialysing the protein into a solution of 2 M GdmCl, the value of $s^0_{20,w}$ dropped to 1 S for three different solute concentrations analysed. Because of the broadness of the boundaries a more accurate $s^0_{20,w}$ value was not obtainable, even with the DCDT software, but the drop in $s^0_{20,w}$ clearly demonstrates a more asymmetric conformation, and/or a change in its solution properties.

The ultracentrifuge results would therefore appear to be in general agreement with the fluorescence data.

The purpose of this study was to probe the model suggested for fragments CR~15–17, 15–16 and 16 in our previous work [3], and also to collect new information about the modules' behaviour in fragment 16–17. The data obtained are consistent with the major conclusion made for the unfolding and stability of modules in the 15–17 and 15–16 fragments. The 16–17 junction unfolds first, followed by module 17 and then by module 16. Unfolding of CR1~16–17 followed by fluorescence revealed a broad profile which actually appears to be similar to that obtained for module 16. This basically supports the conclusions made in Ref. [3] that modules 16 and 17 interact less intensively, or in other words the junction between modules 16 and 17 is less extensive compared to the 15–16 junction.

The axial ratio of about 5:1 obtained for CR1~16–17 in nondenaturing buffer conditions indicates that this fragment adopts an elongated conformation consistent with an end-to-end arrangement of modules. The result obtained for CR1~16–17 at 2 M GdmCl indicates that the fragment adopts a partially unfolded state, which appears to be in agreement with fluorescence data, which also show conformational changes under these conditions. Analysis of the results obtained in our previous study led us to the conclusion that the modules undergo an expanded intermediate, or in other words a partially random form, prior to a random coil state. According to our model at 2 M GdmCl the 16–17 junction is unfolded and modules 17 and 16 are expanded. The value of $s^0_{20,w}$ obtained for CR1~16–17 at 2 M GdmCl agrees well with this suggestion. A combination of ultracentrifugation and fluorescence has allowed us to find differences between the stability and unfolding of modules 16 and 17 when they are in the 16–17 fragment compared to when they are a part of the triple 15–17 fragment of CR1.

Acknowledgements M.D.K. is funded by the Human Frontiers Science program. The Edinburgh Centre for Protein Technology is funded by the Biotechnology and Biological Sciences Research Council (BBSRC). D.T.F.D. is funded by the Royal Society of Great Britain. The NCMH unit at the University of Nottingham is supported by the BBSRC and the Engineering and Physical Sciences Research Council. Work in St. Louis was supported in part by funding from the National Institute of Health (R01 AI41592) and from CytoMed (Cambridge, Mass.). J.P.A. and Washington University have a financial interest in CytoMed.

References

1. Doolittle RF (1985) Trends Biochem Sci 114:233–237
2. Reid KBM, Bentley DR, Campbell RD, Chung LP, Sim RB, Kristensen T, Tack BF (1986) Immunol Today 7:230–234
3. Kirkitadze MD, Krych M, Uhrin D, Dryden DTF, Smith BO, Wang X, Hauhart R, Atkinson JP, Barlow PN (1999) Biochemistry 38:7019–7031
4. Dorling A, Riesbeck K, Warrens A, Lechler R (1997) Lancet 349:867–871
5. Lowell CA, Klickstein LB, Carter RH, Mitchell JA, Ahearn JM (1989) J Exp Med 170:1931–1943
6. Ward T, Pipkin PA, Clarkson NA, Stone DM, Minor PD, Almond JW (1994) EMBO J 13:5070–5074
7. Philo JS (1997) Biophys J 72:435–444
8. Stafford W (1992) In: Harding SE, Rowe AJ, Horton JC (eds) Analytical ultracentrifugation in biochemistry and polymer science. Royal Society of Chemistry, Cambridge, pp 359–393
9. van Holde KE (1985) Physical chemistry. Prentice Hall, Englewood Cliffs
10. Laue TM, Shah BD, Ridgeway TM, Pelletier SL (1992) In: Harding SE, Rowe AJ, Horton JC (eds) Analytical ultracentrifugation in biochemistry and polymer science. Royal Society of Chemistry, Cambridge, pp 90–125
11. Harding SE, Horton JC, Cölfen H (1997) Eur Biophys J 25:347–360
12. Squire PG, Himmel M (1979) Arch Biochem Biophys 196:165–177
13. Harding SE, Cölfen H (1995) Anal Biochem 228:131–142
14. Wiles AP, Shaw G, Bright J, Perczel A, Cambell ID, Barlow PN (1997) J Mol Biol 272:253–265
15. Garcia de la Torre J, Carrasco B and Harding SE (1997) Eur Biophys J 25:361–372

Progr Colloid Polym Sci (1999) 113: 168–175
© Springer-Verlag 1999

BIOLOGICAL SYSTEMS

B. Kernel
G. Zaccai
C. Ebel

Determination of partial molal volumes, and salt and water binding, of highly charged biological macromolecules (tRNA, halophilic protein) in multimolar salt solutions

B. Kernel · G. Zaccai · C. Ebel (✉)
Institut de Biologie Structurale
CEA-CNRS, 41 avenue des martyrs
F-38027 Grenoble Cedex 1
France
e-mail: ebel@ibs.fr
Tel.: +33-4-76889638
Fax: +33-4-76885494

Abstract Highly charged macromolecules such as nucleic acids or halophilic proteins cannot be considered without their environment of ions and water. Density, equilibrium sedimentation and neutron scattering were used to characterize yeast tRNA[Phe] in high concentrations of NaCl or $MgCl_2$ and halophilic malate dehydrogenase from *Haloarcula marismortui* in high-concentration NaCl solutions. Methods were compared and used in a complementary way to determine partial molal volumes and solvent binding parameters, which, when the hypothesis of an invariant particle is valid, provide solvation values for the macromolecule.

Key words Yeast tRNA[Phe] · Malate dehydrogenase from *Haloarcula marismortui* · Partial molal volume · Salt-binding parameter · Equilibrium sedimentation

Introduction

Highly charged macromolecules such as nucleic acids or halophilic proteins cannot be considered without their environment of ions and water. Interactions between the charged macromolecules, ions and water contribute to their folding, stability, activity and capacity to interact with other macromolecules. Proteins and nucleic acids from extremely halophilic bacteria are adapted to an environment nearly saturated in salt [1–3]. Halophilic proteins have a negatively charged surface. They require high concentrations of salt for stability and they have been shown to be solvated by large amounts of salt (KCl or NaCl) [4–7]. The interactions between proteins and nucleic acids are known to be associated with ion and water molecule displacements [8]. They are in general inhibited when increasing the salt concentration. As an exception, they were seen to be reinforced in high salt concentrations in the case of the TATA box binding protein from *Pyrococcus woesei*, whose cytoplasm contains 0.8 M salt [9]. In order to explore the role of concentrated salt environments on nucleic acids, we measured the solvent interactions of yeast Phe tRNA between 1 and 4 M NaCl, and 1 and 3 M $MgCl_2$. We determined for each salt condition the macromolecular molal volume by density measurements at constant composition of the solvent component. The increments of mass or neutron scattering length density were measured by density, ultracentrifugation and neutron scattering experiments performed at a constant chemical potential of the solvent components. Comparing the methods, the best technique appears to be sedimentation equilibrium, from which binding parameters are calculated. From partial molal volumes and binding parameters, and using the invariant particle hypothesis, we were able to determine a solvation in NaCl of 4–5 mol water per mole of nucleotide, with 60% of the counterions dissociated, consistent with previous calculations. The results in $MgCl_2$ have to be confirmed but indicated 6–7 mol water per mole of nucleotide, with about all the counterions dissociated. On halophilic malate dehydrogenase, in 2–5 M NaCl, we determined both the partial molal volume and the salt binding parameter by using complementary density and neutron scattering performed at a constant potential of the solvent components. Using the invariant particle hypothesis, these experiments allowed the protein to be described as solvated by 2000 mol per mole of water and 55 mol per mole of salt.

Theory

Note that the concentrations, c, are expressed in moles per liter. Temperature and pressure are assumed to be constant.

Definitions of components

As explained by Eisenberg [10], in order to be able to consider solution component concentrations as independent thermodynamic variables, they are described as an electrically neutral combination of species. There are different ways to define the components of a system. We consider here the system composed of water (component 1), macromolecule (the charged nucleotide chain or the polypeptide chain plus the number of counterions just required for electroneutrality) (component 2), and salt (component 3). As a consequence, the molar mass, M_2, and the neutron scattering length density, b_2, of component 2 (Table 1), as well as the partial molal volume of the polyelectrolyte, $\bar{V}_2$, depend on the nature of the accompanying counterions.

Partial molal volume

$$\bar{V}_2 = (\partial V_m / \partial m_2)_{m_1,m_3} \ , \tag{1}$$

where $\bar{V}_2$ is the partial molal volume (ml/mol), V_m is the volume of the solution corresponding to 1 kg of principal solvent (water), and m_i is the molality of component i (mol/kg water).

$\bar{V}_2$ measurement by densimetry

The molal volumes of salts can be calculated rigorously from Eq. (1) after trivial conversion from density tables.

Table 1 Molar masses and neutron scattering lengths of yeast tRNA^{Phe}, malate dehydrogenase from *Haloarcula marismortui* (*HmMalDH*), NaCl and MgCl$_2$. Values for tRNA are given per mole of nucleotide. The index i is 2 for the macromolecules and 3 for the salts

	M_i (g/mol)	b_i (cm/mol)
tRNA^{Phe} + 76Na$^+$	349.8	6562 × 10^9
tRNA^{Phe} + 38Mg^{2+}	338.9	
HmMalDH + 156Na$^+$	134138	1.9245 × 10^{15}
NaCl	58.44	7.9546 × 10^{11}
MgCl$_2$	95.218	

In the case of the quite low concentrations that are most often considered for biological macromolecules, the following equation can be derived:

$$(\partial \rho / \partial c_2)_{m_1,m_3} = \left[(\rho - \rho^0)/c_2 \right]_{m_1,m_3} = M_2 - \rho^0 \bar{V}_2 \ . \tag{2}$$

The mass density increment at constant molality of all components except component 2 is measured by the density ρ of a solution at concentration c_2 of component 2 (mol/l), and of that ρ^0, of the solvent. It has to be noted that the experimental determination of $\bar{V}_2$ by densimetry is extremely difficult since it requires the addition to the solvent of just the polyelectrolyte and its counterions (component 2), without any other modification in the solvent composition.

The mass density increment $(\partial \rho / \partial c_2)_{\mu_1,\mu_3}$ and the neutron scattering length density increment $(\partial \rho_N / \partial c_2)_{\mu_1,\mu_3}$

These correspond to the increase in density, ρ, or neutron scattering length density, ρ_N, of the solution due to the addition of 1 mol/l component 2, at constant chemical potential of all components except component 2. This condition can be achieved by dialysis. These are the parameters which determine the ability of a macromolecule to sediment and to scatter, respectively [10–12].

The measurements of $(\partial \rho_N / \partial c_2)_{\mu_1,\mu_3}$ and $(\partial \rho / \partial c_2)_{\mu_1,\mu_3}$

1. Small-angle neutron scattering

$$I(0)/c_2 = N_A (\partial \rho_N / \partial c_2)^2_{\mu_1,\mu_3} \ . \tag{3}$$

$I(0)$ is the forward scattering intensity derived from the Guinier relation at one macromolecule concentration extrapolated to zero macromolecule concentration from a set of measurements. N_A is Avogadro's number [13, 14].

2. Density measurement (after dialysis)

$$(\partial \rho / \partial c_2)_{\mu_1,\mu_3} = \left[(\rho - \rho^0)/c_2 \right]_{\mu_1,\mu_3} \ . \tag{4}$$

3. Equilibrium sedimentation. For an ideal homogeneous macromolecule in solution

$$(\partial \rho / \partial c_2)_{\mu_1,\mu_3} = (2RT/\omega_2) \quad \partial \ln c_2 / \partial r^2 \ . \tag{5}$$

Water and salt binding parameters $(\partial m_1 / \partial m_2)_{\mu_1,\mu_3}$ and $(\partial m_3 / \partial m_2)_{\mu_1,\mu_3}$

$(\partial m_1 / \partial m_2)_{\mu_1,\mu_3}$ (also named the interaction parameter or the preferential interaction parameter with water) represents the hypothetical amount of water (in moles of water per mole of protein) that would have to be added to or subtracted from a protein solution to maintain the values of μ_1 and μ_3 identical to those in pure solvent. $(\partial m_3/\partial m_2)_{\mu_1,\mu_3}$ represents the hypothetical amount of salt (in moles of salt per mole of protein) to be added or subtracted for the same purpose. The two binding parameters are related by the solvent composition:

$$[(\partial m_1/\partial m_2)_{\mu_1,\mu_3}]/[(\partial m_3/\partial m_2)_{\mu_1,\mu_3}] = (w_1/w_3) \ , \qquad (6)$$

where (w_1/w_3) is the molar ratio between water and salt in the solvent.

Relation between $(\partial\rho/\partial c_2)_{\mu_1,\mu_3}$ and $(\partial\rho_N/\partial c_2)_{\mu_1,\mu_3}$ and the binding parameters

The mass density increment and the neutron scattering length density increment can be expressed as functions of $\bar{V}_2$, $\bar{V}_1$, b_2, b_1 (the scattering length densities of the component in centimeters per mole) and $(\partial m_1/\partial m_2)_{\mu_1,\mu_3}$

$$(\partial\rho/\partial c_2)_{\mu_1,\mu_3} = (M_2 - \rho^0\bar{V}_2) + (\partial m_1/\partial m_2)_{\mu_1,\mu_3}$$
$$\times (M_1 - \rho^0\bar{V}_1) \qquad (7)$$

$$(\partial\rho_N/\partial c_2)_{\mu_1,\mu_3} = (b_2 - \rho_N^0\bar{V}_2) + (\partial m_1/\partial m_2)_{\mu_1,\mu_3}$$
$$\times (b_1 - \rho_N^0\bar{V}_1) \qquad (8)$$

or, alternatively, as functions of $\bar{V}_2$, $\bar{V}_3$, b_2, b_3 and $(\partial m_3/\partial m_2)_{\mu_1,\mu_3}$

$$(\partial\rho/\partial c_2)_{\mu_1,\mu_3} = (M_2 - \rho^0\bar{V}_2) + (\partial m_3/\partial m_2)_{\mu_1,\mu_3}$$
$$\times (M_3 - \rho^0\bar{V}_3) \qquad (9)$$

$$(\partial\rho_N/\partial c_2)_{\mu_1,\mu_3} = (b_2 - \rho_N^0\bar{V}_2) + (\partial m_3/\partial m_2)_{\mu_1,\mu_3}$$
$$\times (b_3 - \rho_N^0\bar{V}_3) \qquad (10)$$

The complementarity of Eqs. (7) and (8) or Eqs. (9) and (10) will be developed later.

Structural model for the determination of water and salt binding

For a macromolecule binding (per mole) n_1 mol water and n_3 mol salt

$$(\partial m_1/\partial m_2)_{\mu_1,\mu_3} = n_1 - n_3(w_1/w_3) \qquad (11)$$

$$(\partial m_3/\partial m_2)_{\mu_1,\mu_3} = n_3 - n_1(w_3/w_1) \ . \qquad (12)$$

For a particle whose solvation is constant in a solvent of various compositions (various w_1/w_3 ratios), plots of Eqs. (11) or (12) will give straight lines. Alternatively, if the binding parameters measured at various salt to water ratios satisfies Eqs. (11) or (12) with n_1 and n_3 as

constant values, the particle (solvated macromolecule) can be considered as invariant in composition.

Remark: the Donnan effect

The dissociation of the counterions of a polyelectrolyte induces solvent rearrangements that are equivalent to negative salt binding. For complete counterion dissociation from a negatively charged polyelectrolyte in a solution containing a salt, the values of n_3 corresponding to the Donnan effect are, per mole of charge

$n_3 = -1/2$ for a salt X^+Y^-

$n_3 = -1/3$ for a salt $2X^+Y^{2-}$

$n_3 = -1/6$ for a salt $X^{2+}2Y^-$

$n_3 = -1/4$ for a salt $X^{2+}Y^{2-}$.

Materials and methods

Macromolecule preparation

Yeast tRNA^{Phe}

tRNA^{Phe} was obtained from Sigma. It had a Phe acceptance of 1100 pmol/unit absorbance at 260 nm. We verified that the extinction coefficient (1 unit optical density at 260 nm corresponds to 1.86 nM [15]) is constant in NaCl concentrations above 0.1 M, and in MgCl$_2$ concentrations above 10 mM. Unless otherwise specified, all tRNA buffers contained 10 mM MOPS (4-morpholine-propane-sulfonic acid), pH 7, and 1 mM ethylenediaminetetraacetate.

Malate dehydrogenase from Haloarcula marismortui

Malate dehydrogenase from *H. marismortui* (HmMalDH) was overexpressed in *E. coli* and purified using a classical protocol [16, 17]. It was stocked at 6 °C in 4 M NaCl, passed on a gel filtration column (Superose 12HR10–30 from Pharmacia) and reconcentrated on Centricon 30 (Amicon) before the experiments. All buffers contained 50 mM tris(hydroxymethyl)aminomethane, pH 8.2. The macromolecule content was measured spectroscopically after the density and neutron measurements (1 unit optical density at 280 nm corresponds to 6.51 μM [4]).

$\bar{V}_2$ of tRNA^{Phe} from density measurements

We followed the protocol of Cohen and Eisenberg [18]. tRNA^{Phe} (6–8 mg) was dissolved in 1 ml 0.1 M NaCl or MgCl$_2$, heated at 65 °C for 5 min, filtered, and extensively dialyzed at 4 °C against water. The measurement of the density of this solution (A) allowed its RNA concentration to be determined. On a high-precision Mettler balance (1/100 mg), in a 4-ml glass bottle, were added (by weight) desiccated salt and solution (A), from which the weight of water and salt molarity of the final solution (B) are known precisely. A set of five solvents of similar molarity were prepared. The densities of all solutions were measured on a Paar DMA 602 M regulated at 20 °C. The measurement of the reference solvents allowed the density of the solvent in solution (B) to be determined exactly. $\bar{V}_2$ was then determined according to Eq. (2).

Measurements by density of the mass density increments at constant chemical potential of diffusible components

tRNA[Phe] was dissolved in 0.1 M NaCl, heated for 5 min at 65 °C, and filtered before extensive dialysis against the appropriate salt solution at room temperature. HmMalDH was dialysed against the appropriate salt solution, directly from its stock solution. Two bags were also filled with the buffer in order to check the condition of equilibrium dialysis. To avoid solvent evaporation, samples were taken from the dialysis bags in a cold room using a syringe and pricking the bag when inside the bath.

Sedimentation equilibrium

tRNA (5 μl) in 0.1 NaCl or MgCl$_2$ was dissolved in 135 μl of appropriate buffer. Experiments were performed on an Optima XL-A ultracentrifuge (Beckman) at 15 000 and 23 000 rpm and analysis was performed using Beckman software, EQASSOC4, using an experimental baseline determined after ultracentrifugation at 42 000 rpm.

Neutron scattering experiments

Neutron scattering experiments were performed at room temperature on the D11 instrument at the ILL, Grenoble [19], using a wavelength of 6 Å (tRNA[Phe]) and 10 Å (HmMalDH), a collimation of 2.5 m, and sample detector distances of 2.8 and 1.2 m. Usual normalization of the data was performed by using the scattering of 1.00 mm water [13]. Samples were prepared as for density experiments with a tRNA[Phe] concentration between 3 and 8 g/l, and a HmMalDH concentration between 7 and 21 g/l, in order to extrapolate parameters to zero concentration of macromolecule [20].

Results

tRNA

Partial molal volumes

The values of partial molal volumes measured for Na-tRNA and Mg-tRNA are compared in Fig. 1. When increasing the salt concentration, $\bar{V}_2$ increases due to the fact that water molecules are electrostricted around charged groups in the solvent. The difference of about 10 cm^3/mol between $\bar{V}_2$ of tRNA in NaCl and in MgCl$_2$

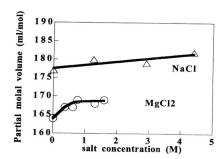

Fig. 1 Partial molal volumes measured for yeast Na-tRNA[Phe] and Mg-tRNA[Phe]

can be related to the reported difference between the partial molal volume of the counterions (−1.2 cm^3/mol for Na$^+$, −10.6 cm^3/mol for 1/2 Mg^{2+} at infinite dilution [21]). When expressed in cubic centimeters per gram, the partial specific volume increases for Na-tRNA from 0.505 to 0.52 cm^3/g (the values found for Na-DNA in the same range of salt were 0.50 and 0.54 cm^3/g [18]) and for Mg-tRNA from 0.48 to 0.50 cm^3/g. Note that these values are different from the operational apparent specific volumes, ϕ'_2, used for the interpretation of sedimentation experiments. The ϕ'_2 values are obtained at constant chemical potential of water and salt and are not thermodynamic quantities. They are significantly larger than the partial specific volumes [22]. Combined ϕ'_2 from density and equilibrium sedimentation experiments were successfully used to determine the molecular weight and dimerization of 16S ribosomal RNA [23].

Neutron scattering experiments

Neutron scattering increments for tRNA in various salts up to 0.9 M have been published [24, 25]. We attempted to extend these data to 1–4 M NaCl conditions [26]. The accuracy in the absolute values of $I(0)/c_2$ was insufficient for the analysis, however, because of the small mass of the macromolecule and the slight polydispersity in the solutions. These data were therefore not used for further treatment.

Mass density increment at constant potential of diffusible components

From Table 2, it can be seen that above 1 M NaCl data obtained from density and ultracentrifugation experiments are in good agreement. Ultracentrifugation appears clearly as the most powerful technique, since it needs much less material, and requires much less caution in the preparation of the sample for the equivalent precision. The curves (not shown) of the density increments obtained from ultracentrifugation as functions of density for Na-tRNA and Mg-tRNA are very slightly shifted with respect to each other.

Interaction parameters and structural model of solvation for tRNA

The salt binding parameters calculated from density and/or equilibrium sedimentation according to Eq. (7) using $\bar{V}_2$ from density measurements ($\bar{V}_2$ was extrapolated as parallel to the NaCl curve at high MgCl$_2$ concentrations) are plotted in Fig. 2. Above 1 M NaCl, the curve can be considered to be a straight line, and thus the data can be interpreted in terms of a macromolecule with constant solvation (Eq. 11). The intercept

Table 2 Mass density increments determined for tRNA[Phe] in NaCl and MgCl$_2$. Salt concentrations (mol/l), density ρ^0 (g/ml), density increments $(\partial\rho/\partial c_2)_{\mu_1,\mu_3}$ (g/mol nucleotide), for which errors have been determined from sets of experiments to be ± 2.6 g/mol

Salt concentration	Densimetry in NaCl		Equilibrium sedimentation			
			in NaCl		in MgCl$_2$	
	ρ^0	$(\partial\rho/\partial c_2)_{\mu_1,\mu_3}$	ρ^0	$(\partial\rho/\partial c_2)_{\mu_1,\mu_3}$	ρ^0	$(\partial\rho/\partial c_2)_{\mu_1,\mu_3}$
0.1	1.003	165	1.002	170	1.006	163
0.48			1.020	160		
0.96			1.036	154	1.073	140
1	1.040	149				
1.92			1.078	142	1.136	121
2	1.077	137				
2.87			1.114	129	1.190	106
3.7	1.142	121				
3.83			1.145	118		
4.74			1.176	112		

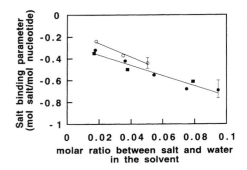

Fig. 2 Salt binding parameters calculated for yeast Na-tRNA[Phe] and Mg-tRNA[Phe] from density and/or equilibrium sedimentation. *Open circles*: in MgCl$_2$ from equilibrium sedimentation; *filled circles*: in NaCl from equilibrium sedimentation; *filled squares*: in NaCl from density measurements

of the curve gives a value of $n_3 = -0.3$ mol associated salt to 1 mol nucleotide, and the slope has a value of $n_1 = 5$ mol associated water. This means that the Donnan effect dominates this term, since negative values of binding are associated with dissociation of counterions. These values are very close to those found for DNA in the same range of salt concentrations ($n_1 = 4$, $n_3 = -0.3$ [27]).

The modeling of the salt binding related to the Donnan effect for a regularly charged cylinder (schematizing B-DNA) [28] has shown that salt binding decreases when increasing the salt concentration. At 0 M monovalent salt, the salt binding parameter is -0.06, a value calculated from Manning condensation theory [29]; it decreases strongly between 0 and 1 M salt, reaching a value of -0.3. Between 1 and 4 M salt it decreases only slightly, reaching a value of -0.34. These last values correspond to the experimental value of n_3. At low salt concentration, the salt binding parameters of a cylinder and a sphere of the same effective charge density differ, but above 1 M monovalent salt, the calculated values are similar and nearly constant, in agreement with our results on tRNA.

For tRNA in MgCl$_2$, the invariant particle hypothesis gave values of $n_1 = 6$ mol water per mole of nucleotide, and $n_3 = -0.14$ mol of salt per mole of nucleotide, a value closed to the Donnan value of -0.16 (corresponding to the total dissociation of the counterions). This suggests that in MgCl$_2$ at high salt concentration, tRNA is more hydrated and with its counterions more dissociated than in NaCl; however, these experimental results have to be completed by more measurements around 1 M MgCl$_2$.

Halophilic malate dehydrogenase

Complementarity between neutron and mass experiments (equilibrium or density), in H$_2$O and D$_2$O, for the determination of both the partial molal volume and the solvent binding parameter of biological polyelectrolytes

The protocol described for the determination of the partial molal volume by densimetry for tRNA cannot be applied to HmMalDH. During the extensive dialysis against water, the protein would denature in an essentially irreversible way. We thus used the complementarity of Eqs. (9) and (10) to determine both $\bar{V}_2$ and $(\partial m_3/\partial m_2)_{\mu_1,\mu_3}$ from $(\partial\rho/\partial c_2)_{\mu_1,\mu_3}$ and $(\partial\rho_N/\partial c_2)_{\mu_1,\mu_3}$. Inspection of Table 3 allows us to understand how this complementarity can be used in the best way for the four experiments: mass/H$_2$O, mass/D$_2$O, neutron/H$_2$O and neutron/D$_2$O.

The first comment concerns the contrast variation capability of each "technique" when increasing the salt concentration. From column 7 of Table 3, it can be seen that the density or neutron scattering length density varies significantly between 4 M NaCl and water. From this fact, each of the four techniques has a good potential to determine macromolecule solvation at multimolar salt concentrations (combining Eqs. 9 or 10 and 12). The comparison and complementarity of these techniques in this frame was developed in a previous paper [12].

Table 3 Mass and neutron scattering properties in water and 4 M NaCl solutions with hydrogenated or deuterated water of a protein of 100 000 g/mol. Columns 1–6 concern the properties in water; column 7 considers the differences in the properties between 4 M NaCl (*) and water. The solvent densities, ρ^0, or neutron scattering length densities, ρ_N^0, the molar masses, M_2, M_3, and neutron scattering lengths, b_2, b_3, of the protein (index 2) and salt (index 3) are indicated. We consider a partial molal volume for the protein of 73 000 cm^3/mol and for the salt of 20 cm^3/mol. The units for columns 1 and 7 are: g/ml or 10^{-10} cm^{-2}; for columns 2, 3, 4 and 6: 10^{-5} g/mol or 10^{-15} cm/mol; for column 5: g/mol or 10^{-10} cm/mol

	1 ρ^0 or ρ_N^0	2 M_2 or b_2	3 $\rho^0\bar{V}_2$ or $\rho_N^0\bar{V}_2$	4 $(M_2-\rho^0\bar{V}_2)$ or $(b_2-\rho_N^0\bar{V}_2)$	5 M_3 or b_3	6 $100(M_3-\rho^0\bar{V}_3)$ or $100(b_3-\rho_N^0\bar{V}_3)$	7 $(\rho^{0*}-\rho^0)$ or $(\rho_N^{0*}-\rho_N^0)$
Mass/H$_2$O	1.00	1.00	0.73	0.27	58	0.04	0.16
Mass/D$_2$O	1.11	1.01	0.81	0.20	58	0.04	0.14
Neutron/H$_2$O	−0.56	1.45	1.86	1.86	79.5	0.09	0.36
Neutron/D$_2$O	6.40	2.16	−2.15	−2.51	79.5	−0.05	0.22

The comparison of the relative weights of columns 2 (M_2 or b_2), 3 ($\rho^0\bar{V}_2$ or $\rho_N^0\bar{V}_2$) and 6 [($100(M_3-\rho^0\bar{V}_3)$ or $100(b_3-\rho_N^0\bar{V}_3)$)] gives an indication of the relative weights of the protein, of the excluded solvent mass, and of the perturbed solvent, in the mass density or the neutron scattered length density of the solvated protein. It appears that mass/H$_2$O plus mass/D$_2$O are only slightly complementary, compared to each of the other combinations of techniques. For example, in the neutron/D$_2$O technique, the influence of solvent interactions is negligible, and that of the excluded solvent mass considerable, and in the neutron/H$_2$O technique excluded solvent does not have a large weighting.

However, it appears that the measurements performed in D$_2$O are not usable for the precise determination of both $\bar{V}_2$ and the binding parameters because M_2 and b_2 cannot be precisely defined: they depend on the amount of hydrogen exchanged by deuterium in the protein, a quantity hardly defined since it depends on sample preparation (and note that in the case of polyelectrolytes in the presence of salts mass spectroscopy, a powerful technique to determine hydrogen/deuterium exchange, is not able to give good results). We calculate that a 10% error in the estimation of the amount of exchangeable exchanged hydrogen will provide a variation of about 5% in either M_2 or b_2, and will lead to an error in partial specific volumes of about 0.1 cm^3/g.

Thus we used mass density and neutron scattering experiments performed in H$_2$O to determine simultaneously both $\bar{V}_2$ and $(\partial m_3/\partial m_2)_{\mu_1,\mu_3}$. Because HmMalDH can be obtained as a very stable and homogeneous material in large amounts, as it is a protein of quite large molar mass, a very precise determination of $(\partial\rho_N/\partial c_2)_{\mu_1,\mu_3}$ can be obtained from the extrapolation to zero protein concentration of the forward intensities measured in a range of 5–25 g/l protein concentration for a given salt condition [20]. The determination of $(\partial\rho/\partial c_2)_{\mu_1,\mu_3}$ was performed by density measurements, since this tetrameric protein was seen by ultracentrifugation to dissociate at low protein concentration [4]. As a consequence, the determination of $(\partial\rho/\partial c_2)_{\mu_1,\mu_3}$ by equilibrium sedimentation is difficult since it is coupled to the determination of an association constant, while density measurements are performed at protein concentrations at which dissociation does not occur, and furthermore, they would be insensitive to the protein dissociation if the interaction with the solvent of dissociated and associated protein were not significantly different [12].

A graphical determination of both $\bar{V}_2$ and $(\partial m_3/\partial m_2)_{\mu_1,\mu_3}$

From $(\partial\rho/\partial c_2)_{\mu_1,\mu_3}$ measured at each salt condition, we calculated a series of solutions to Eq. (9): [$\bar{V}_2$; $(\partial m_3/\partial m_2)_{\mu_1,\mu_3}$]. In the same way, from $(\partial\rho_N/\partial c_2)_{\mu_1,\mu_3}$ measured at the same salt concentrations, we calculated a series of solutions to Eq. (10). These solutions, when plotted on the same graph, provided the solution which satisfies both Eqs. (9) and (10). This determination for HmMalDH in 4 M NaCl is shown in Fig. 3. The complementarity can be seen from the fact that the two lines cross. We note that mass/D$_2$O would give a line with a slope similar to that of mass/H$_2$O attesting to the poor complementarity of these two techniques, and neutron/D$_2$O a nearly vertical line. However, as previously mentioned, the uncertainties in hydrogen/deuterium exchange in the protein make the

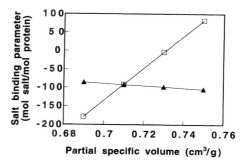

Fig. 3 Graphical determination of $\bar{V}_2$ and $(\partial m_3/\partial m_2)_{\mu_1,\mu_3}$ for malate dehydrogenase from *Haloareula marismortui* (*HmMalDH*) in 4 M NaCl. Pairs of solutions [$\bar{V}_2$; $(\partial m_3/\partial m_2)_{\mu_1,\mu_3}$] from $(\partial\rho/\partial c_2)_{\mu_1,\mu_3}$ and $(\partial\rho_N/\partial c_2)_{\mu_1,\mu_3}$ are plotted on the same graph, the crossing point giving the solution

techniques with D_2O solvents unsuitable for the $[\bar{V}_2; (\partial m_3/\partial m_2)_{\mu_1,\mu_3}]$ determination: in the graphical representation, depending on the hydrogen/deuterium exchange, lines are significantly shifted.

Partial specific volumes of HmMalDH in 2–5 M NaCl

The values calculated for partial specific volumes of HmMalDH from $\bar{V}_2$ in 2, 3, 4 and 5 M NaCl are 0.700, 0.709, 0.700 and 0.708 cm^3/g, respectively. We note that as expected, they are nearly constant or perhaps increase slightly, a feature which can be related to limited electrostriction at higher salt concentrations.

From the values and formulae given by Harpaz et al. [30] and neglecting the participation of counterions we calculated a partial specific volume of 0.713 cm^3/g for HmMalDH, derived from either the volumes of the amino acids in water corrected by the volume of the peptide bond, or the volumes obtained from crystallographic studies with correction for electrostriction around charged residues. A limiting value of $\bar{V}_2 = 0.735$ cm^3/g is calculated considering no electrostriction. Partial specific volumes of an ion in aqueous solutions can be attributed to two major components, one being its intrinsic volume (which corresponds to a radius slightly larger than the crystallographic one), and the other its electrostatic volume [18]. For HmMalDH associated with its counterions we calculated limiting values of 0.684 cm^3/g and 0.725 cm^3/g, corresponding to maximum and zero electrostriction. The experimental values are found within this range.

Solvation of HmMalDH in the invariant particle hypothesis

From the slope and intercept of the line presented in Fig. 4, we found using the invariant particle hypothesis

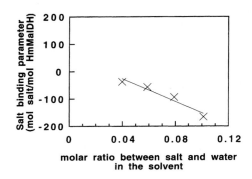

Fig. 4 Salt binding parameters calculated for HmMalDH from density and neutron scattering experiments in NaCl

a solvation of 55 mol salt and 2000 mol water in addition to the 156 counterions. The validity of the hypothesis will be discussed in subsequent work. The results are similar while not identical to those found previously by Bonneté et al. [4], due not only to the more rigorous formalism used here and to the use of different values for $\bar{V}_2$, but also to some discrepancy in the measurements (which were not performed at the same temperature and pH): 1 g protein was described as solvated by 0.08 g NaCl and 0.2 g water; here we find, by adding 55 mol salt and the 156 counterions, 0.05 g salt and 0.27 g water per gram of protein (considered without its counterions). Preliminary results suggest that the solvation and hydration varies depending on the nature of the salt present in the solvent [31]. The respective role of the cations and anions on the stabilization of HmMalDH was also explored in order to discriminate specific and nonspecific effects [7]. The solvent structure around HmMalDH is currently being studied by X-ray crystallography [32].

References

1. Lanyi JK (1974) Bacteriol Rev 38:272–290
2. Jaenicke R (1981) Annu Rev Biophys Bioeng 10:1–67
3. Eisenberg H, Mevarech M, Zaccaï G (1992) Adv Protein Chem 43:1–62
4. Bonneté F, Ebel C, Zaccaï G, Eisenberg H (1993) J Chem Soc Faraday Trans 89:2659–2666
5. Ebel C, Guinet F, Langowski J, Urbanke C, Gagnon J, Zaccaï G (1992) J Mol Biol 223:361–371
6. Ebel C, Altekar W, Langowski J, Urbanke C, Forest E, Zaccaï G (1995) Biophys Chem 54:219–227
7. Ebel C, Faou P, Kernel B, Zaccaï G (1999) Biochemistry 38:9039–9047
8. Record MT Jr, Ha J-H, Fischer MA (1991) Methods Enzymol 208:291–343
9. O'Brien R, DeDecker B, Fleming KG, Sigler PB, Ladbury JE (1998) J Mol Biol 279:117–125
10. Eisenberg H (1976) In: Biological macromolecules and polyelectrolytes in solution. Clarendon, Oxford, pp 29–63
11. Eisenberg H (1981) Q Rev Biophys 14:141–172
12. Ebel C (1995) Prog Colloid Polym Sci 99:17–23
13. Jacrot B, Zaccaï G (1981) Biopolymers 20:2413–2426
14. Jacrot B, Zaccai G (1983) Annu Rev Biophys Bioeng 12:139–157
15. Gueron M, Leroy JL (1978) Anal Biochem 91:691–695
16. Cendrin F, Chroboczek J, Zaccaï G, Eisenberg H, Mevarech M (1993) Biochemistry 32:4308–4313
17. Madern D, Pfister C, Zaccai G (1995) Eur J Biochem 230:1088–1095
18. Cohen G, Eisenberg H (1968) Biopolymers 6:1077–1100
19. Ibel K (1976) J Appl Cryst 9:296–309
20. Ebel C, Faou P, Zaccaï G (1999) J Cryst Growth 196:395–402
21. Millero FJ (1971) Chem Rev 71:147–175
22. Eisenberg H (1990) In: Saenger W (ed) Landolt-Bornstein new series biophysics-nucleic acids, vol VII. Springer,

Berlin Heidelberg New York, pp 257–276

23. Pearce TC, Rowe AJ, Turnock G (1975) J Mol Biol 97:193–205
24. Li ZQ, Giegé R, Jacrot J, Oberthur R, Thierry JC, Zaccaï G (1983) Biochemistry 22:4380–4388
25. Zaccai G, Xian SY (1988) Biochemistry 27:1316–1320
26. Kernel B (1997) Thesis. Université J. Fourier, Grenoble, France
27. Eisenberg H (1994) Biophys Chem 53:57–68
28. Sharp KA (1995) Biopolymers 36:227–243
29. Manning GS (1978) Q Rev Biophys 11:179–246
30. Harpaz Y, Gerstein MC, Chothia C (1994) Structure 2:641–649
31. Ebel C, Faou P, Franzetti B, Kernel B, Madern D, Pascu M, Pfister C, Richard S, Zaccai G (1998) In: Oren A (ed) Microbiology and biogeochemistry of hypersaline environments. CRC Press, Boca Raton, pp 227–237
32. Richard S (1998) Thesis. Université J. Fourier, Grenoble, France

Progr Colloid Polym Sci (1999) 113 : 176–181
© Springer-Verlag 1999

Studying membrane proteins in detergent solution by analytical ultracentrifugation: different methods for density matching

G. Mayer
B. Ludwig
H.-W. Müller
J.A. van den Broek
R.H.E. Friesen
D. Schubert

G. Mayer (✉) · J.A. van den Broek
D. Schubert
Institut für Biophysik, JWG-Universität
D-60590 Frankfurt am Main, Germany
e-mail: mayer@biophysik.uni-frankfurt.de
Tel.: +49-69-63015837
Fax: +49-69-63015838

B. Ludwig · H.-W. Müller
Institut für Biophysikalische Chemie
und Biochemie, Abt. Molekulare Genetik
JWG-Universität
D-60439 Frankfurt am Main, Germany

R.H.E. Friesen
Department of Microbiology
University of Groningen
9751 NN Haren
The Netherlands

Abstract Determining the complex size and the association behaviour of intrinsic membrane proteins by analytical ultracentrifugation is usually performed in aqueous solutions containing nonionic, nondenaturing detergents. In sedimentation equilibrium experiments, the contribution of the protein-bound detergent to the quantities of interest, as well as the contribution of free detergent micelles to the sedimentation profiles, can be eliminated by performing the measurements at a solvent density which equals that of the detergent ("density matching"). In the past, density matching had been almost exclusively done by substituting H_2O by appropriate H_2O/D_2O mixtures. We have now applied sucrose or glycerol to blank out the contribution of either nonaethylene glycol lauryl ether ($C_{12}E_9$) or Triton X-100 (reduced form). In addition, we adjusted the density of a N,N-dimethyllaurylamine N-oxide/$C_{12}E_9$ mixture to that of the solvent. The model protein used in our study was cytochrome c oxidase from *Paracoccus denitrificans*. We found that all approaches described work well: the results obtained were virtually identical to those determined in H_2O/D_2O mixtures. The detergent densities to be matched, however, were lowered by both sucrose and glycerol. Sucrose and glycerol are well known for their ability to stabilize protein conformation; thus, applying these reagents for density matching, instead of D_2O, could be helpful in ultracentrifuge studies on more labile membrane proteins.

Key words Membrane proteins · Association behaviour · Detergents · Density matching · Sedimentation equilibrium

Introduction

Solubilization and purification of intrinsic membrane proteins in aqueous solutions containing suitable nonionic detergents apparently does not disturb the proteins' secondary, tertiary and quaternary structure [1, 2]. As a consequence, molar mass determination of monomers or stable complexes of such proteins, as well as studies on their homologous or heterologous protein–protein associations, can be performed on the detergent-solubilized proteins [2, 3]. Analytical ultracentrifugation is considered to be the best technique available for these purposes [2, 3].

In ultracentrifuge experiments on membrane proteins in detergent solutions, the protein-bound detergent contributes to both the molar mass, M, and the partial specific volume, $\bar{v}$, of the protein particles. It also influences their kinetic parameters: sedimentation, diffusion and frictional coefficient. In addition, free detergent micelles may contribute to the measured absorbance-versus-radius, $A(r)$, or refractive index-versus-radius distributions. In sedimentation equilibrium experiments, however, these effects in many cases do not represent a serious problem:

1. Detergent contributions to the proteins' kinetic parameters are not relevant.

2. With a large number of detergents, the other contributions can be nearly completely eliminated by performing the experiments at a solvent density, ρ, which equals that of the detergent [2, 4].

The latter technique is called "density matching"; it is analogous to "contrast variation", as applied in small-angle X-ray and neutron scattering [5].

In the past, density matching has been done almost exclusively by substituting H_2O by appropriate H_2O/D_2O mixtures [2, 4]. However, it was clear from the beginning that other choices do exist, such as the use of sucrose, glycerol, salts or mixtures of these additives with D_2O to increase buffer density to that of the detergent [6–10]. In addition, an alternative procedure was suggested and applied, namely the use of a detergent mixture "tailored" to match the density of dilute aqueous buffers [11]. Unfortunately, it turned out that this detergent was only useful with a very limited number of membrane proteins.

Two of the additives applicable for increasing solvent density mentioned above, sucrose and glycerol, are well known for their ability to stabilize protein conformation [12–15]. In the course of our work on a very labile membrane protein, the erythrocyte anion exchanger (band 3), we became interested in utilizing these additives for stabilizing the protein in its native conformation. We have now systematically studied whether they can be applied to the matching of the density of two popular nonionic detergents, nonaethylene glycol lauryl ether ($C_{12}E_9$) and Triton X-100, reduced form (rTX-100). In addition, in analogy to Ref. [11] we adjusted the density of a detergent mixture to that of the solvent. In contrast to Ref. [11], we started with two detergents of widespread use in membrane research, namely N,N-dimethyllaurylamine N-oxide (LDAO), one of the rare detergents with a density less than 1 g/ml [1, 2], and $C_{12}E_9$. We applied the approaches outlined to cytochrome c oxidase from *Paracoccus denitrificans*, a stable protein/pigment complex consisting of four different protein subunits and two haem a molecules, which can easily be isolated as a homogeneous complex [16, 17]. Particular consideration was given to the questions of whether the additives lead to nonidealities in the sedimentation behaviour and to changes in the partial specific volume of the protein. We also tried to apply density matching by sucrose or glycerol to dodecyl maltoside (DDM), a frequently used detergent of relatively high density.

Materials and methods

Materials

$C_{12}E_9$ ("Thesit") was purchased from Boehringer (Mannheim), rTX-100 from Sigma-Aldrich (Deisenhofen), LDAO from Fluka (Buchs), and DDM from Biomol (Hamburg). 1,6-Diphenyl-1,3,5-hexatriene (DPH) was obtained from Serva (Heidelberg).

Cytochrome c oxidase from wild-type *P. denitrificans* (strain ATCC 13543) was isolated as described earlier [16] in the detergent DDM, and was transferred into the detergent solutions described below by ion-exchange chromatography on a (8×1)-cm column of DEAE-Sepharose CL 6B (Serva). The final protein concentration (in the buffer described below) was approximately 0.2 g/l, equivalent to an absorbance, at 425 nm for pathlength of 1 cm, of 0.30–0.33. The isolated protein complex did not contain any contaminants absorbing around 425 nm, i.e., in the absorption range of the attached pigment.

Methods

In order to find out the concentrations of D_2O, sucrose or glycerol matching the density of $C_{12}E_9$ micelles, the effective molar mass of the micelles, $M_{eff} = M(1 - \bar{v}\rho)$, was determined as a function of the concentration of the additives by sedimentation equilibrium analysis [18]. The $C_{12}E_9$ micelles were labelled with the dye DPH, which does not significantly influence their density [18]. Density matching of rTX-100 and DDM by sucrose and glycerol was investigated similarly, but without DPH labelling. Analogous experiments were performed with LDAO/$C_{12}E_9$/DPH mixtures in order to determine the weight ratio of the two detergents which yields, in aqueous buffer, micelles with $M_{eff} = 0$. Mixtures of $C_{12}E_9$ and DPH, and of LDAO, $C_{12}E_9$ and DPH, both with a DPH content of 0.5 mg/g detergent, as well as mixtures of LDAO and $C_{12}E_9$ were initially prepared in chloroform/methanol (2:1 v/v). The organic solvents were then removed in a rotary evaporator. Afterwards, the detergents, at a total concentration of 0.1–0.5% (w/v), were resolubilized in 10 mM tris(hydroxymethyl)amino-methane/HCl (pH 8.0), 50 mM NaCl and 0.5 mM ethylene-diaminetetraacetate containing the required concentrations of D_2O, sucrose or glycerol. rTX-100, unstained $C_{12}E_9$ or DDM were directly dissolved in the aqueous solvent.

Sedimentation equilibrium analysis of $C_{12}E_9$, rTX-100 and LDAO/$C_{12}E_9$ micelles and complexes of cytochrome c oxidase with these detergents, in the buffers described, was performed using a Beckman Optima XL-A analytical ultracentrifuge, an An-60 Ti or an An-50 Ti rotor, and standard Epon 6-channel centrepieces. The sample volume was 135 μl. DPH-stained detergent micelles were studied at a wavelength of 357 nm, the (unstained) micelles of rTX-100 at 275 nm and samples of cytochrome c oxidase at 425 nm. The step mode ($\Delta r = 0.002$ cm) and 20–50 replicates were used to scan the cells. The rotor speed was 13 000, 20 000 and 40 000 rpm for the detergent micelles, and 12 000 rpm in the studies on the enzyme; the latter was followed by a few hours at 40 000 rpm to allow baseline position and the quality of density matching to be checked. Studies on DDM used an Optima XL-I ultracentrifuge, a rotor speed of 20 000 rpm and unstained micelles, under otherwise identical conditions. The rotor temperature was 4 °C.

Solution densities were measured in a Paar DMA 02 density meter (Anton Paar, Graz), and kinematic viscosities in a Ubbelohde viscometer, ref. no. 50101 (Schott, Hofheim a. Ts.), at a temperature of 4 °C.

Results

The exact density of a detergent micelle depends not only on its chemical structure but also on other parameters: on solvent composition (see later), solvent temperature and, for chemically heterogeneous detergents (such as $C_{12}E_9$ or rTX-100), on the manufacturer and, even with one manufacturer, on the detergent batch. Thus, for each experimental condition, the exact density matching conditions have to be established by experiment. In view of the large number of detergents (and manufacturers) as well as the broad spectrum of experimental conditions applied in biomembrane research, we did not attempt to establish a compilation of exact density matching conditions for a large number of detergents and conditions (for approximate figures see Refs. [8, 19]). Instead, it was our aim to demonstrate the general feasibility of a few selected approaches which seemed to be promising. Application of the methods described to other systems or conditions should be straightforward.

Establishing the conditions for density matching

For a series of concentrations of the additives sucrose and glycerol, sedimentation equilibrium experiments were performed on the detergents $C_{12}E_9$ and rTX-100 in order to determine the additive concentrations which cancel the sedimentation of the detergent micelles. The resulting data "detergent M_{eff} versus solvent density" for all additives are shown in Fig. 1 (for additional data on the system $C_{12}E_9/H_2O/D_2O$ see Ref. [18]). The solvent densities that match the micelle density under the given conditions correspond to 6.4% (w/v) sucrose, 13.2% (w/v) glycerol or 46% (v/v) D_2O for $C_{12}E_9$ and 9.7% (w/v) sucrose or 23.0% (w/v) glycerol for rTX-100. As already noted, these figures can vary slightly with the batch and the temperature. They were found to be virtually independent of detergent concentration, c_d, [0.1% < c_d < 0.5% (w/v)]. It is obvious from Fig. 1 that the matching density is strongly dependent on the solvent system and is lower for glycerol and even more so for sucrose solutions than for H_2O/D_2O mixtures. Thus, the apparent partial specific volume of the detergent micelles distinctly depends on the type of additive used. This effect is also responsible for the nonlinearity of the curves in Fig. 1A.

We performed analogous measurements on DDM, a detergent frequently applied for the solubilization of labile membrane proteins. When the solution density was increased by addition of $D_2O/D_2{}^{18}O$, the density of this detergent (at 6 °C) was found to be 1.253 g/ml by extrapolation [18]. Using sucrose and glycerol, actual density matching could be obtained, the additive concentrations required being approximately 39% (w/v) and 76% (v/v) (958 g/l), respectively. Since we consider these additive concentrations as too high for use in sedimentation equilibrium studies, we did not pursue this approach (see later).

Experimental data on the effective molar mass of mixed $LDAO/C_{12}E_9$ micelles as a function of the LDAO content are shown in Fig. 2, for a total detergent concentration of 0.2% (w/v). According to the figure,

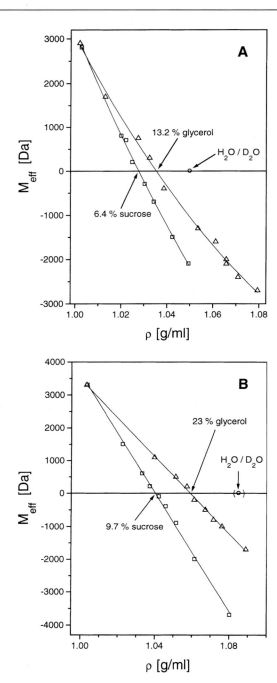

Fig. 1A,B Dependency of the effective molar mass of micelles of **A** nonaethylene glycol lauryl ether ($C_{12}E_9$) [0.3% (w/v)] and **B** Triton X-100, reduced form (*rTX-100*) [approximately 0.33% (w/v)] on solvent density, varied by sucrose (□), glycerol (△) or D_2O (○). Rotor speed: 40 000 rpm. The figure for rTX-100 in H_2O/D_2O, measured at 6 °C, is taken from Ref. [26]

the $A(r)$ profiles should be parallel to the baseline at an LDAO content of the detergent mixture of 43 % (w/w). This was in fact found to be the case (see insert in Fig. 2). The insert also shows that no separation occurs

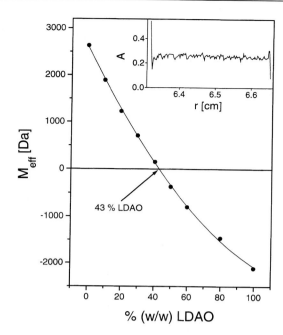

Fig. 2 Dependency of the effective molar mass of mixed *N,N*-dimethyllaurylamine *N*-oxide (*LDAO*)/C$_{12}$E$_9$ micelles on the relative LDAO content. *Insert*: Typical *A(r)* profile (at 357 nm) under the condition of density matching. Total detergent concentration: 0.2% (w/v); rotor speed: 40 000 rpm

into LDAO-rich and LDAO-poor micelles (otherwise, the *A* values would increase both towards the meniscus and the bottom). For total detergent concentrations, c_{dt}, below 0.3% (w/v), the density of the mixed micelles was found to increase upon dilution, probably due to the different critical micelle concentrations of the two detergents [20]: for example, for $c_{dt} = 0.3\%$ or 0.5% (w/v), 40% (w/w) LDAO was required for density matching, and for $c_{dt} = 0.1\%$ (w/v) approximately 45% (w/w) LDAO was required.

Two effects could influence the results described:

1. The formation of a density gradient due to the additives (especially sucrose).
2. The dependency of micelle density on hydrostatic pressure.

Both effects are more pronounced at higher rotor speeds, and under adverse conditions they could make true density matching impossible. However, in our experiments the effects (presumably mainly 1) were visible only in the presence of sucrose (6 or 10%) at a rotor speed of 40 000 rpm (under these conditions, the solution density increases between the meniscus and the bottom of a sector in a six-channel centrepiece by approximately 1.4%, and the pressure at the bottom reaches 40–45 atm). They are far below the limit of detection at rotor speeds below 20 000 rpm, as used in sedimentation equilibrium experiments on membrane proteins.

Sucrose and glycerol addition lead to an increase in the viscosity of the solvents. This in turn will increase the time required to reach sedimentation equilibrium, which counteracts the stabilizing effects of the two additives on the conformation of labile proteins. Under the conditions of density matching, the relative increase in the kinematic viscosity of the solution at 4 °C, as compared to water, was 86% for rTX-100/glycerol, 44% for C$_{12}$E$_9$/glycerol, 36% for rTX-100/sucrose and 22% for C$_{12}$E$_9$/sucrose.

Application to cytochrome *c* oxidase

The intrinsic membrane protein used in our sedimentation equilibrium experiments, cytochrome *c* oxidase from *P. denitrificans*, is a uniform protein/pigment complex (molar mass 128 500 Da; amino acid composition see Refs. [21, 22]) under all experimental conditions applied by us. Its $\bar{v}$ value, calculated from the known amino acid and pigment composition according to the method of Cohn and Edsall [23, 24] and from additional data applicable to the pigment [25], is 0.753 ml/g at 25 °C and 0.732–0.748 ml/g at 4 °C, correcting for temperature according to Ref. [24]. Testing the usefulness of the different approaches of density matching described above, we asked the following questions:

1. Can the experimental *A(r)* data obtained be fitted by a single exponential describing the equilibrium sedimentation distribution of a uniform protein particle showing ideal sedimentation behaviour?
2. Assuming these protein particles represent monomeric cytochrome *c* oxidase, does the calculated $\bar{v}$ correspond to the value calculated by the Cohn–Edsall method and/or determined in H$_2$O/D$_2$O mixtures?

We considered a positive answer to both questions as obligatory for the applicability of the method studied. Fitting the data according to question 1 showed that the sample molecules behaved as ideally sedimenting uniform particles in all density matching variants tested. This is shown, for the system requiring the highest additive concentration of all experiments and for the one with the detergent "blend", in Fig. 3. The effective molar masses determined from the fits, the solvent densities, and the $\bar{v}$ values calculated from these data (applying the known monomer molar mass) are shown in Table 1. It is obvious that a close agreement exists between the different $\bar{v}$ values, all figures determined being in the range (0.770 ± 0.005) ml/g (the uncertainty of the individual figures is between ±1.0 and ±1.5% of $\bar{v}$). In contrast, all $\bar{v}$ values in Table 1 are slightly but significantly higher than the one calculated by the Cohn–Edsall method, approximately 0.740 ml/g (see above); however, this is also true for $\bar{v}$ determined by

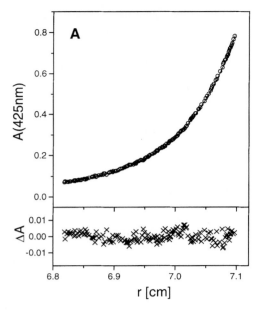

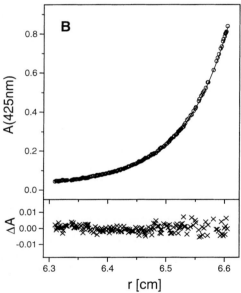

Fig. 3A,B *Top*: Experimental sedimentation equilibrium data $A(r)$ for cytochrome c oxidase in **A** glycerol/rTX-100 and **B** LDAO/$C_{12}E_9$ (o) under the conditions of density matching, and curves fitted to the data assuming the presence of uniform macromolecules showing ideal sedimentation behaviour (——). *Bottom*: Residuals of the fits (x). (Total) detergent concentration: 0.3% (w/v); rotor speed: 12 000 rpm

density matching by H_2O/D_2O. Apart from that, the reputation of the Cohn–Edsall method is based almost exclusively on its success with water-soluble proteins, but not with intrinsic membrane proteins in detergent solutions. We therefore consider our experimental figures the more reliable ones. It should be noted that $\bar{v}$ values for two other pigment-containing intrinsic

Table 1 Solvent density, ρ, and measured effective molar mass, M_{eff}, and resulting partial specific volume, $\bar{v}$, of cytochrome c oxidase, determined in solvents of different composition under conditions of density matching; (total) detergent concentration: 0.3% (w/v)

Additive/detergent	ρ [g/ml]	M_{eff} [kDa]	$\bar{v}$ [ml/g]
$D_2O/C_{12}E_9$	1.050	26.1	0.766
Sucrose/$C_{12}E_9$	1.028	27.3	0.766
Glycerol/$C_{12}E_9$	1.033	26.8	0.766
Sucrose/rTX-100	1.043	24.6	0.775
Glycerol/rTX-100	1.054	24.3	0.769
Detergent/detergent			
LDAO/$C_{12}E_9$	1.003	28.9	0.773

$\bar{v}$ ($D_2O/C_{12}E_9$) is corrected for H/D exchange according to Ref. [28]

membrane proteins determined by our group [18, 26] are very close to those in Table 1.

Discussion

In this paper, we have explored the potential of sucrose and glycerol as alternatives to D_2O in "density matching", i.e. in blanking out the contributions of protein-bound detergent and of free detergent micelles to the concentration-versus-radius profiles in sedimentation equilibrium studies on intrinsic membrane proteins. It was shown that for cytochrome c oxidase and two commonly used detergents, $C_{12}E_9$ and rTX-100, the results obtained are virtually identical to those found with density matching by addition of D_2O. In particular, there was no indication for additive-induced nonideal sedimentation behaviour of the protein, nor could we find distinct additive-induced changes in its $\bar{v}$ value.[1] What we found, however, was a distinct additive-induced increase in apparent micelle $\bar{v}$, which was larger in sucrose than in glycerol. When using these additives for density matching, this behaviour is advantageous since it reduces the necessary additive concentration. The additive-induced increase in micelle $\bar{v}$ also allows the range of detergent densities which can be matched to be extended to higher values. This seemed of particular interest with respect to DDM, a powerful but still mild detergent. However, the additive concentrations required for matching the DDM density were found to be very high: 39% (w/v) for sucrose and 76% (v/v) for glycerol. This leads to a distinct density gradient (in the case of sucrose) and to solvent viscosities which drastically increase the time required to reach sedimentation equilibrium. In addition, it seems highly questionable whether proteins still show ideal sedimentation behaviour at the high

---

[1] An analogous observation was made in the past with several water-soluble proteins [12, 27].

additive concentrations required, and whether appropriately corrected $\bar{v}$ values can be obtained. In our opinion, it is preferable to exchange DDM (by ion-exchange chromatography) against another mild detergent, the density of which can be matched easily, as described in this study.

An alternative to density matching using high additive concentrations is to perform a series of sedimentation equilibrium experiments at different solvent densities and to extrapolate the results to the matching density of the detergent [2, 3]. With H_2O/D_2O mixtures, this procedure works well even for proteins in DDM [18, 26], but even this system seems to be applicable to homogeneous samples only.

Using detergent mixtures and adjusting the density of the mixed micelles to that of the solvent seems to be an interesting alternative to classical density matching [2, 3]. The principle of this procedure was first suggested by Ludwig et al. [11]. Their detergent of fixed composition, however, turned out to be not applicable to most membrane proteins. On the other hand, the procedure suggested here not only uses detergents which have been successfully applied to a large number of membrane proteins but is also very versatile: at least for the high-density component of the mixture, a large number of other detergents can be selected, for example, poly(ethylene glycol) derivatives with longer fatty acid chains, Triton X-100 or even DDM. Thus, the properties of the detergent mixture, with respect to both solubilizing power and lack of influence on protein conformation, could be "tailored" so as to obtain an acceptable compromise between the high solubilizing power of LDAO and the "mildness" of the polyether detergents. It should, however, be noted that the composition of the protein-bound detergent may differ from that of the starting material. This would lead to incomplete matching of the bound detergent and thus to an error in the molar mass determination of the protein component. On the other hand, a mismatch by ± 0.02 g/ml, at 0.5 g detergent bound per gram of protein, would influence the apparent protein mass by only approximately $\pm 4\%$.

Acknowledgements The work of B. Ludwig was supported by the Deutsche Forschungsgemeinschaft (SFB 472).

References

1. Helenius A, Simons K (1975) Biochim Biophys Acta 415:29
2. Tanford C, Reynolds JA (1976) Biochim Biophys Acta 457:133
3. Schubert D, Schuck P (1991) Prog Colloid Polym Sci 86:12
4. Reynolds JA, Tanford C (1976) Proc Natl Acad Sci USA 73:4467
5. Cantor CR, Schimmel PR (1980) Biophysical Chemistry, part II. Freeman, San Francisco, p 831
6. Schubert D, Boss K, Dorst HJ, Flossdorf J, Pappert G (1983) FEBS Lett 163:81
7. Schubert D, Boss K (1985) Z Naturforsch Teil C 40:908
8. Reynolds JA, McCaslin DR (1985) Methods Enzymol 117:41
9. Ralston G (1993) Introduction to analytical ultracentrifugation. Beckman Instruments, Fullerton, USA, p 58
10. Tsiotis G, Psylinakis M, Woplensinger B, Lustig A, Engel A, Ghanotakis D (1999) Eur J Biochem 259:320
11. Ludwig B, Grabo M, Gregor I, Lustig A, Regenass M, Rosenbusch JP (1982) J Biol Chem 257:5576
12. Lee JC, Timasheff SN (1981) J Biol Chem 256:7193
13. Chang BS, Beauvais RM, Arakawa T, Narhi LO, Dong A, Aparisio DI, Carpenter JF (1996) Biophys J 71:3399
14. Jarabak J, Seeds AE Jr, Talalay P (1966) Biochemistry 5:1269
15. Bradbury SL, Jacoby WB (1972) Proc Natl Acad Sci USA 69:2373
16. Hendler RW, Pardhasaradhi K, Reynafarje B, Ludwig B (1991) Biophys J 60:415
17. Iwata S, Ostermeier C, Ludwig B, Michel H (1995) Nature 376:660
18. Schubert D, Tziatzios C, van den Broek JA, Schuck P, Germeroth L, Michel H (1994) Prog Colloid Polym Sci 94:14
19. Durchschlag H, Zipper P (1995) J Com Esp Deterg 26:275
20. Helenius A, McCaslin DR, Fries E, Tanford C (1979) Methods Enzymol 56:734
21. Saraste M (1990) Q Rev Biophys 23:331
22. Witt H, Ludwig B (1997) J Biol Chem 272:5514
23. Cohn EJ, Edsall JT (1943) In: Cohn EJ, Edsall JT (eds) Proteins, amino acids and peptides as ions and dipolar ions. Hafner, New York, pp 370–381
24. Durchschlag H (1986) In: Hinz H-J (ed) Thermodynamic data for biochemistry and biotechnology. Springer, Berlin Heidelberg New York, pp 45–128
25. Durchschlag H, Zipper P (1994) Prog Colloid Polym Sci 94:20
26. Tziatzios C, Schuck P, Schubert D, Tsiotis G (1994) Z Naturforsch Teil C 49:220
27. Gekko K, Morikawa T (1981) J Biochem 90:39
28. Edelstein SJ, Schachman HK (1967) J Biol Chem 242:306

Progr Colloid Polym Sci (1999) 113:182–184
© Springer-Verlag 1999

J. Behlke
O. Ristau

Analytical ultracentrifugation of the nitrogenase of *Azotobacter vinelandii* under anaerobic conditions

J. Behlke (✉) · O. Ristau
Max Delbrück Center
for Molecular Medicine
Robert-Rössle-Strasse 10
D-13092 Berlin, Germany
e-mail: behlke@mdc-berlin.de
Tel.: +49-30-94062205
Fax: +49-30-94062802

Abstract Nitrogenase (*Azotobacter vinelandii*) is a high-molecular-mass enzyme complex responsible for the fixation and metabolization of nitrogen. The complex consists of two different moieties, a larger MoFe protein (240 kDa) and a smaller Fe protein (60 kDa). The stoichiometry and affinity of both components were studied in solution by sedimentation equilibrium in an XL-A analytical ultracentrifuge. Because both components are highly sensitive to oxygen, the experiments were carried out in an argon atmosphere. Data analysis performed using the program Polymole yielded a 2:1 stoichiometry of Fe protein to MoFe protein. Assuming two independent binding sites on the MoFe protein, the association constant for the first Fe protein bound was $(1.99 \pm 0.48) \times 10^7 \ \mathrm{M}^{-1}$.

Key words Nitrogenase · Sedimentation equilibrium · Complex formation · Association constants · Argon

Introduction

The methods of analytical ultracentrifugation have experienced a remarkable renaissance in recent years with the introduction of the Beckman XL-A ultracentrifuge [1, 2]. This applies to sedimentation velocity as well as to the sedimentation equilibrium techniques that are widely used to determine the size and shape of macromolecules or to study homologous or heterologous associations with respect to the affinity and stoichiometry of interacting systems. This field is of interest in understanding regulatory principles of biological macromolecules. Significant methodical knowledge is required to analyze macromolecules under aerobic conditions; however, there are also substances of importance that are stable only under anaerobic conditions, and experiments to characterize such compounds require somewhat more effort than usual.

Our experience with the nitrogenase complex is presented in this communication. It is an important,

large enzyme responsible for biological nitrogen fixation and is composed of Fe protein and MoFe protein components [3]. Both metalloproteins are extremely oxygen-sensitive. The Fe protein is a homodimer of about 64 kDa. Each subunit contains a 4Fe:4S cluster and can bind one MgADP or MgATP. The MoFe protein is a heterotetramer ($\alpha_2\beta_2$) of about 240 kDa with two different metallocenters. Whereas the so-called P clusters (8Fe:7S) are located between α and β chains, the MoFe cofactors (1Mo:7Fe:9S:1homocitrate) are found only in the α subunits where the substrate reduction takes place. The P clusters are involved in the electron transfer between the Fe protein and the MoFe protein. The complex of both proteins is essential for the fixation and metabolism of nitrogen. It is important to know whether, in solution, the MoFe protein binds one or two Fe protein dimers. Sedimentation equilibrium experiments were carried out in order to answer this question. The experiments were performed in an argon atmosphere in order to avoid inactivation of the nitrogenase by oxygen.

Materials and methods

The nitrogenase (*Azotobacter vinelandii*), a mixture of two parts of the Fe protein and one part of the MoFe component, was kindly provided by Hermann Schindelin (Cal-Tech). The sample and buffer were stored under argon. In order to study the complex formation between both molecules, sedimentation equilibrium experiments were carried out using an XL-A analytical ultracentrifuge (Beckman, Palo Alto, USA). The double-sector cells used for these experiments were assembled as usual and put on the bottom of a glove box containing argon. After extended gas exchange through the holes of the centerpiece, the cells were closed by plugs. About 70 μl of the protein samples and buffer was injected into the cells using a syringe needle piercing the plugs. Afterwards, the cells were closed with gaskets and screws as usual. Sedimentation equilibrium was reached after a 2 h overspeed at 10 000 rpm followed by an equilibrium speed of 8000 rpm for 30 h. Concentration distribution curves were recorded at 390, 400 and 410 nm using the millimolar absorbance coefficients 112.6, 100.1 and 91.08 for the MoFe protein and 12.61, 10.86 and 9.71 for the Fe protein, respectively, for the 12-mm optical path length. We used our computer program Polymole [4] for the analysis of complex formation and stoichiometry. This fits the concentration distribution curves which can be described as radial absorbance with the indices A for receptor molecule (MoFe protein) and L for ligand molecule (Fe protein) by Eq. (1):

$$Abs_r = \varepsilon_A c_{0A} \exp(F_A M_A G) + \varepsilon_L c_{0L} \exp(M_L G)$$

$$+ c_{0A} \times \sum_{j=1}^{n} \frac{1}{n^j} \binom{n}{j} \times (\varepsilon_A + j\varepsilon_L) \times (c_{0L} k_1)^j$$

$$\times \exp[(F_A M_A + jM_L)G] \tag{1}$$

with

$$G = \frac{(1 - \rho\bar{v})\omega^2}{2RT}(r^2 - r_0^2) . \tag{2}$$

Here the sum of the free reactants (first two terms) together with the terms representing the complexes are given. ε_A and ε_L are the extinction coefficients of the respective reactants and F_A takes into account the deviation in buoyancy between A and L. In Eq. (2) ρ denotes the density of the solvent, $\bar{v}$ the partial specific volume of the Fe protein, ω the angular velocity, R the gas constant and T the absolute temperature. To reduce the number of estimated parameters in Eq. (1), we fixed the molecular masses of the reactants A and L at 232.4 kDa and 63.48 kDa. Furthermore, the initial concentrations of the reactants, c_A and c_L, as total concentrations can be analyzed by integration of the areas below the concentration distribution curve (Eq. 1). The total concentration of A is given by

$$c_{At} = \frac{2c_{0A}}{(r_b^2 - r_m^2)} \int_{r_m}^{r_b} \exp(F_A M_A G) r \, dr$$

$$+ \frac{2c_{0A}}{(r_b^2 - r_m^2)} \int_{r_m}^{r_b} \sum_{j}^{n} \frac{1}{n^j} \binom{n}{j}$$

$$\times (c_{0L} k_1)^j \times \exp[(F_A M_A + jM_L)G] r \, dr . \tag{3}$$

The total concentration of L is given by

$$c_{Lt} = \frac{2c_{0L}}{(r_b^2 - r_m^2)} \int_{r_m}^{r_b} \exp(M_L G) r \, dr$$

$$+ \frac{2c_{0A}}{(r_b^2 - r_m^2)} \int_{r_m}^{r_b} \sum_{j}^{n} \frac{j}{n^j} \binom{n}{j}$$

$$\times (c_{0L} k_1)^j \times \exp[(F_A M_A + jM_L)G] r \, dr . \tag{4}$$

The partial specific volume used for the Fe protein has a value of 0.739 ml/g and its ratio to the MoFe protein (factor F_A in Eq. 1) was calculated to be 0.9906.

Results and discussion

Concentration distribution curves obtained under equilibrium conditions are shown in Fig. 1. By performing global fitting using data from three different wave lengths the results obtained are subject to more rigorous constraints. When determining only the molecular mass of this mixture an average value of 318 kDa was obtained indicating a complex of somewhat higher than 1:1 stoichiometry. However, in solutions of micromolar concentrations, complexes of moderate affinity contain a significant amount of free reactants. Because their size is clearly smaller compared to that of the complex, one has to expect a smaller average molecular mass for the mixture containing both complex and reactants. Therefore, we have fitted the concentration distribution curves using different compositions of the complex. As previously mentioned a 1:1 stoichiometry can be excluded. The best data fit was obtained assuming a binding of two Fe protein molecules to one MoFe protein molecule. Assuming two different binding sites on the MoFe protein, we can calculate the association constants using the statistical binding model. The affinity for the binding of the first Fe protein was determined to be (1.99 $\pm$ 0.48) $\times 10^7$ M^{-1}. For the second Fe protein we have to assume an affinity that amounts to a quarter of the value

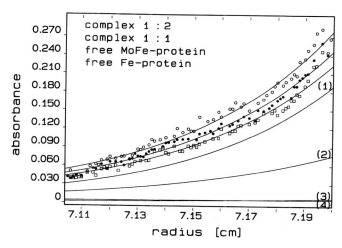

Nitro d: \C31-C40\C31\C31-10_1.R03 2x

Fig. 1 Radial absorbance scans at 390 nm (○), 400 nm (●) and 410 nm (□) from a mixture containing 0.98 μM MoFe protein and 1.96 μM Fe protein in 0.1 M cacodylate buffer pH 6.5 at 20 °C. The *solid curves* represent the best simultaneous fit of the 1:2 association (*1*), and a 1:1 association complex (*2*) of the MoFe protein/Fe protein system as well as the free reactants MoFe protein (*3*) and Fe protein (*4*)

derived for the first. These values are reflected in the optimum fit of the curves presented in Fig. 1.

Recently, the complex formation between the MoFe protein and the Fe protein in crystals has been analyzed by other techniques [5]. From the X-ray diffraction data, a 2:1 complex was determined as was demonstrated here for the behavior in solution. By means of sedimentation equilibrium, using our program Polymole, we are able to determine not only the stoichiometry of complex formation but also the affinity of reactants for dissolved components. The data presented here demonstrate the possibility of determining the affinities of reactants that are extremely sensitive to oxygen. Thus, special precautions are required in order to carry out such experiments. The curve-fitting analysis has demonstrated that these very sensitive components of nitrogenase are stable over a long time period when working in an argon atmosphere.

Acknowledgement The authors are grateful to Hermann Schindelin, Cal-Tech, for providing the nitrogenase.

References

1. Harding SE, Rowe AJ, Horton JC (eds) (1992) Analytical ultracentrifugation in biochemistry and polymer science. Royal Society of Chemistry, Cambridge

2. Schuster TM, Laue TM (eds) (1994) Modern analytical ultracentrifugation. Birkhäuser, Boston

3. Schlesinger WH (1991) Biochemistry, an analysis of global change. Academic Press, San Diego

4. Behlke J, Ristau O, Schoenfeld HJ (1997) Biochemistry 36: 5149

5. Schindelin H, Kisker C, Schlessman JL, Howard JB, Rees DC (1997) Nature 387: 370

Progr Colloid Polym Sci (1999) 113:185–191
© Springer-Verlag 1999

BIOLOGICAL SYSTEMS

T. Tongdang
H.F.J. Bligh
K. Jumel
S.E. Harding

Combining sedimentation velocity with SEC-MALLS to probe the molecular structure of heterogeneous macromolecular systems that cannot be preparatively separated: application to three rice starches

T. Tongdang · K. Jumel
S.E. Harding (✉)
NCMH Unit, University of Nottingham
School of Biological Science
Sutton Bonington
Leics. LE12 5RD, UK

H.F.J. Bligh
Division of Plant Science
University of Nottingham
School of Biological Science
Nottingham NG7 2RD, UK

Abstract A combined approach using sedimentation velocity in the Beckman Optima XL-I ultracentrifuge together with high performance size exclusion chromatography coupled to multiangle laser light scattering (SEC-MALLS) was used to assess the molecular structure of rice amylose from three different rice starches without pre-separation by preparative means. The amylose content of the rice starches was analysed by the iodo-colorimetric method. We found for the three species of amylose starch studied (*Thaibonnet*, *Cypress* and *Dixiebell*): (1) the weight average molecular weights, M_w, (from SEC-MALLS) for the amyloses were similar (4.9, 4.8, 5.1 × 10^5 g/mol); (2) the sedimentation coefficients, $s^\circ_{20,w}$ (from apparent distribution of sedimentation coefficient analysis) were also similar (5.7, 5.3, 7.1 S), despite the fact that the amylose content (relative to amylopectin) for the three rice starches was quite different. The study also shows that attempts to obtain molecular weight by sedimentation velocity (s-D and s-k_s) on the unseparated materials are not useful: SEC-MALLs is the method of choice for these substances. For heterogeneous materials where preparative separation is not possible, the utility of using SEC-MALLs to characterize component molecular weights and velocity sedimentation to characterize sedimentation coefficients is thus indicated.

Key words Rice amylose · Sedimentation velocity · Size exclusion chromatography-multiangle laser light scattering

Introduction

Rice starch, as other starches, is composed of two polymeric forms of the same glucose monosaccharide unit – amylose and amylopectin – although structurally these forms are quite different. Amylose contains linear $\alpha(1$–$4)$ linked chains (with some slight branching) whereas amylopectin has extensive $\alpha(1$–$6)$ linked branching of much higher molecular weight, 0.45–2.5 × 10^5 Da for amylose and 60–110 × 10^6 Da for amylopectin [1]. Despite many attempts, and the desire to study the molecular properties of each component, effective separation has proved extremely difficult [2]. Banks and

Greenwood [3] dissolved starch in dimethyl sulfoxide (DMSO) and precipitated with butanol and the component taken to be amylose was then separated; this method had been used by many workers [4–7]. Concanavalin A has also been used as precipitant for amylopectin [7]. However, these methods are very susceptible to contamination by intermediates and also anomalous amylose [8]. Strategies that can be developed which can analyse the molecular properties of either component without relying on preparative separation would therefore appear to be of considerable value.

Relatively recently, high performance size exclusion chromatography (SEC) has been employed on whole, unfractionated starch that has been debranched with isoamylase enzyme [9], as a means for determining the molecular weight distribution of amyloses and as a means for investigating the fine structure of amylopectin. With amylopectin this approach appears to work only with debranched material [10]. However, amylose is a different proposition and it is possible to (1) analytically separate the amylose from amylopectin and (2) characterise the molecular weight of the amylose component, providing the SEC is coupled on-line to multi-angle laser light scattering (SEC-MALLS [11]). Recent advances in velocity sedimentation analysis, in terms of distribution of sedimentation coefficient analysis, has meant that ultracentrifugation can provide an additional and independent route for studying the molecular properties of the amylose component of starch, and, just as with SEC-MALLS, without the requirement for preparative separation. Such developments have taken advantage of the ability to capture large amounts of data from the refractometric (Rayleigh interferometric) optical records from the ultracentrifuge directly via a digital camera into a PC.

Material and methods

Rice starch

Rice starches were isolated by a modified alkaline extraction method [12]. With this method, rice grains (three varieties, *Thaibonnet*, *Cypress* and *Dixiebelle*) were soaked in six times their weight of 0.2% NaOH (w/v) for 12 h. After draining off the supernatant, the soaked grains were blended at low speed for ~5 min. The blended sample was passed through a muslin cloth and then centrifuged at 2000 rpm for 15 min. The supernatant was drained and the yellowish upper part of the sediment was scraped off slightly. The sediment was continually centrifuged and re-separated with fresh NaOH solution until a clear supernatant was obtained or a negative response in the Biuret test was observed. An excess of distilled water was added to the purified starch and hydrochloric acid (1 M) was used to neutralise the alkaline. The starch slurry was passed through a sintered funnel (size 3) and then dried at room temperature. This material was then defatted by the soxhlet extraction method for 18 h using methanol (85% v/v) as the solvent. The starch was then dried in air.

Amylose content measurement by iodo-colorimetry

The procedure followed was based on the International Standard ISO 6647 method [13] for amylose determination of rice flour (approx. 90% starch). Defatted rice starch samples (100 mg) were dispersed with 1 ml of ethyl alcohol and suspended in 9 ml of 1 M NaOH solution. The suspensions were heated for 30 min at 95 °C and then left overnight. The solution was diluted to 100 ml and a 5 ml aliquot was pipetted into a 50 ml volumetric flask with 25 ml of distilled water, 0.5 ml of 1 M acetic acid together with 1 ml of iodine solution (0.2% iodine in 2% potassium iodide) were added and the volume was diluted to 50 ml with distilled water. The solution was well mixed before leaving for ~20 min for the colour to develop in a dark room. The optical absorbance was recorded at 620 nm in a spectrophotometer (Ultraspec 4050, LKM Biochrom Croydon, UK). A sample blank was prepared for the reference cell

simultaneously. Each measurement was performed in triplicate and the mean value calculated.

Standard curve

For calibration purposes (amylose/amylopectin ratio), pure potato amylose and pure corn amylopectin (ICN Biomedicals, USA) solutions (1 mg/ml) were prepared in a similar manner to those from the rice starch samples. The potato amylose and amylopectin solutions were mixed together in a proportion of amylose (by volume) ranging from 0 to 40%. Aliquots (5 ml) of the mixtures were reacted with iodine solution and the absorbance was read in the same manner as the sample. The standard curve was then plotted as absorbance at a wavelength of 620 nm against amylose content.

Sedimentation velocity

Sample preparation

Sedimentation velocity ultracentrifugation work was performed on an Optima XL-I Analytical Ultracentrifuge (Beckman, Palo Alto, USA) equipped with scanning interference optics [14]. A suspension of the purified and defatted rice starch in 90% DMSO (Sigma, Poole, UK) was heated at 95 °C for 15 min and allowed to cool to ambient temperature and then stirred at for 24 h under a nitrogen atmosphere. The starch solution was passed through a sintered glass funnel (no. 3) and subsequently diluted for the analyses. Amylose concentrations were then obtained from total starch concentrations using the amylose/amylopectin ratio.

Ultracentrifugation

The solution (400 μl, 3–8 mg/ml) and the solvent (90% DMSO) were injected into the sample and solvent channels, respectively, of 12-mm optical pathlength ultracentrifuge cells before loading into a 4-hole titanium rotor. The samples were run on the XL-I at 50 000 rpm at 25.0 °C. The Rayleigh interference optical system was used to scan the movement of the sedimenting boundaries at 5 min intervals. Standard Fourier transform software converted the fringe profiles into plots of fringe displacement, j (relative to the meniscus), versus radial position, r [14]. The sedimentation coefficient was calculated by using the XL-I software [sedimentation time derivative $g(s^*)$ analysis], calculating the apparent distribution $g^*(s^*)$ [written for convenience as just $g(s^*)$] of the apparent sedimentation coefficients s^* from the time derivative, $(\partial c/\partial t)_t$, of the concentration profile in the cell [15]. The analysis procedure was essentially as described in Stafford [15] and the instrument manual of the Optima XL-I ultracentrifuge [16]. The apparent sedimentation coefficient, s^*, of a boundary for each loading concentration of amylose, c (obtained from the total loading concentration adjusted by the amylose/amylopectin ratio), was obtained from the peak of the $g(s^*)$ profile (fitted to a Gaussian) and then converted to standard conditions of viscosity and density of water at 20.0 °C, $s_{20,w}$ [17]. The partial specific volume, $\bar{v}$ of amylose and amylopectin required for this correction were 0.643 ml/g [18]. The corrected $s_{20,w}$ values were then plotted against concentration. By extrapolation to zero concentration, the sedimentation coefficient at infinite concentration ($s^{\circ}_{20,w}$) was obtained. According to Gralen [19]:

$$s_{20,w} = s^{\circ}_{20,w}(1 - k_s c) \tag{1}$$

where $s^{\circ}_{20,w}$ is the sedimentation coefficient at infinite dilution (in Svedbergs, S = 1×10^{-13} s) and k_s (ml/g), from the limiting slope at $c = 0$, is the "Gralen" parameter. An alternative representation has been given in the case of more severe forms of concentration dependence: $(1/s_{20,w}) = (1/s_{20,w}^{\circ})(1 + k_s c)$. However, in such cases, non-linear extrapolation of $s_{20,w}$ to $c = 0$ still usually gives a better estimate for $s_{20,w}^{\circ}$.

SEC-MALLS method

Sample preparation

The procedure essentially followed that described in [9]. It differs from that used for the DMSO-based procedure followed for sedimentation velocity essentially because the SEC columns available to us were not suited to DMSO. To solubilise in a purely aqueous solvent, the amylopectin component had to be debranched instead, following the method described in [20]. Ethanol (1 ml) and 1 M NaOH (10 ml) were added to the purified, defatted rice starch (100 mg). The suspension was left at room temperature for 24 h and then heated in boiling water for 15 min. After cooling, the solution was neutralized with 1 M HCl. Then 1 ml of sodium acetate (30 mM, pH 3.5) and 20 μl of isoamylase were added and mixed. The sample solution (2 ml) was added, well mixed and then incubated at 40 °C for 24 h. Enzyme activity was terminated by boiling in a water bath for 5–10 min. The column eluent was a pH 8.6 buffer containing Na_2HPO_4 (0.1 M), NaH_2PO_4 (0.05 M) and NaN_3 (0.02%) and was filtered under vacuum through a 0.2-μm nylon membrane filter (Gelman Sciences, 47 mm) in a glass Millipore solvent filtration system.

The debranched sample was filtered through a 0.45 μm syringe filter (Whatman, 13 mm, PVDF) and injected into the sample loop via a Rheodyne injection valve (Rheodyne 7125) fixed on the column oven (Anachem, Luton, UK). The sample with eluent was then pumped (Waters 590 HPLC pump) into a guard column (TSK GEL PWXL: 6×40 mm) followed by five analytical columns (TSK GEL G3000 PWXL, Asahipak GS-320H (2×), TSK G2500 PWXL and G-Oligo PWXL) connected in that order, in the column oven at 30.0 °C. The eluate was analysed in a Dawn F multi-angle laser photometer (Wyatt Technology, Santa Barbara, USA) [11], linked with a He-Ne laser operating at 632 nm and equipped with 18 detectors at angles ranging from 22 to 160°, and subsequently (on-line) by a refractive index concentration detector (Wyatt Optilab) and a refractive index increment, $dn/dc = 0.152$ ml/g [21]. The responses from the detectors were transmitted to a personal computer, and the light-scattering and the RI chromatograms were displayed on the screen during fractionation. The column eluent (phosphate buffer, pH 8.6) was degassed with an on-line degasser (Degasys, DG1200, uniflow from HPLC Technology, UK). The pump flow rate was set at 0.5 ml/min.

Data analysis

The absolute weight average molecular weight, M_w, of amylose was extracted from Zimm-Debye plots [22]:

$$R_\theta / K^* c = M_w P(\theta)(1 - 2A_2 c M_w P(\theta)) \tag{2}$$

where R is the excess Rayleigh ratio or excess scattering of the polymer in solution when compared to that of the solvent in solution at a scattering angle of θ; K^* is the optical constant = $4\pi n_0^2 (dn/dc)^2 \lambda_0^{-4} N_A^{-1}$, where n_0 is the refractive index of the solvent at the incident radiation (vacuum) wavelength, λ_0 (cm), N_A is Avogadro's number (6.02×10^{23} mol^{-1}) and c is the concentration (g/ml); A_2 is the second thermodynamic virial coefficient (because of the low concentrations used this was assumed to be $\sim$0), and $P(\theta)$, the scattering function, is the theoretically derived form factor, approximately equal to

$$1 - 2\mu^2 \langle r^2 \rangle / 3! + \cdots \tag{3}$$

where $\mu = (4\pi/\lambda)\sin(\theta/2)$, and $\langle r^2 \rangle$ is the mean square radius of gyration R_g (cm). M_w values were extracted using ASTRA analysis software (Wyatt Technology, Santa Barbara, USA); this software also estimates the number average molecular weight, M_n [11]. Corresponding degrees of polymerization (weight average $DP_w = M_w/162$ and number average $DP_n = M_n/162$) were obtained.

Results

Amylose/amylopectin ratio

The iodo-colorimetric method showed a clear difference in the relative amylose content for the three rice starches. Further studies were then performed to determine if this difference is matched by a difference in physical properties.

Sedimentation behaviour of rice starch components

Figure 1a shows the progress of the sedimenting boundary for rice starch (in 90% DMSO) from *Dixiebell* at a total starch loading concentration of 8 mg/ml. The rapid sedimentation of the amylopectin component (moving to the base within 20 min of starting) can be clearly seen (first scan). The single amylose boundary follows, which is fairly homogeneous. From analysis of the change in the concentration distribution with time, the corresponding $g(s^*)$ versus s^* plots could be generated (Fig. 1b–d) for the amylose.

Sedimentation coefficients

After correction of the sedimentation coefficient to the standard solvent conditions, a plot of $s_{20,w}$ against amylose concentration could then be constructed, and Fig. 2 shows the characteristic drop in $s_{20,w}$ with increase in c, caused by non-ideality effects for all three samples: *Thaibonnet* (Fig. 2a), *Cypress* (Fig. 2b) and *Dixiebell* (Fig. 2c). The corresponding "infinite dilution" ($s_{20,w}^{\circ}$) values are 5.7 ± 0.3, 5.3 ± 0.3 and 7.1 ± 0.2 S. The same results were obtained from reciprocal plots of $(1/s_{20,w})$ versus c. These results are also within the range of 4–14 S reported for amylose from rice starches [23].

Molecular weights from SEC-MALLS analysis

SEC elution profiles for the three rice starches (after debranching of the amylopectin) are shown in Fig. 3, and show that the amylose component is essentially homogeneous, consistent with the $g(s^*)$ profiles. Two further peaks in the refractive index profile at $\sim$32 and 36 ml correspond to debranched amylopectin, but are

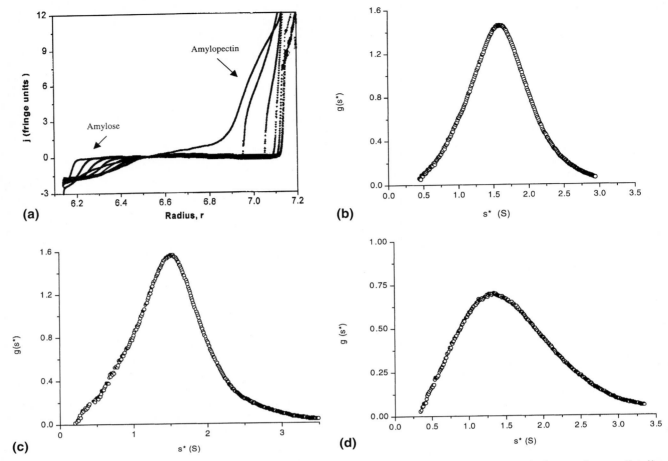

Fig. 1 a Fringe displacements, $j(r)$, relative to the meniscus of rice starch components from *Dixiebell* as a function of radial position, r. Optima XL-I ultracentrifuge used at 25.0 °C and 50 000 rpm rotor speed; scan interval: 20 min; solvent: 90% DMSO, 10% water; loading amylose concentration = 2.68 mg/ml. **b** Corresponding $g(s^*)$ versus s^* plot. **c** $g(s^*)$ versus s^* plot for *Thaibonnet* amylose (loading amylose concentration = 1.73 mg/ml). **d** $g(s^*)$ versus s^* plot for *Cypress* amylose (loading amylose concentration = 1.38 mg/ml)

too small to show a light scattering signal and hence cannot be analysed.

Corresponding weight average molecular weights for the amylose component eluting at ∼25 ml show clear light-scattering signals and can be analysed. A comparison of the results for the three rice starches is given in Table 1, and, as with the sedimentation coefficients, show the amyloses to be very similar also in this regard: 4.8, 4.9 and 5.1×10^5 g/mol respectively for *Thaibonnet*, *Cypress* and *Dixiebell*. It should be stressed that these molecular weights correspond to debranched material. Although debranching primarily affects the amylopectin, it may also have some small effect on the amyloses, so the values in Table 1 could be underestimates. The Wyatt ASTRA software also yields estimates for the number average molecular weight (from the form of the molecular weight versus elution volume distribution); these too are very similar for the three amyloses; corresponding "polydispersity indices" (i.e. M_w/M_n) are therefore also very similar and quite high at ∼1.4.

Further attempts to obtain molecular weights from manipulations of the sedimentation velocity data were then performed as follows.

1. *s-D* method. From the boundary spreading with time through diffusion it is possible also to estimate (in addition to the sedimentation coefficient) the translational diffusion coefficient (further, from combination of s and D, M can then be obtained [17]); for monodisperse systems such as proteins this can be done successfully and commercial packages such as DCDT [15] or SVEDBERG [24] are very useful. However, for polydisperse systems – the hallmark of amylose and other polysaccharides – the values returned are not reliable, and this was confirmed for our materials with widely ranging estimates for different loading concentrations.

2. *s-k_s* method. It is also possible in principle to estimate M from a combination of s with the Gralen coefficient, k_s [25]. Again this appears to work well with proteins, but for polysaccharides the limitation is that an

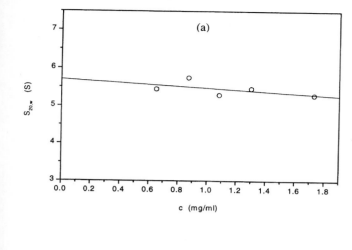

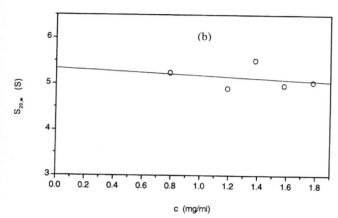

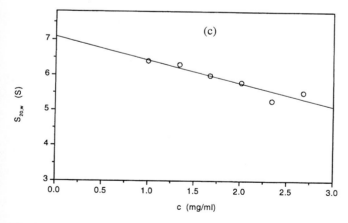

Fig. 2 Plot of $s_{20,w}$ against amylose concentration for **a** *Thaibonnet*, **b** *Cypress* and **c** *Dixiebell*. Amylose concentrations estimated from total loading concentrations adjusted by the amylose/amylopectin ratio

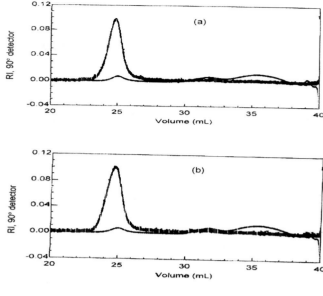

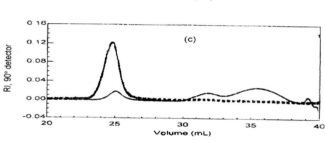

Fig. 3 SEC elution profiles from the concentration detector (refractive index detector) (*the lower peak*) and 90° light scattering profiles (*the higher peak*) (for clarity only this angle is shown) for **a** *Thaibonnet*, **b** *Cypress* and **c** *Dixiebell* rice starches (after debranching) in pH 8.6, $I = 0.1$ phosphate buffer. Consistent with the $g^*(s)$ plots (Fig. 2), single peaks were seen for the amylose component

g often suffices for proteins, it is impossible to pin down with any precison for polysaccharides: changes in v_s from 1 to 50 yield enormous changes in the estimated M (for the amyloses, from 7000 to 4.5×10^5).

So, although methods (1) and (2) are also not requiring preparative separation of the amylose, the SEC-MALLS procedure seems the only realistic procedure at present for molecular weight analysis.

Frictional ratio f/f_0

The conformation/molecular hydration parameter f/f_0, the translational frictional ratio (ratio of the frictional coefficient of the amylose to the frictional coefficient of a spherical particle of the same anhydrous mass), can also be obtained by combining $s^\circ_{20,w}$ with the molecular weight (see, for example [26]):

$$f/f_0 = M(1 - \bar{v}\rho_0)/([(N_A 6\pi\eta_0 s^\circ_{20,w})(4\pi N_A/3M)^{1/3}] \quad (4)$$

estimate for the "swollen specific volume", v_s (volume of the macromolecule swollen through solvation per unit anhydrous mass) is required. Whereas a value of ~ 1 ml/

Table 1 Physicochemical parameters for amyloses rice starches (*Thaibonnet*, *Cypress* and *Dixiebell*) from SEC-MALLS

Properties	*Thaibonnet*	*Cypress*	*Dixiebell*
Amylose content (% d.w.b)[a]	20.6 ± 0.3	19.8 ± 0.7	33.5 ± 0.4
$s^{\circ}_{20,w}$ (S)	5.7 ± 0.3	5.3 ± 0.3	7.1 ± 0.2
$M_w \times 10^{-5}$ (g/mol)	4.9 ± 0.3	4.8 ± 0.2	5.1 ± 0.3
DP_w	3000	2900	3100
$M_n \times 10^{-5}$ (g/mol)	3.8 ± 0.2	3.5 ± 0.3	3.7 ± 0.2
DP_n	2300	2200	2300
f/f_0	5.3	5.6	4.4
Polydispersity index (M_w/M_n)	1.3	1.4	1.4

[a] d.w.b.: dry weight basis

with N_A Avogadro's number and ρ_0 and η_0 the density and viscosity of the solvent (which will be water at 20.0 °C since the $s_{20,w}$ value has been corrected to this); again, similarly high values ($\sim$5) are obtained for the three rice starches. Unfortunately, it is not possible to unequivocally comment on the conformation or hydration based on this ratio alone: it will be a function of both, as manifested by the Perrin relation:

$$f/f_0 = P(v_s/\bar{v})^{1/3} \qquad (5)$$

with P the shape parameter and v_s the swollen specific volume as noted above. For a spherical particle irrespective of its size or hydration, the shape function $P = 1$. To take the extremes:

1. Spherical domain. $P = 1$, and hence $v_s \sim 80$ ml/g. Although this is huge compared to proteins, it is similar to estimates for other polysaccharides and heavily glycosylated mucin systems (see, for example [27, 28]).
2. Hydration similar to that of globular proteins. $v_s \sim 1$ ml/g and hence $P \sim 4$; for a rigid macromolecule this corresponds to an ellipsoid of revolution of axial ratio $\sim$100:1 (see [29]), an unlikely scenario for an uncharged single-stranded polysaccharide (even if there is significant helical content).

Conclusions

The main purpose of this paper is not to give a detailed description of the molecular structure of starch polysaccharides but to show how it is possible to extract useful structural information on a heterogeneous macromolecular system where the components cannot be easily isolated preparatively prior to analysis; starch is an outstanding example of one such system. In that regard, starch presents us with an additional problem in that the amylopectin component in its intact form is poorly soluble in aqueous solvents; for sedimentation analysis this is not a problem so long as other (non-aqueous) solvents can be used since no separation "medium" is required. For SEC-MALLS analysis, however, a separation medium (the SEC column material) is required with appropriate assumptions over chemical inertness (although columns are indeed available where chemical inertness can be at least assumed, in this study we have followed the simpler route using an aqueous solvent and solubilising the amylopectin by debranching). Our structural calculations based on the frictional ratio, however, have assumed that the properties of amylose in DMSO and aqueous buffer are essentially similar; although this assumption may be open to question, we have at least shown how both molecular weight and conformation/hydration information can be extracted.

This combination can be made even more powerful if we can add further information from additional hydrodynamic parameters, for example the intrinsic viscosity $[\eta]$ (possible from viscometers that can be also be coupled downstream from the SEC separation system – see e.g. [30]). The MALLS can in principle at least also provide R_g information [cf. Eq. (4)]. Both $[\eta]$ and R_g, like $s^{\circ}_{20,w}$, can provide – *when combined together* – conformational and hydration information without ambiguity.

References

1. Fronimos P (1990) MPhil Dissertation. University of Leicester, UK; see also Aberle T, Berchard W, Vorwerg W, Radota S (1994) Starch/Starke 46:329
2. Hizukuri S, Takeda Y, Abe J, Hanashio I, Matsunobu G (1997) In: Frazier PJ, Richmond P, Donald AM (eds) Starch structure and functionality. Royal Society of Chemistry, Cambridge, pp 121–128
3. Banks W, Greenwood CT (1975) Starch and its components. Edinburgh University Press, Edinburgh
4. Takeda Y, Hizukuri S, Juliano BO (1986) Carbohydr Res 48:299
5. Gilbert ML, Gilbert AG, Spragg PS (1964) In: Whistler LR (ed) Methods in carbohydrate chemistry, vol 4. Academic Press, New York, pp 25–27
6. Jane JL, Chen JF (1992) Cereal Chem 69:60
7. Metheson KN, Welsh AL (1988) Carbohydr Res 180:301

8. Kringler WR, Zimbalski M (1992) Starch/Starke 44:414
9. Ong MH, Jumel K, Tokarczuk FP, Blanshard MVJ, Harding SE (1994) Carbohydr Res 260:99
10. Fernandez A (1996) PhD Thesis. University of Nottingham, UK
11. Wyatt PJ (1992) In: Harding SE, Sattelle DB, Bloomfield VA (eds) Laser light scattering in biochemistry. Royal Society of Chemistry, Cambridge, pp 35–58
12. Juliano BO (1984) In: Whistler RL, Bemiller JN, Paschall EF (eds) Starch chemistry and technology, 2nd edn. Academic Press, London, pp 507–525
13. ISO (1987) Rice—determination of amylose content. ISO 6647 (E) International Organization for Standardization, Switzerland
14. Furst A (1997) Eur Biophys J 35:307

15. Stafford W (1992) Anal Biochem 203:295
16. Beckman Instruments (1997) The instruction manual of the Optima XL-I analytical ultracentrifuge with integrated optical systems. The Spinco Business of Beckman Instruments, Palo Alto, Calif
17. Tanford C (1961) Physical chemistry of macromolecules. Wiley, New York
18. Fujii M, Honda K, Fujita H (1973) Biopolymers 12:1177
19. Gralen N (1944) PhD Dissertation. University of Uppsala, Sweden
20. Ramesh M, Ali SZ, Bhattacharya KR (1999) Carbohydr Polym 38:337
21. Banks W, Greenwood CT, Sloss J (1969) Carbohydr Res 11:399
22. Zimm BH (1948) J Chem Phys 16:1093

23. Juliano OB (1985) In: Juliano OB (ed) Rice: chemistry and technology. American Association of Cereal Chemists, St Paul, Minn pp 59–175
24. Philo J (1994) In: Schuster TM, Laue TM (eds) Modern analytical ultracentrifugation. Birkhauser, Boston, pp 156–170
25. Rowe AJ (1977) Biopolymers 16:2595
26. Harding SE (1995) Biophys Chem 55:69
27. Harding SE, Rowe AJ, Creeth JM (1983) Biochem J 209:893–896
28. Harding SE, Day K, Dhami R, Lowe PM (1997) Carbohydr Polym 32:81–87
29. Harding SE, Horton JC, Colfen H (1997) Eur Biophys J 25:347–359
30. Harding SE (1997) Prog Biophys Mol Biol 68:207–262

Progr Colloid Polym Sci (1999) 113:192–200
© Springer-Verlag 1999

BIOLOGICAL SYSTEMS

P. Lavrenko
O. Okatova
H. Dautzenberg

Inhomogeneity and conformational parameters of low-substituted carboxymethyl cellulose from analytical ultracentrifugation data

P. Lavrenko (✉) · O. Okatova
Institute of Macromolecular Compounds
Russian Academy of Sciences
Bolshoy pr. 31, 199004 St. Petersburg
Russia
e-mail: lavrenko@mail.macro.ru

H. Dautzenberg
Universität Potsdam
WIP-Forschungsgruppe für
Polyelektrolytkomplexstrukturen
Kantstrasse 55, D-14513 Teltow
Germany

Abstract The sedimentation velocities of carboxymethyl cellulose molecules in different solvents have been studied and compared with those of other cellulose derivatives to analyze a function of the substituent chemical structure, the degree of substitution, and the degree of polymerization. Complexation and charge effects are briefly discussed. The degree of coiling and draining of the macromolecule chains in dilute solution is characterized by the Kuhn segment length and the hydrodynamic diameter. The x-spectrum of the macromolecules in the range of the solvent–solution sedimentation boundary is analyzed to determine the inhomogeneity of the samples. Special attention is paid to problematic questions of the concentration effects in the sedimentation analysis of the polymers in dilute solutions.

Key words Carboxymethyl cellulose · Analytical ultracentrifugation · Conformation · Inhomogeneity

Introduction

Analytical ultracentrifugation is a well-known method with great potential for precise measurements of polydisperse systems of macromolecules. Characteristics such as molecular weight, chain dimensions, structural parameters, inhomogeneity, and others, may be evaluated [1]. Many experimental results obtained with this physical method during the last 40–50 years in different laboratories are available for different polymers, for instance, in the *Polymer Handbook* [2].

Macromolecules of cellulose derivatives are very difficult to analyze because of their inhomogeneity not only with respect to molecular weight but also with respect to composition (degree of substitution, DS, and regularity in structure). The analysis is further complicated by the potential of the macromolecules of these compounds to self-associate in solution, and their polyelectrolyte characteristics. All these properties are typical for carboxymethylated cellulose, which was widely studied with different methods in water, salt water, and cadoxen solutions [3–7].

Here, we report the characterization of low-substituted carboxymethyl cellulose (CMC) in dilute solutions in different solvents. Results are compared with the data obtained for some other cellulose derivatives via analytical ultracentrifugation during the last few years. Special attention is paid to the properties of cellulose products with low DS [8, 9] because they may be considered as models suitable for the cellulose chain. We also discuss some problematic questions of the method when applied to the analysis of cellulose and cellulose derivatives.

In many cases satisfactory information can only be achieved by combining results from different techniques; therefore, diffusion and viscosity measurements were also performed.

Experimental

Samples and solutions

All Na-CMC samples were prepared based on linters as cellulose starting material. We followed alkalization by carboxymethylation with monochloroacetate in a medium composed of sodium hydroxide, water, and ethanol or isopropanol [8]. Samples thus prepared were purified according to standard procedures prior to use. The fractions of the samples were taken as a soluble part after cutting CMC soluble in water/ethanol (15/85) as well as CMC insoluble in water/ethanol (80/20) mixtures. The DS was determined by elemental analysis and is indicated in the following in parentheses. A cadoxen solution (6 wt% Cd, 28 wt% ethylenediamine, no sodium hydroxide) was prepared as described earlier [10] and subsequently denoted as "cadoxen". Dilute water cadoxen (DWC) (10 parts water per 1 part cadoxen, by weight) was used as a solvent. The solvent density was 1.0030 g/ml, the viscosity 1.02×10^{-2} g cm^{-1} s^{-1} and the refractive index 1.3370 at 26 °C. The preparation of solutions and experimental details were described earlier [11]. Dialysis of CMC solutions was performed simultaneously for 7 days against the solvent using a cellophane membrane as described in Ref. [12].

Methods

Analytical ultracentrifugation was performed in a 3180 analytical ultracentrifuge from the Hungarian Optical Works MOM [13] equipped with interference optics [14]. The rotor speed was 50 000 rpm with an accuracy of ± 20 rpm. The temperature of the rotor (26.0 °C) was maintained with an absolute accuracy of ± 0.1 °C and with a constancy of regulation of ± 0.05 °C. A polyamide single-sector cell with a 12-mm light path and quartz windows was used. The base line was determined in a separate experiment with solvent alone.

Free diffusion of the polymer molecules in solution was investigated using a gradient method in which a sharp boundary was formed between solvent and solution in a Teflon cell [15] with $h = 2.0$-cm light path. A Tsvetkov interferometric diffusometer with spar twinning ($a = 0.10$ cm) and compensator fringes' interval ($b = 0.15$ cm) was used [16].

Viscosity measurements were performed in an Ostwald-type viscometer at 26 °C with an average velocity gradient of 360 s^{-1}. The kinetic energy correction was negligible.

All data were evaluated via nonweighted linear regression analysis and refined by least squares.

Results and discussion

Pycnometry

The densities of CMC(0.6) (fraction no. 27)/DWC solutions of different solute concentrations were determined in a pycnometer (54.575 ml at 26 °C) before and after dialysis. Results in terms of ρ(g/ml)/c (wt%) as obtained for both dialyzed (a) and nondialyzed (b) solutions of CMC(0.6) are as follows.

(a) 1.00712/0.99, 1.00643/0.796, 1.00541/0.591, 1.00466/0.390, 1.00373/0.199

(b) 1.00725/0.99, 1.00645/0.796, 1.00553/0.591, 1.00470/0.390, 1.00378/0.199.

Figure 1 shows that points a and b form a linear dependence of $\rho(c)$ with slope $\partial\rho/\partial c = 0.42 \pm 0.02$ (curve 1). No curvature of the $\rho(c)$ dependence was detected; therefore, we can accept the slope, $\partial\rho/\partial c$, as equal to the buoyancy factor, $(1 - \bar{v}\rho_0)$, with $\bar{v}$ being the partial specific volume of CMC(0.6) in DWC solution.

For CMC(0.9) results are presented by points 2a and 2b in Fig. 1. We have

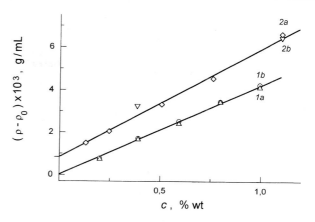

Fig. 1 Concentration dependence of **a** dialyzed and **b** nondialyzed solution density for (1) carboxymethyl cellulose (CMC)(0.6) and (2) CMC(0.9) in dilute water cadoxen (DWC) at 26 °C. For convenience, curve 2 is shifted up to the marked value

(a) 1.00846/1.10, 1.00640/0.765, 1.00516/0.508, 1.00390/0.245, 1.00335/0.130

(b) 1.00810/1.10, 1.00490/0.385

and $\partial\rho/\partial c = 0.52 \pm 0.02$.

It is important to note that for CMC with both DS = 0.6 and 0.9, points a in Fig. 1 coincide with points b, and dialyzed solutions give the same results as non-dialyzed ones. We can conclude that selective sorption of the solvent components by macromolecules is unnoticeable, and a complexation phenomenon does not occur in the CMC–DWC system.

Velocity sedimentation

A series of experiments with a freely sedimenting boundary between the solvent and the solution was performed at several different dilutions (with decreasing value of c down to the minimum possible value) and under identical experimental conditions (frequency, temperature, cell). The contour of the interference boundary was symmetrical in form. The first moment of the sedimentation curve was therefore used as the boundary position, x_m, counted from the axis of rotation of the rotor. The sedimentation coefficient, s, was determined as usual from the time dependence of x_m by $s = \omega^{-2} (\mathrm{d}\ln x_m / \mathrm{d}t)$. Here ω is the angular rotor rate, $\omega = 2n/60$, n is the frequency of the rotor rotation ($n = 50\,000$ rpm was used), and t is the sedimentation time. A typical dependence of $\ln x_m$ on t as obtained at several different dilutions is illustrated in Fig. 2. The experimental points fall on a straight line with a slope of $s\omega^2$. Pressure effects did not exceed the experimental uncertainty (3%) in the s determination and were, therefore, ignored.

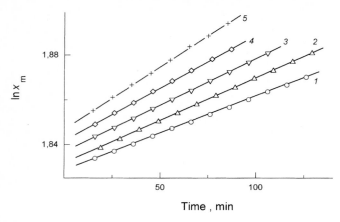

Fig. 2 ln x_m versus time for CMC(0.6), fraction 19 in DWC at 26 °C. The solute concentrations, c, are 2.05 (*1*), 1.52 (*2*), 1.22 (*3*), 0.89 (*4*), and 0.56 g/l (*5*). For convenience, curves *2–5* are shifted up to values of 0.024 (*2*), 0.050 (*3, 4*), and 0.052 (*5*)

Figure 3 shows that the reciprocal sedimentation coefficient, $1/s$ (points) depends linearly on solute concentration and that the slope of the plot decreases as the intercept increases. The plot was approximated by a linear function (solid curves) described by the expression $1/s = (1/s_0)(1 + k_s c)$ where $s_0 = \lim s_{c \to 0}$ and k_s is the concentration parameter. Results are listed in Table 1. The k_s parameter was approximated by a power function of s_0: $k_s = 0.12 (s_0 \times 10^{13})^{2.6 \pm 0.2}$. The average value of the relation $k_s/[\eta]$ was found to be 0.90 ± 0.09, which is typical for many cellulose substances [17].

The unanswered question here remains the obvious inversion of the $s(c)$ functions at $c > 1.5$ g/l.

The time dependence of the diffusion curve dispersion, $\overline{\sigma^2}$ (the second central moment) is shown in Fig. 4. $\overline{\sigma^2}$ was calculated by the height–area method [16]:

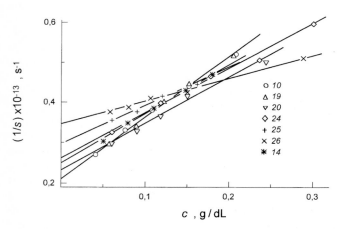

Fig. 3 $1/s$ versus solute concentration for CMC(0.6) in DWC at 26 °C. Here and in Figs. 4 and 5 the numbers of the curves correspond to the numbers of the samples in Table 1

$\overline{\sigma^2} = (a^2/2^3)/[\text{argerf}(aH/Q)]^2$, where H and Q are the curve height and area, respectively, and "argerf" means the argument of the error function. The diffusion coefficient, D, was evaluated from the slope of the $\overline{\sigma^2}(t)$ dependence by $D = (1/2) \partial\overline{\sigma^2}/\partial t$, where t represents time. The D value was extrapolated to vanishing concentration. The correction to the concentration effect did not exceed 14%.

The fringe area, Q, was used to estimate the specific refractive increment by $dn/dc = (Q/c)(\lambda/abh)$, where λ is the wavelength of light. The average value $(dn/dc)_{546} = 0.15 \pm 0.02$ ml/g thus obtained agrees well with those known for molecular solutions of CMC(0.9) in different solvents [11] if we take into account differences in the refractive indices of the solvents. We may conclude that CMC(0.6) as well as CMC(0.9) forms a real solution in DWC.

Molecular weight

The molecular weight of the samples was calculated from sedimentation and diffusion data as obtained for CMC in DWC using the value of $(1 - \bar{v}\rho_0) = 0.42$ in combination with the Svedberg equation

$$M_{SD} = [RT/(1 - \bar{v}\rho_0)](s_0/D) . \qquad (1)$$

Here R is the gas constant and T the temperature in Kelvin. The degree of polymerization (DP) was calculated from DP $= M_{SD}/M_0$ where M_0, is the mass of a monomer unit. M_0 was evaluated by taking into account the difference in the DS values for the Na-CMC samples using $M_0 = 162.16 + 80 \times$ DS. Results are listed in Table 1.

Viscometry

The intrinsic viscosity, $[\eta]$, was obtained from the extrapolation of the relative viscosity increment, $\eta_i \equiv \eta_{sp}/c = (\eta - \eta_0)/\eta_0 c$, to vanishing solute concentration, $[\eta] = \lim(\eta_{sp}/c)_{c \to 0}$ as made according to the Huggins equation $\eta_{sp}/c = [\eta] + [\eta]^2 k_H c$, where k_H is the Huggins constant. A plot of η_{sp}/c versus c is shown in Fig. 5. Experimental data are well approximated by a linear function with the intercept $[\eta]$ and the slope which yields k_H. The average values of k_H are 0.43 ± 0.12 and 0.47 ± 0.12 for samples and fractions of CMC(0.6), respectively.

The intrinsic viscosity, $[\eta]$, was also determined in 0.5 M NaNO$_3$, where k_H was found to be 0.64 ± 0.15. The results are illustrated in Fig. 6. The $[\eta]$ values for CMC(0.6) as well as for CMC(0.9) obtained in DWC are obviously close to those obtained in salt water (points are close to the solid line drawn with a unit slope). We may conclude that CMC forms real molecular solutions

Table 1 Sedimentation properties of carboxymethyl cellulose (*CMC*) (0.6) in dilute water cadoxen (*DWC*) at 26 °C

Sample	Degree of substitution	Degree of polymerization	s_0 (S)	k_s (ml/g)	$[\eta]$ (dl/g)	$D \times 10^7$ (cm²/s)	dn/dc (ml/g)	$A_0 \times 10^{10}$ (erg K⁻¹ mol⁻¹/³)	σ_s/s_0
Samples									
10	0.57	2060	4.75	7.3	8.55	0.69	0.14	3.6	0.14
14	0.54	1700	4.40	5.0	6.55	0.78	0.14	3.4	0.08
11	0.59	1480	4.36	–	7.3	0.88	0.15	3.9	–
15	0.58	910	3.95	–	4.2	1.3	0.13	4.0	–
17	0.60	530	2.87	–	2.5	1.6	0.13	3.5	–
18	0.67	440	2.85	–	2.5	1.9	0.17	3.9	–
12	0.66	350	2.63	–	2.15	2.2	0.16	4.0	–
Fractions[a]									
19	0.64	1600	4.76	7.2	7.0	0.87	0.15	3.9	0.14
20	0.65	1440	4.35	5.0	5.5	0.87	0.13	3.5	0.08
22	0.69	1230	4.20	–	5.3	1.00	0.15	3.7	–
23	0.56	1050	3.82	–	4.7	1.09	0.16	3.7	–
24	0.54	990	3.70	3.9	4.0	1.13	0.12	3.5	0.15
25	0.71	750	3.39	3.25	3.2	1.3	0.15	3.5	0.16
21	0.68	570	3.03	–	2.4	1.53	0.14	3.4	–
27	0.65	540	2.95	–	2.25	1.6	0.15	3.4	–
26	0.63	500	2.90	1.76	2.3	1.7	0.16	3.6	0.23

[a] Polymers 19, 20, 21, 23, 24, 26, and 27 are fractions of the Na-CMC samples 10, 11, 12, 14, 15, 17, and 18, respectively
The values were obtained with using $s_0 = s(1 + 0.9[\eta]c)$

in both solvents used. An insignificant change in the size of the macromolecules (or low aggregation of the CMC molecules in salt water) may be responsible for the somewhat higher average $[\eta]$ value obtained for CMC in this solvent.

Hydrodynamic invariant

Sedimentation–diffusion properties of CMC(0.6) in DWC were used in calculating the hydrodynamic parameter, A_0, defined by [18]

$$A_0 = (D\eta_0/T)^{2/3}(R[s][\eta]/100)^{1/3} . \qquad (2)$$

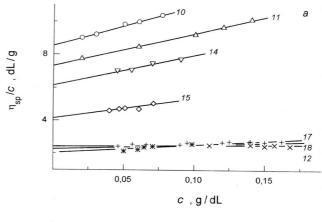

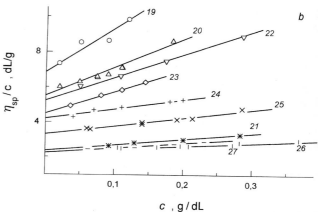

Fig. 5 Concentration dependence of a solution relative viscosity increment η_{sp}/c as measured for **a** samples and **b** fractions of CMC(0.6) in DWC at 26 °C

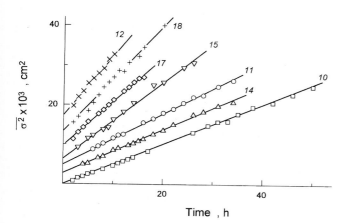

Fig. 4 Dispersion of the diffusion boundary, $\overline{\sigma^2}$, versus time for CMC(0.6) in DWC at 26 °C

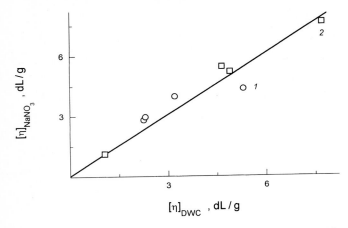

Fig. 6 Relation between the $[\eta]$ values as measured in 0.5 M NaNO$_3$ and in DWC for (*1*) CMC(0.6) and (*2*) CMC(0.9) at 26 °C

Here [s] is the characteristic value of the sedimentation coefficient defined as $[s] = s_0\eta_0/(1 - \bar{v}\rho_0)$. The average value of A_0 thus obtained for the CMC fractions is $(3.6 \pm 0.2) \times 10^{-10}$ erg K^{-1} mol$^{-1/3}$: it falls into the range $(3.5 \pm 0.8) \times 10^{-10}$ erg K^{-1} mol$^{-1/3}$ as reported for cellulose derivatives in water and in aqueous buffers [18].

Molecular-weight dependences

Molecular-weight dependences of hydrodynamic properties are usually described by the empirical Mark–Kuhn equations

$$s_0 = K_s(\mathrm{DP})^{1-b} \ , \tag{3}$$

$$[\eta] = K_\eta(\mathrm{DP})^a \ . \tag{4}$$

A logarithmic plot of these functions for the CMC fractions in DWC is presented in Fig. 7, showing a linear dependence over the DP range investigated. Thus, Eqs. (3) and (4) can be applied with the scaling exponents $1 - b$ and a given in Table 2. The values of $K_s = 22 \times 10^{-15}$ (if s is given in sin sec) and $K_\eta = 0.54$ (if $[\eta]$ is given in milliliters per gram) are valid for the CMC(0.6) fractions in DWC.

For unfractionated CMC(0.6) samples, one can see a wider scatter of the experimental points (marked as 1 in Fig. 7). The values for $1 - b$ and a of 0.36 ± 0.03 and 0.85 ± 0.07, respectively, and the coefficients $K_s = 31 \times 10^{-15}$ and $K_\eta = 1.4$ are determined here with higher error.

In every case the values of M_{SD} as obtained from sedimentation–diffusion data in DWC were used.

Inhomogeneity

Inhomogeneity was evaluated by the absolute method of analytical ultracentrifugation. The dispersion, $\overline{\sigma^2}$, of the

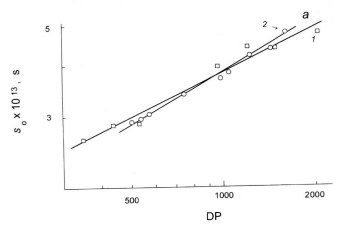

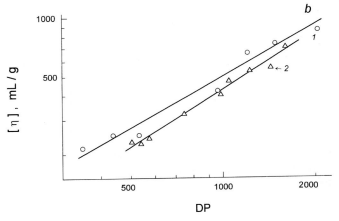

Fig. 7 Log-log dependence of **a** s_0 and **b** $[\eta]$ on the degree of polymerization (DP) for (*1*) samples and (*2*) fractions of CMC(0.6) in DWC at 26 °C

x spectrum was calculated at different times using the boundary profile as described previously [11]. A typical change in $\overline{\sigma^2}$ with time is shown in Fig. 8. This dependence is obviously not linear, the rate of change being profoundly affected by solute concentration. Note that in experiments with solute concentration exceeding 1.5 g/l, the $\overline{\sigma^2}$ points fall below the straight line (curve 6) describing the diffusion-spreading of the concentration boundary. Hence, polydispersity and diffusion-dispersion are both suppressed by concentration effects at these c values, and extrapolation of the data to infinite

Table 2 Numerical parameters of the Mark–Kuhn equations for the CMC fractions at 26 °C. The Cd content of the initial cadoxen of the DWC (10 + 1) was 6 wt%

Polymer	Solvent	$1 - b$	a	Ref.
CMC(0.6)	DWC	0.41 ± 0.02	0.96 ± 0.04	This work
CMC(0.9)[a]	DWC	0.40 ± 0.02	0.86 ± 0.07	11
CMC(0.9)[a]	Cadoxen	0.42 ± 0.04	0.84 ± 0.09	11

[a] Unfractionated samples

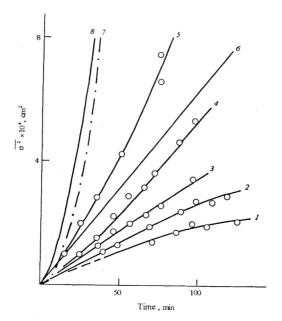

Fig. 8 Dispersion of the concentration boundary, $\overline{\sigma^2}$, during sedimentation run versus time for CMC(0.6), fraction 19 in DWC at 26 °C. Here and in Fig. 9 the solute concentrations are 2.05 (*1*), 1.52 (*2*), 1.22 (*3*), 0.89 (*4*), and 0.56 g/l (*5*). The *solid lines* guide the eye. Curve *6* presents diffusion-spreading of the boundary as calculated by $\overline{\sigma^2} = 2Dt$. Curves *7* and *8* present the hypothetical $\overline{\sigma^2}(t)$ dependences as found from the conditions $t \to \infty$ and $c \to 0$, respectively

time does not lead here to the elimination of diffusion effects, in contrast to the generally accepted opinion.

To eliminate concentration and diffusion effects with the aid of a recently proposed method [19, 20], the standard deviation, σ, of the x spectrum is plotted in Fig. 9 against the boundary position, x_m. As earlier [20],

we can approximate them with linear functions. The slope, $\partial\sigma/\partial x_m$, was determined and then extrapolated to vanishing concentration in the semilogarithmic coordinate axes of $\partial\sigma/\partial x_m$ versus Δs, where $\Delta s \equiv s_0 - s$ (Fig. 10). The result is $(\partial\sigma/\partial x_m)_0 = \lim (\partial\sigma/\partial x_m)_{\Delta s \to 0}$. The time dependence of $\overline{\sigma^2}_{c \to 0}$ at long times was approximated in terms of radial distance, x, by

$$\left(\overline{\sigma^2}_{c \to 0}\right)_{t \to \infty} = (\partial\sigma/\partial x_m)_0^2 (x_m - x_0)^2 \ , \tag{5}$$

where x_o is equal to x_m at the start time, $t = 0$. By using the definition of the sedimentation coefficient, s, given by $x_m = x_0 \exp(\omega^2 s_0 t)$, Eq. (5) was, rewritten in terms of the sedimentation time as

$$\left(\overline{\sigma^2}_{c \to 0}\right)_{t \to \infty} = A_1 [\exp(A_2 t) - 1]^2 \ . \tag{6}$$

Equation (6) obviously reflects the concentration-free dependence of $\overline{\sigma^2}$ on time provided by both diffusion and inhomogeneity effects.

On the other hand, during concentration-independent sedimentation, the dispersion of the x spectrum is known [17] to satisfy equation

$$\overline{\sigma^2} = 2Dt + x_m^2 \omega^4 \overline{\sigma_s^2} t^2 \ , \tag{7}$$

where $\overline{\sigma_s^2}$ is the dispersion of the s spectrum (distribution of the macromolecules with respect to sedimentation coefficients). To combine Eqs. (6) and (7) we have to find a power series (Eq. 8) which approaches Eq. (7) at the start time (mainly diffusion effects) and which, in turn, approaches Eq. (6) at long times when the boundary-spreading is mainly provided by heterogeneity effects.

$$\overline{\sigma^2}_{c \to 0} = A_3 t + A_4 x_m^2 \omega^4 t^2 \tag{8}$$

Here A_3 and A_4 are numerical coefficients. Comparison of Eqs. (7) and (8) leads to the conclusion that $A_3 = 2D$ and $\overline{\sigma_s^2} = \lim(t \to \infty) A_4$, where

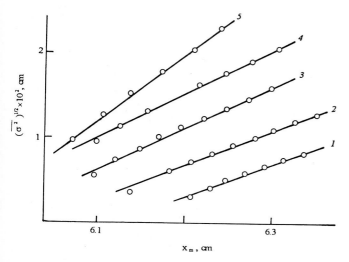

Fig. 9 Standard deviation, σ, of the x spectrum plotted against the boundary position, x_m, for CMC(0.6), fraction 19 in DWC at 26 °C and at the same solute concentrations as in Fig. 8

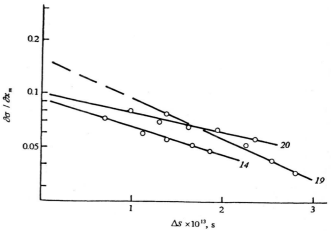

Fig. 10 $\partial\sigma/\partial x_m$ versus Δs for CMC(0.6) in DWC at 26 °C. The numbers of the curves correspond to the numbers of the samples in Table 1

$$A_4 = [(\overline{\sigma^2}_{c \to 0})_{t \to \infty} - 2Dt]/[x_0 \omega^2 t \, \exp(\omega^2 s_0 t)]^2 \ . \qquad (9)$$

The time-dependence of the A_4 value at long times was used to estimate the maximum value of A_4 that was accepted as $\overline{\sigma_s^2}$.

The results so obtained are given in Table 1 in terms of the relative standard deviation, σ_s/s_0, of the s spectrum. One can see that σ_s/s_0 varies from 0.08 to 0.23. Hence, according to sedimentation velocity data, inhomogeneity of the CMC(0.6) fractions and samples is low: this is typical for cellulose. Fractionation of the CMC(0.6) samples leads to residues with slightly higher DS values (by about 8%) but with the same inhomogeneity parameters with respect to the sedimentation coefficients. Nevertheless, study of the CMC fractions yields more precise molecular-weight functions of hydrodynamic properties and conformational parameters of the CMC macromolecules.

Degree of chain-coiling

A cadoxen content of 1 part per 10 parts water proved to be sufficient to suppress charge effects [7] and to secure a molecular state of dispersion of CMC in solution [11]. After assessing the conditions for a valid evaluation of sedimentation, diffusion and viscosity measurements on CMC solutions without disturbing charge effects, the dependence of the hydrodynamic parameters of CMC in DWC solutions on the DP of the polymer will now be discussed with respect to chain conformation.

The rather high values of b and a, exceeding 0.5, reflect noticeable draining of the CMC macromolecules in solution; this is confirmed by some other experimental observations [11]. Thus, for assessing the dimensions of the CMC molecules in DWC, the theory of the wormlike chain [22, 23] seems to be adequate, as it implies a draining effect. According to this theory, the $[s]$ value can be expressed by a linear function of the square root of the DP according to

$$[s] = (M_0/N_A P_\infty)(\lambda A)^{-1/2} DP^{1/2}$$
$$+ (M_0/3\pi\lambda N_A)[\ln(A/d) - \gamma] \ . \qquad (10)$$

Here $P_\infty = 5.11$, $\gamma = 1.056$ [23], N_A is Avogadro's number, and λ is the length of a monomer unit along a chain contour.

The slope of the dependence of $[s]$ on $(DP)^{1/2}$ is used to obtain the Kuhn statistical segment length, A, and the hydrodynamic chain diameter, d. The linearity of this dependence is demonstrated in Fig. 11. The A and d values thus obtained are listed in Table 3. The d values are compatible with the chemical structure of the polymer chain. The degree of coiling of the CMC(0.6) molecules in DWC solution is characterized by a Kuhn segment length of 15 ± 1 nm. This value agrees well

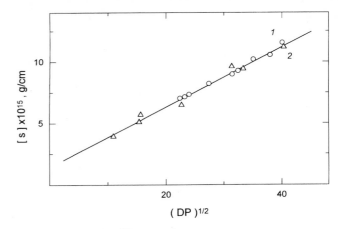

Fig. 11 $[s]$ versus $(DP)^{1/2}$ as obtained for the CMC(0.6) fractions (1) and for CMC(0.9) (2) in DWC at 26 °C

with the viscometric data but is somewhat lower than that for CMC(0.9), also obtained in DWC, $A = 19 \pm 2$ nm (Table 3).

Inhomogeneity of the CMC samples was taken into account as follows. According to Eq. (13) as given in Ref. [31] for heterogeneous samples, P_∞ in Eq. (10) should be replaced with $P_\infty \times q_{\text{corr}}^{-}$. The polymolecularity correction factor, q_{corr}, for CMC(0.6) samples (with the apparent value for M_z/M_w of 1.05–1.4) was evaluated from Table 9.1 in Ref. [31] to be less than or equal to 1.05. Hence, this correction does not exceed the experimental uncertainty (5%) in the A determination and can therefore be ignored. Nevertheless, treatment of the $[s]$-$(DP)^{1/2}$ data obtained for unfractionated samples by Eq. (10) leads to an unrealistically low d value (0.16 nm); therefore, the data obtained for fractions were considered here to be preferable to those of unfractionated samples.

It seems reasonable to compare the data above with those of unsubstituted cellulose on the one hand and those of highly substituted cellulose derivatives on the other, as obtained via analytical ultracentrifugation. Table 3 shows that the equilibrium rigidity of the cellulose chain is characterized by a segment length close to 10 nm, whereas for other compounds the A value is much higher.

The experimental values of the Kuhn segment length, A, obtained here can be compared with theoretical ones calculated on the assumption of a completely free rotation around the covalent links of the main chain. A parameter of hindrance to intramolecular rotation, σ, is defined as $\sigma = (A/A_{\text{fr}})^{1/2}$, where A_{fr} is the value of A for a completely free rotation within the polymer chain, amounting to 1.21 nm for a cellulose chain [32]. As can be seen from Table 3, the σ values for CMC as well as for other cellulose derivatives are rather high, significantly exceeding a value of 3, which is considered as an upper limit for flexible macromolecules even in the

Table 3 Conformational parameters of some cellulose esters in solutions

Polymer (degree of substitution)	Solvent	A (nm)	d (nm)	σ	Ref.
CMC(0.9)	DWC	19 ± 2	1.1 ± 0.5	4.0	This work
CMC(0.6)	DWC	15 ± 1	0.7 ± 0.2	3.5	This work
CMC(0.9)	Cadoxen	13 ± 2	0.6 ± 0.2	3.3	24
Cellulose	Cadoxen	11–14	0.1–1	2.7–3.4	25–27
Cellulose nitrate (2.3)	Ethylacetate	20.0	1.0	4.1	28
Cyanoethyl cellulose (2.6)	Acetone	35.0	1.4	5.4	29
Cellulose carbanilate (2.67)	Ethylacetate	30.0	1.1	5.0	30

presence of bulky side groups. Considering these results, it may be assumed that with CMC as well as with other cellulose derivatives, additional intramolecular interactions, especially the formation of hydrogen bonds, are effective in enhancing the equilibrium rigidity of the macromolecules.

DS and temperature effects

It is hardly possible to discuss the DS effects in sedimentation properties of the CMC samples studied because of their small differences in DS values (0.6 and 0.9). A more significant effect was observed, for example, in the transport properties of cellulose nitrates in ethylacetate. Here, the A value increases 2.5 times along the change in the DS value from 2 to 3 (open squares, Fig. 12). For cellulose carbanilate (CC) molecules with DS = 2.1, 2.2, and 2.67, the segment length, A, is 16, 19, and 30 nm, respectively [30, 37, 38]. The same tendency is noticeable for CMC (Table 3).

A change in solvent and temperature may also significantly affect the sedimentation behavior of the cellulose ester molecules in solution [33]. For example, the equilibrium rigidity of the CC(2.67) molecules in dilute solution in ethylacetate increases two fold with a temperature decrease from 50 to 2.5 °C (filled squares, Fig. 12).

Hence, the sedimentation behavior of the cellulose derivatives in dilute solutions depends strongly on the chemical structure of the substituents, the degree of substitution, and the nature of the solvent. It is also affected by temperature and inhomogeneity (origins) of the samples. Particularly strong interactions between substituents lead to 3–4 times higher A values (CN, CC, cyanoethyl cellulose). Weakening of these interactions (provided by a lower DS) leads to lower A values approaching values of native cellulose.

Conclusions

In accordance with sedimentation velocity data, complexation of CMC (DS = 0.6–0.9) molecules in cadoxen is similar to that for native cellulose [39]. In a solution of CMC in DWC, complexation and polyelectrolyte effects are practically absent, and sedimentation analysis of the inhomogeneity of the samples and the evaluation of the conformation of the CMC chain seem to be most successful in this solvent [40]. The Kuhn segment length of 15 nm as found here for CMC(0.6) in DWC is close to that of unsubstituted cellulose and somewhat lower than that obtained for CMC(0.9) in the same solvent. Obviously, within the range of molecular weights investigated here, the molecular conformation of both cellulose and CMC complies with the model of a draining Gaussian coil with a sufficiently large number of statistical segments in the equivalent chain. On the other hand, the equilibrium rigidity of CMC(0.6) molecules is definitely lower than that of highly substituted cellulose derivatives. This difference is caused by the still comparatively low DS as well as by the rather low degree of bulkiness of the carboxymethyl substituents.

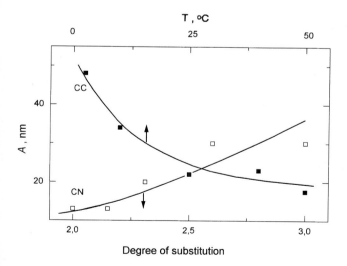

Fig. 12 Temperature (filled squares) and degree of substitution effects (open squares) in equilibrium rigidity (segment length A) of cellulose carbanilate (*CC*) (2.67) [33] and cellulose nitrate (*CN*) [28, 34–36] molecules in solution as evaluated from sedimentation data

References

1. Harding SE, Rowe AJ, Horton JC (eds) (1992) Analytical ultracentrifugation in biochemistry and polymer science. Royal Society of Chemistry, Cambridge
2. Brandrup J, Immergut EH (eds) (1989) Polymer handbook. Wiley-Interscience, New York
3. Sitaramaiah G, Goring DAJ (1962) J Polym Sci 58:1107
4. Dautzenberg H, Dautzenberg H, Linow K-J (1978) Faserforsch Textiltech 29:593
5. Gelman RA (1982) J Appl Polym Sci 27:2957
6. Yamamoto T, Mori Y, Ookubo N, Hayakawa R, Wada Y (1982) Colloid Polym Sci 260:20
7. Okatova OV, Lavrenko PN, Tsvetkov VN, Dautzenberg Ho, Philipp B (1989) Acta Polym 40:297
8. Philipp B, Dautzenberg H, Lukanoff B (1979) Vysokomol Soedin Ser A 21:2579
9. Borsa J, Reicher J, Ruzznak I (1992) Cell Chem Technol 26:261
10. Jayme G, Neuschaeffer K (1957) Naturwissenschaften 44:62
11. Lavrenko PN, Okatova OV, Tsvetkov VN, Dautzenberg H, Philipp B (1990) Polymer 31:348
12. Vink H (1957) Ark Kemi 11:29
13. Görnitz E, Linow K-J (1992) In: Harding SE, Rowe AJ, Horton JC (eds) Analytical ultracentrifugation in biochemistry and polymer science. Royal Society of Chemistry, Cambridge, pp 26
14. Tsvetkov VN, Skazka VS, Lavrenko PN (1971) Polym Sci USSR 13:2530
15. Lavrenko PN, Okatova OV, Khokhlov KS (1977) Instrum Exp Tech (USSR) 20:1477
16. Tsvetkov VN (1951) J Exp Theor Phys 21:701
17. Lavrenko PN, Linow K-J, Goernitz E (1992) In: Harding SE, Rowe AJ, Horton JC (eds) Analytical ultracentrifugation in biochemistry and polymer science. Royal Society of Chemistry, Cambridge, pp 517
18. Tsvetkov VN, Lavrenko PN, Bushin SV (1984) J Polym Sci Polym Chem Ed 22:3447
19. Lavrenko PN, Okatova OV (1994) Polymer 35:2137
20. Lavrenko PN, Okatova OV, Philipp B, Dautzenberg H (1995) In: Kennedy JF, Phillips GO, Williams PO, Piculell L (eds) Cellulose and cellulose derivatives. Woodhead, Cambridge, pp 177
21. Baldwin RL, Williams JW (1950) J Am Chem Soc 72:4325
22. Hearst J, Stockmayer WH (1962) Chem Phys 37:1425
23. Yamakawa H, Fujii M (1973) Macromolecules 63:407
24. Lavrenko PN, Okatova OV, Filippova NV, Shtennnikova IN, Tsvetkov VN, Dautzenberg H, Philipp B (1986) Acta Polym 37:663
25. Brown W, Wikström R (1965) Eur Polym J 1:1
26. Henley D (1962) Ark Kemi 18:327
27. Lubina SY, Klenin SI, Strelina IA, Troitskaya AV, Khripunov AK, Urinov EU (1977) Vysokomol Soedin Ser A 19:244
28. Pogodina NV, Lavrenko PN, Pozhivilko KS, Melnikov AB, Kolobova TA, Marchenko GN, Tsvetkov VN (1982) Polym Sci USSR 24:358
29. Tsvetkov VN, Lavrenko PN, Andreeva LN, Mashoshin AI, Okatova OV, Mikriukova OI, Kutsenko LI (1984) Eur Polym J 20:823
30. Andreeva LN, Urinov E, Lavrenko PN, Linow K-J, Dautzenberg H, Philipp B (1977) Faserforsch Textiltech 28:117
31. Bareiss RE (1989) In: Brandrup J, Immergut EH (eds) Polymer handbook. Wiley-Interscience, New York, pp 149
32. Holtzer AM, Benoit H, Doty PJ (1954) J Phys Chem 58:624
33. Lavrenko PN, Urinov EU, Andreeva LN, Linow K-J, Dautzenberg Ho, Philipp B (1976) Polym Sci USSR Ser A 18:2948
34. Pogodina NV, Melnikov AB, Mikryukova OI, Didenko SA, Marchenko GN (1984) Vysokomol Soedin Ser A 26:2515
35. Pogodina NV, Melnikov AB, Yevlampieva NP (1984) Khim Drev 6:3
36. Newman S, Loeb L, Conrad CM (1953) J Polym Sci 10:463
37. Lavrenko PN, Riumtsev EI, Shtennikova IN, Andreeva LN, Pogodina NV, Tsvetkov VN (1974) J Polym Sci Symp 44:217
38. Andreeva LN, Lavrenko PN, Urinov E, Kutsenko LI, Tsvetkov VN (1975) Vysokomol Soedin Ser B 17:326
39. Huglin MB, O'Donohue SJ, Sasia PM (1988) J Polym Sci Part B Polym Phys 26:1067
40. Lavrenko PN, Okatova OV, Dautzenberg H, Philipp B (1993) Cellul Chem Technol 27:468

Progr Colloid Polym Sci (1999) 113 : 201–204
© Springer-Verlag 1999

G.A. Morris
A. Ebringerova
S.E. Harding
Z. Hromadkova

UV tagging leaves the structural integrity of an arabino-(4-O-methylglucurono)xylan polysaccharide unaffected

G.A. Morris (✉) · S.E. Harding
NCMH Unit,
School of Biological Sciences,
University of Nottingham,
Sutton Bonington,
Leicestershire LE12 5RD, UK
e-mail: scxgam@szn1.agric.nottingham.
ac.uk
Tel.: +44-115-9516197
Fax: +44-115-9516142

A. Ebringerova · Z. Hromadkova
Institute of Chemistry
Slovak Academy of Sciences
SK-842 38 Bratislava, Slovakia

Abstract The ultracentrifuge is a useful tool for probing the effects of chemical or other types of substitution/mutation in macromolecules, although very few studies have been performed on polysaccharides. In this study we demonstrate that the substitution of hydroxyl groups on the polysaccharide xylan by p-carboxybenzyl (CB) bromide groups has little observed effect on molecular weight and little apparent effect on conformation (as monitored by sedimentation velocity and intrinsic viscosity). Corn-cob arabino-(4-O-methylglucurono)xylan (AGX) and its chemically modified derivative CB-AGX were found to have intrinsic viscosities of (82.8 ± 0.4) ml/g and (77.6 ± 0.5) ml/g, respectively. Sedimentation coefficients, $s_{20,w}^0$, were calculated to be (1.72 ± 0.06) S and (1.77 ± 0.06) S and (weight-average) molecular weights, M_w, from sedimentation equilibrium were found to be (37 000 ± 1500) g/mol and (40 000 ± 3000) g/mol. From these results we concluded that the UV tagging of arabinoxylan AGX does not significantly change the structural integrity of the xylan molecule; however, the decrease in intrinsic viscosity may indicate a very slight conformational change.

Key words UV tagging · Arabinoxylans · Carboxybenzyl · Analytical ultracentrifugation

Introduction

A lot of interest has been directed towards monitoring the effect of substitution or mutation of groups or residues on proteins using the ultracentrifuge; however, the ultracentrifuge can be equally useful in monitoring such effects on polysaccharides. In this study we aim to demonstrate whether the substitution of hydroxyl groups by p-carboxybenzyl (CB) bromide has any observed effect on molecular weight and/or conformation (as monitored by sedimentation velocity and intrinsic viscosity).

Xylan, a structural polysaccharide of the hemicellulose class, is the second most abundant naturally occurring organic compound [Ref. [1] and references therein]. Xylan is a major component of plant cell walls and can be found at levels up to 30% in some hardwoods. Xylan is often covalently linked to other cell wall components, such as lignin, and is thought to form an interface between cellulose and lignin in the plant cell wall [1] (Fig. 1).

The arabinoxylans of wheat grains compose approximately 70% of the endosperm cell wall and are therefore the principle source of fibre in flour. Water-soluble arabinoxylans leached from the endosperm form high-viscosity solutions able to hold 10 times their weight in water [3].

The xylan backbone is built up essentially by $(1 \rightarrow 4)$ linked β-D-xylopyranose units:

$$\cdots \rightarrow 4)\beta\text{-D- xyl}p\ (1 \rightarrow \cdots$$

O-acetyl groups substitute many of the OH groups at C2 and C3 of the xylans in their native state [1].

In this study the particular xylan examined was known as AGX from corn cob: it is of the arabino-(4-O-

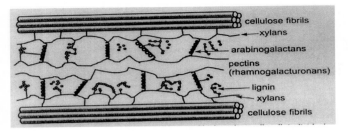

Fig. 1 Simplified diagrammatic representation of a typical higher plant cell wall. (From Ref. [2])

methylglucurono)xylan type typical of softwoods and annual plants. The chemical modification is achieved by reaction with excess CB bromide in the presence of aqueous sodium hydroxide [4].

Materials and methods

Samples

The heteroxylan, AGX was isolated from an alkaline extract of the mill corn cob by ethanol precipitation [5]. Moisture contents were calculated to be 9.96% and 10.68% for AGX and CB-AGX, respectively. Sample purity was investigated spectrophotometrically by the measurement of the UV absorbance spectra in the range 200–320 nm on an LKB Biochrom Ultraspec 4050 (Fig. 2); this was also used to distinguish spectral differences between native and chemically modified samples. The degree of substitution was calculated (by photometric titration of H^+ [6]) to be approximately 24% of the monomeric sugars. Prior to the hydrodynamic analysis it was necessary that both the sample and the buffer were in identical ionic environments, and it was therefore imperative that the sample was dialysed against the appropriate buffer. This process of dialysis against a large excess of buffer allows the sample to be treated as a simple two-component system and therefore simplifies calculations accordingly [7]. In general 24 h is considered

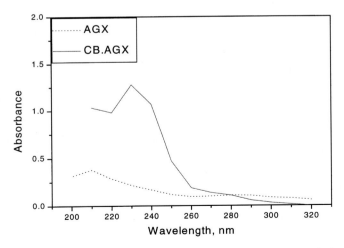

Fig. 2 UV absorption profiles for corn-cob arabino-(4-O-methylglucurono)xylan (*AGX*) and *p*-carboxybenzyl (*CB*)-AGX in a phosphate–chloride buffer, pH 7.5, $I = 0.1$ mol/l. Concentrations both 0.1 mg/ml

to be long enough for the sample and dialysate to reach isoionic equilibrium. All samples were dialysed for approximately 24 h against an ionic strength of 0.1, pH 7.5 phosphate–chloride buffer [8] of composition $Na_2 HPO_4 \cdot 12 H_2O$ (5.444 g); KH_2PO_4 (0.589 g) and NaCl (2.923 g).

Viscometry

Solutions and reference solvents were analysed using a 2-ml automatic Oswald viscometer (Schott-Geräte, Mainz, Germany), under precise temperature control (25.00 ± 0.02 °C).

Sedimentation velocity

Arabinoxylan samples of different concentrations were prepared for use in both Beckman XL-A and XL-I (Beckman Instruments, Palo Alto) analytical ultracentrifuges. A rotor speed of 50 000 rpm and a 4-mm column length in 12-mm optical path length double-sector cells were used together with an accurately controlled temperature of 20.0 °C. We can take advantage of the fact that both AGX and CB-AGX absorb in the UV (Fig. 2) and therefore both UV absorption and interference optics were used to record solute distributions. The $g(s^*)$ (sedimentation-time-derivative) method was used with the program DC/DT to determine apparent sedimentation coefficients ($s_{T,b}$) at each concentration by the procedure described in Ref. [9]. $s_{T,b}$ values were corrected to standard conditions of viscosity and temperature of water at 20 °C to give $s_{20,w}$ [10].

$$s_{20,w} = s_{T,b}[(1 - \bar{v}\rho_{20,w})\eta_{T,b}]/[(1 - \bar{v}\rho_{T,b})\eta_{20,w}] \ , \qquad (1)$$

where $s_{20,w}$ is the sedimentation coefficient expressed in terms of the standard solvent of water at 20.0 °C, $s_{T,b}$ is the measured sedimentation coefficient in the experimental solvent at the experimental temperature, T, $\eta_{T,b}$ and $\eta_{20,w}$ are the viscosities of the solvent and water at the experimental temperature and at 20.0 °C, respectively, and $\rho_{20,w}$ and $\rho_{T,b}$ are the corresponding densities. A partial specific volume, $\bar{v}$, of (0.625 ± 0.006) ml/g was used for both native and modified samples [11]. $s_{20,w}$ values were extrapolated to zero concentration using the standard equation [12]

$$s_{20,w} = s_{20,w}^0(1 - k_s c), \qquad (2)$$

where the Gralen parameter [13], k_s (ml/g), is a measure of concentration dependence.

Sedimentation equilibrium

Cells were scanned every 4 h at 18 000 rpm until two consecutive scans gave identical optical records. The centrifuge was then accelerated to 55 000 rpm in order to obtain a baseline required for data analysis using the optical absorption system. All absorbance and interference data sets were analysed using the MSTARA and MSTARI programs [14], respectively.

Results and discussion

The UV absorption profiles (Fig. 2) show that both native and modified samples appear to be free from any peptide contaminants (or at least those containing aromatic groups). It is also clear that the introduction of the CB group causes a marked change in the spectophotometric properties of the samples.

Reduced viscosity versus concentration plots shown in Fig. 3 reveal only a small decrease (about 6%) in

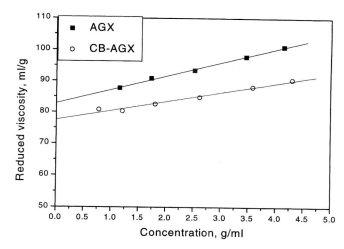

Fig. 3 Reduced viscosity versus concentration for AGX and CB-AGX. Buffer conditions as for Fig. 2

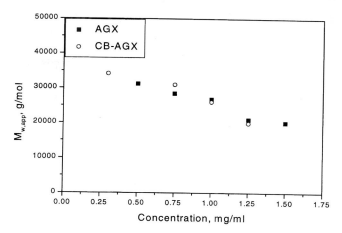

Fig. 5 Apparent molecular weight (from low-speed sedimentation equilibrium) versus cell loading concentration for AGX and CB-AGX

intrinsic viscosity, from (82.8 ± 0.4) ml/g to (77.6 ± 0.6) ml/g, caused by chemical substitution. It is worth pointing out that these values may be partially affected by possible systematic dilution errors caused by dialysis, although we would like to stress that both samples received identical dialysis protocols. (Such dialysis errors would not affect $s_{20,w}^0$ or M discussed later, since concentration is not an integral part of the function, unlike reduced viscosity, and concentration errors do not affect the extrapolated $c = 0$ values).

Sedimentation coefficients (Fig. 4) for both native AGX and CB-AGX do not show any dramatic decrease

with increase in concentration, and follow almost identical behaviour, with $s_{20,w}^0 = (1.72 \pm 0.06)$ S and (1.77 ± 0.06) S, respectively.

Molecular weight behaviour (Fig. 5) is also indistinguishable between the native and substituted xylans, showing the same decrease in $M_{w,app}$ with concentration through non-ideality. Extrapolated "ideal" values are also very similar: $M_w = (37\,000 \pm 1500)$ g/mol and $(40\,000 \pm 3000)$ g/mol, respectively.

Conclusions

As there appear to be no significant differences in the sedimentation coefficients or molecular weights on chemical substitution it would appear that the chemical substitution of the aromatic tag CB has no effect on the physical integrity of the arabinoxylan AGX, with no major conformational change or no induced self-association. The small decrease in intrinsic viscosity may possibly indicate a slight conformational change. It is possible, of course, that increasing the degree of substitution could accentuate this minor affect (or conversely, reducing the degree of substitution could remove it completely). Our conclusions are nonetheless similar to those of Errington et al. [15] for "blue dextran" and [16] for di-iodotyrosine labelled dextran, viz. labelling of polysaccharides for visualisation purposes does, at least in these cases, not seriously affect their structural integrity.

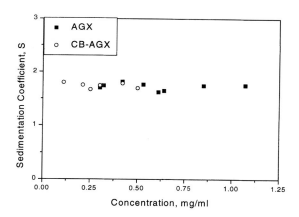

Fig. 4 Apparent sedimentation coefficient versus sedimenting concentration for AGX and CB-AGX. Buffer conditions as in Fig. 2

References

1. de Gruyter G (ed) (1984) Wood chemistry, ultrastructure, reaction. Springer, Berlin Heidelberg, New York
2. Tombs MP, Harding SE (1998) An introduction to polysaccharide biotechnology. Taylor and Francis, London
3. Stephen AM (ed) (1995) Food polysaccharides and their applications. Dekker, New York
4. Ebringerova A, Novotna M, Kacurakova M, Machova E (1996) J App Polym Sci 6:1043
5. Ebringerova A, Simkovic I, Hromadkova H, Toman R (1988) Czech Patent CS 244591; Chem Abs 109:8277k
6. Kohn R, Hromadkova H, Ebringerova A, Toman R (1986) Collect Czech Chem Commun 51:2243
7. Cassassa EF, Eisenberg H (1964) Adv Protein Chem 19:287
8. Green AA (1933) J Am Chem Soc 55:2331
9. Ralston G (1993) Introduction to analytical ultracentrifugation. Beckman Instruments, Fullerton, USA
10. Dhami R, Harding SE, Elizabeth NJ, Ebringerova A (1995) Carbohydr Polym 28:113
11. Stafford WF (1992) In: Harding SE, Rowe AJ, Horton JC (eds) Analytical ultracentrifugation in biochemistry and polymer science. Royal Society of Chemistry, Cambridge, pp 359–393
12. Pavlov GM (1997) Eur Biophys J 25:385
13. Gralen N (1944) PhD dissertation. University of Uppsala, Sweden
14. Cölfen H, Harding SE (1997) Eur Biophys J 25:333
15. Errington NC, Harding SE, Rowe AJ (1992) Carbohydr Polym 17:151
16. Errington NC, Harding SE, Illum L, Schacht EH (1992) Carbohydr Polym 18:289

Progr Colloid Polym Sci (1999) 113: 205–208
© Springer-Verlag 1999

G.A. Morris
S.N.G. Butler
T.J. Foster
K. Jumel
S.E. Harding

Elevated-temperature analytical ultracentrifugation of a low-methoxy polyuronide

G.A. Morris (✉) · S.N.G. Butler
K. Jumel · S.E. Harding
NCMH Unit
School of Biological Sciences
University of Nottingham
Sutton Bonington, Leics. LE12 5RD, UK
e-mail: scxgam@szn1.agric.notting-
ham.ac.uk
Tel.: +44-115-9516197
Fax: +44-115-9516142

T.J. Foster
Product Microstructure
Unilever Research, Colworth House
Sharnbrook, Beds. MK44 1LQ, UK

Abstract Relatively little has been published on the ultracentrifuge behaviour of macromolecular solutions at elevated temperature (> 40 °C). In this study we look at the sedimentation velocity behaviour of one particular food grade polyuronide – pectin – from 20 °C to 60 °C in a specially adapted Model E ultracentrifuge. Reduced specific viscosity measurements were also determined over the same temperature range. A small decrease in the reduced viscosity, together with a similar increase in $s_{20,w}$, suggests that the pectin chain is more flexible at elevated temperatures, but that the overall molecular integrity remains intact.

Key words Beckman Model E adaptation · Viscometry · Size exclusion chromatography-multi-angle laser light scattering

Introduction

Relatively little has been published on the ultracentrifuge behaviour of macromolecular solutions at elevated temperature (> 40 °C). This is surprising, considering that, in many food or pharmaceutical processes, macromolecules are exposed to elevated temperatures. Attempts to increase our knowledge of the behaviour of these molecules, in terms of structural integrity and state of oligomerisation, would therefore be useful.

In this short study we look at the behaviour of one particularly important food grade polyuronide type of polysaccharide – pectin. Pectins are major components of plant cell walls and are composed of $1 \rightarrow 4$ linked galacturonic acid residues. Many of these acidic residues are substituted with methyl groups, forming methyl esters (methoxy esters); the degree of esterification (DE)/degree of methoxylation (DM) plays an important part in the properties of pectins. Pectins of DE $> 50\%$ are regarded as high methoxy (HM) and DE $< 50\%$ as low methoxy (LM) pectins. Galacturonic acid chains are often interrupted with rhamnose sugars which result in a "kink" in the chain. There are quite often so-called "hairy regions" which contain abnormally large amounts of arabinose and galactose side chains [1].

Materials and methods

Materials

The low methoxy pectin (degree of esterification $\sim 38\%$) of high degree of purity was a gift from Citrus Colloids (Hereford, UK). The material was solubilised for 24 h in pH 6.8, ionic strength 0.1, standard phosphate/chloride buffer [2], of the following composition: $Na_2HPO_4 \cdot 12H_2O$, 4.595 g; KH_2PO_4, 1.561 g; and NaCl, 2.923 g, all made up to 1 l [2] and then dialysed against this buffer.

Viscometry

Solutions and reference solvents were analysed using a 2 ml Ostwald viscometer (Schott-Geräte, Mainz, Germany) under precise temperature control (± 0.01).

Elevated temperature sedimentation velocity

The high temperature unit (Fig. 1) for the Beckman Model E analytical ultracentrifuge allows centrifugation at temperatures up to 125 °C; however, for aqueous solutions, temperatures > 80 °C are impractical. Since parallel measurements were being made with size exclusion chromatography-multi-angle laser light scattering (SEC-MALLS), and the latter set-up has a high temperature limit of 60 °C, we restricted our ultracentrifuge studies to this upper limit for temperature. Both the rotor and sample cells are pre-heated to the required temperature prior to sample injection. Samples were run at 52 640 rpm and Schlieren images were captured semi-automatically onto photographic film. $s_{T,b}$ values were calculated from Schlieren images using a graphics digitising tablet. $s_{20,w}$ values were then calculated from $s_{T,b}$ values (sedimentation coefficients at temperature, T and in buffer, b) using the standard equation (see, for example, [3]):

$$s_{20,w} = s_{T,b}\left[(1 - \bar{v}\rho)_{20,w}\eta_{T,b}\right] / \left[(1 - \bar{v}\rho)_{T,b}\eta_{20,w}\right] \qquad (1)$$

where $s_{20,w}$ is the sedimentation coefficient of the macromolecular component expressed in terms of the standard solvent density and viscosity of water at 20.0 °C; $s_{T,b}$ is the measured sedimentation coefficient in the experimental solvent at the experimental temperature, T; $\eta_{T,b}$, $\eta_{20,w}$ are the viscosities of the solvent and water at the experimental temperature and 20.0 °C, respectively; $\rho_{20,w}$, $\rho_{T,b}$ are the corresponding densities. A partial specific volume (for the macromolecular component) of 0.63 ± 0.01 was used and assumed to be constant with temperature.

There are many additional problems in the use of elevated temperature analytical ultracentrifugation. At higher temperatures the maximum rotor speed must be reduced by ~1% per 10 °C above 40 °C. Care has to be taken in the choice of gaskets, centrepieces and windows: centrepieces used were of the KEL-F type. Oil condensation on the chamber lenses is a major problem, which results in poor image resolution.

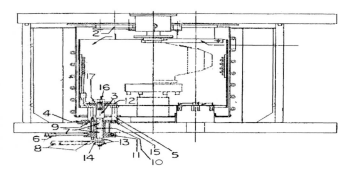

Fig. 1 The radiation can for the Model E high temperature unit. *1* can assembly, *2* baffle assembly, *4* feed through drain assembly. All others are minor components [4]

Size exclusion chromatography coupled to multi-angle laser light scattering

SEC-MALLS allows on-line light scattering of a heterogeneous solute fractionated by size exclusion chromatography. This results in absolute molecular weights (molar masses) and molar mass distributions [5–7]. The Wyatt Technology (Santa Barbara, USA) Dawn F multi-angle laser light scattering photometer (Wyatt, 1992) [8] was coupled to PSS (Mainz, Germany) Hema Bio linear (300 mm × 7.5 mm) and PSS Hema Bio 40 (300 mm × 7.5 mm) columns, both protected by a similarly packed guard column. The eluent was the standard pH 6.8, $I = 0.1$ buffer and the injection volume was 100 μl. The pectin refractive increment (dn/dc) of 0.15 ml/g was assumed to be independent of temperature.

Results and discussion

From Table 1 and Fig. 2 we see that increase in temperature does not affect the unimodality of the sedimenting boundary, as seen in earlier studies [9]. Although $s_{T,b}$ shows a marked increase with increase in temperature (1.57 S to 4.22 S from 20 °C to 60 °C, respectively), after correction to standard conditions of solvent viscosity and density [Eq. (1)] the change is not spectacular (1.57 S to 1.92 S) (Fig. 3). This corresponds to a small decrease in reduced specific viscosity η_{red} from 268 ± 1 ml/g to 241 ± 5 ml/g (Fig. 4). An increase in sedimentation coefficient together with a decrease in η_{red} would suggest a slight change in conformation to a more compact form.

It is important also to establish the state of degradation/association of the material before conclusions over conformation can be drawn, since both the sedimentation coefficient and intrinisc viscosity will be affected by mass as well as shape/hydration considerations; SEC-MALLS provided this. Over the temperature range 30–60 °C the molar mass as studied by SEC-MALLS remains relatively constant at approximately $115\,000 \pm 15\,000$ g/mol (Fig. 5), with no sign of dissociation. This suggests that molecular breakdown is not occurring and therefore any changes in sedimentation coefficient and reduced specific viscosity are the result of conformational changes.

Conclusions

The apparent uniformity in molar mass over the temperature range together with the slight increase of $s_{20,w}$ and

Table 1 The effect of temperature on the sedimentation coefficient and reduced viscosity for LM pectin HL7192

Temperature	$s_{T,b}$ (S)	$s_{20,w}$ (S)	η_{red} (ml/g)	M_w (g/mol)
20	1.57 ± 0.03	1.59 ± 0.03	268 ± 1	–
30	2.05 ± 0.03	1.62 ± 0.03	263 ± 2	$107\,900 \pm 10\,000$
40	2.54 ± 0.04	1.64 ± 0.04	257 ± 2	$113\,900 \pm 10\,000$
50	3.31 ± 0.04	1.78 ± 0.04	243 ± 1	$112\,100 \pm 10\,000$
60	4.22 ± 0.05	1.92 ± 0.05	241 ± 5	$124\,800 \pm 10\,000$

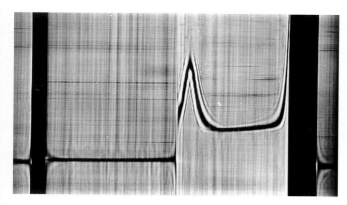

Fig. 2 Schlieren peak for LM pectin HL7192, temperature 60 °C; concentration 2.3 mg/ml; rotor speed 52 640 rpm; $t = 2220$ s, $s_{obs} = 4.22$ S and $s_{20,w} = 1.92$ S

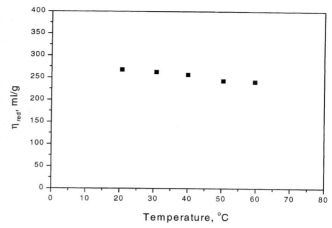

Fig. 4 The effect of increased temperature on the reduced viscosity for LM pectin HL7192

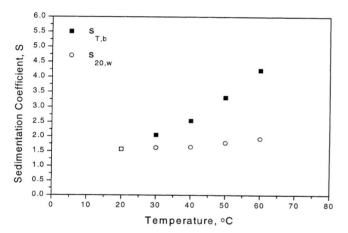

Fig. 3 The effect of increased temperature on s_{obs} and $s_{20,w}$ for LM pectin HL7192

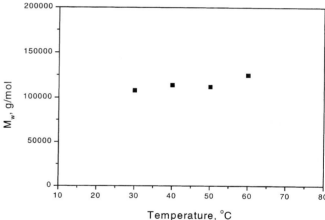

Fig. 5 The effect of increased temperature on the reduced viscosity for LM pectin HL7192

the small decrease in reduced specific viscosity at increased temperatures suggests only a small conformational change. [Because measurements were made at only a single concentration, there is the possibility that this conclusion may be coloured by the fact we have assumed that (1) changes in reduced viscosity reflect changes in intrinsic viscosity (2) the change in $s_{20,w}$ at this concentration reflects changes in $s^{\circ}_{20,w}$ (i.e. at $c = 0$). As noted above, we have also assumed the partial specific volume has not significantly changed at elevated temperature.]

A low-methoxy pectin molecule at room temperature can be thought of as a rod/semi-flexible rod type molecule. At elevated temperatures, increased chain flexibility is apparent; this is most likely due to increased mobility with the input of thermal energy, together with the partial breakdown of inter- and intra-molecular forces. Our observation that pectin undergoes a partial transition but remains intact at increased temperatures, is important as pectin is often used as a stabiliser in heat treatment processes. This can possibly give some further insight into the complex chemistry involved.

Acknowledgements We would like to thank Mr. Les Sarcoe and Mr. Pete Husbands for their expert technical assistance, particularly with installation of the high temperature unit, and to the B.B.S.R.C. and Unilever Research for their financial assistance.

References

1. Tombs MP, Harding SE (1998) An introduction to polysaccharide biotechnology. Taylor and Francis, London
2. Green AA (1933) J Am Chem Soc 25: 2331
3. Pavlov GM (1997) Eur Biophys J 25: 385
4. Beckman Spinco Service Manual (1974). Beckman Instruments, Palo Alto, Calif
5. van Holde KE (1985) Physical biochemistry, 2nd edn. Prentice-Hall, Englewood Cliffs, NJ
6. Jumel K, Browne P, Kennedy JF (1992) In: Harding SE, Sattelle DB, Bloomfeld VA (eds) Laser light scattering in biochemistry, chap 2. Royal Society of Chemistry, Cambridge
7. Jumel K (1994) Ph D Thesis, University of Nottingham
8. Wyatt PJ (1992) In: Harding SE, Sattelle DB, Bloomfeld VA (eds) Laser light scattering in biochemistry, chap 3. Royal Society of Chemistry, Cambridge
9. Harding SE, Berth G, Ball A, Mitchell JR, Garcia de la Torre J (1991) Carbohydr Polym 16:1

Progr Colloid Polym Sci (1999) 113:209–211
© Springer-Verlag 1999

D.J. Scott
S. Leejeerajumnean
J.A. Brannigan
R.J. Lewis
A.J. Wilkinson
J.G. Hoggett

Towards deconvoluting the interaction of the *Bacillus subtilis* sporulation proteins SinR and SinI using tryptophan analogue incorporated proteins

D.J. Scott (✉) · J.G. Hoggett
Department of Biology
University of York
Heslington, York YO10 5DD, UK

S. Leejeerajumnean · J.A. Brannigan
R.J. Lewis · A.J. Wilkinson
York Structural Biology Laboratory
Department of Chemistry
University of York,
York YO10 5DD, UK

Abstract Sporulation in *Bacillus subtilis* is used as the strategy of last resort for survival of the organism and it is a very tightly controlled developmental process. One of the control checkpoints that must be overcome for sporulation to occur is the repression of sporulation genes by the protein SinR (13.5 kDa). This is done by binding of the anti-repressor SinI (6.5 kDa) to form a tightly bound heterodimer. To investigate the interaction of SinI with SinR in solution, an analytical ultracentrifuge study was undertaken. SinR was found to be a tetramer, whereas SinI was in a monomer/dimer equilibrium.

Derivatives of both SinI and SinR, where the native tryptophan was replaced by the analogue 7-aza-tryptophan (7AW), were expressed and found to be as active as the wild-type proteins. The 7AW proteins have the property of having significant absorbance at 315 nm, thus allowing them to be monitored even in the presence of tryptophan containing proteins, making them ideal for studying protein/protein interactions.

Key words Sporulation proteins · *Bacillus subtilis* · Tryptophan · 7-Azatryptophan · Analytical ultracentrifugation

Introduction

The way in which an organism controls its development involves networks of protein/protein and protein/nucleic acid interactions. The regulatory mechanism employed by organisms to coordinate growth and development have been intense topics of research. Sporulation, particularly in the model organism *Bacillus subtilis*, is a much studied example of developmental control: under normal conditions the cell divides in a symmetric manner; however, under certain environmental conditions the cell undergoes asymmetric division, with the final formation of a spore and decay of the mother cell. As such a process is a strategy of last resort for the cell, it is tightly regulated via a series of controls or checkpoints which must be overcome for sporulation to occur [1, 2].

The proteins SinR and SinI are involved in the early decision to proceed to sporulation in *B. subtilis*. SinR (monomer M_r = 13.5 kDa) is a repressor protein which inhibits the transcription of genes required for sporulation, by binding in a cooperative fashion to DNA. SinI (monomer M_r = 6.5 kDa) is its agonist, which binds to SinR forming a tightly bound heterodimer. The SinR/SinI complex can no longer bind DNA in a cooperative fashion; as a consequence, sporulation genes are expressed. Hence, understanding the interaction between SinR and SinI will shed light on our understanding of the decision making process occuring in the cell on whether to proceed to sporulation [3, 4].

The crystal structure of the SinR/SinI complex has been solved recently in our group [5]; it shows a tight interaction in which the two proteins interpenetrate, much in the manner of a handshake (see Fig. 1). There is a range of complementary interactions occurring between the proteins, and indeed it is possible to place a second molecule of SinR in place of SinI and still preserve the same contacts [5].

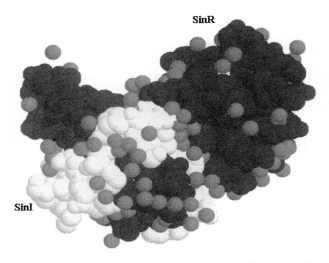

SinR

SinI

Fig. 1 A view of the crystal structure of the SinR/SinI heterodimer displayed as a space filling model. SinI is shown in *white*, while SinR is shown in *dark grey*. Water molecules that were resolved in the crystal structure are shown in *mid-grey*. It can be seen that the proteins form a tight interlocking complex, in the manner of a handshake

Thus, it is easy to see how a molecule of SinI could replace one of SinR in a SinR dimer. To provide information about the behaviour of SinI and SinR in solution, we are currently investigating their interaction using a range of biophysical techniques. We report here how the use of tryptophan analogues incorporated into proteins can simplify the interpretation and analysis of analytical ultracentrifuge results.

Materials and methods

Analytical ultracentrifugation

Sedimentation equilibrium experiments on SinR and SinI were carried out in Tris (pH 7.5) and 150 mM NaCl at 20 °C. Loading concentrations were 0.1, 0.25 and 0.5 mg/ml of each protein. Rotor speeds for SinR were 12 000, 16 000, 22 000 and 28 000 rpm. SinI was subjected to sedimentation equilibrium at 22 000, 26 000, 28 000 and 32 000 rpm.

Results

Equilibrium took around 12 h in the case for SinR and 8 h for SinI. At the end of the run the rotor speed was increased to 36 000 rpm to obtain a true optical baseline. Analysis of the data obtained from these experiments found that SinR had an apparent molecular weight of 53 000 ($\pm$500) kDa, corresponding to a tetrameric protein, whereas SinI was found to be in a reversible monomer/dimer equilibrium with a dissociation constant (K_d) of 4 $\pm$ 2 μM. These data correspond very well with that found by gel filtration [3]. The results are summarised in Table 1.

Table 1 Summary of the sedimentation equilibrium results on SinR and SinI

Species	Mol. wt. (Da)	Oligomeric state
SinR	53 000	Single species (tetramer)
7AWSinR	53 000	Single species (tetramer)
SinI	6500	2M $\leftrightharpoons$ M$_2$

Discussion

Analysis of the interaction of SinI and SinR is complicated by two factors: firstly, that each species is involved in higher order oligomeric states; and secondly, that the dimeric form of SinI has about the same molecular weight as monomeric SinR. In order to deconvolute these complex binding processes, it is desirable to have information about the distribution of the two separate species, SinI and SinR, in the centrifuge cell. This can be accomplished, as detailed in an accompanying paper [6], by biosynthetically incorporating the tryptophan analogue, 7-azatryptophan (7AW), into proteins. This has the property of red-shifting the absorbance spectra some 20 nm; therefore by using wavelengths of around 310–315 nm it is possible to observe the 7AW labelled protein, while any native tryptophan containing proteins will be transparent [7–13]. We have produced 7AWSinR in appreciable amounts and found that it functions in a native fashion [13]. Figure 2 shows a diagram of the absorbance spectra for native SinR and 7AWSinR; it can be seen that there is significant absorbance of the 7AWSinR in the wavelength range of 300–320 nm.

Additionally, it was found by sedimentation equilibrium that this protein was tetrameric (see Table 1), showing that the 7AW incorporated protein had the same

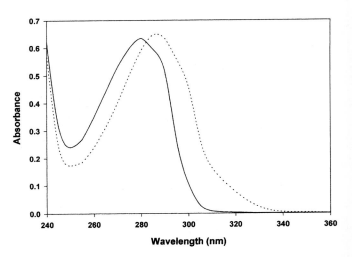

Fig. 2 Ultraviolet absorbance spectra of native and 7AW incorporated SinR. The sample of 7AWSinR was found to have >95% incorporation of the analogue, by examination of its spectra

oligomeric state as the native protein. As only 7AWSinR has significant absorbance at 315 nm, it is possible to monitor it in the analytical ultracentrifuge using absorbance optics even in the presence of excess SinI.

A constant concentration of 7AWSinR (60 μM monomer) was incubated with increasing concentrations of native SinI and analysed by sedimentation equilibrium. Each mixture was scanned at 280, 315 and 400 nm. Each of these wavelengths provides important related information on the system. The data at 280 nm contain information about the whole of the system under study, as both SinR and SinI absorb at this wavelength. The data at 315 nm provide information about the distribution of free 7AWSinR and bound 7AWSinR. Therefore, data obtained at these two wavelengths provide complementary information on the system, and one of its components that can be used to deconvolute the details of the binding process. The scan at 400 nm is used to check that Weiner skewing is not occurring [15]. Details of this are in our accompanying paper [6].

To analyse the data, we calculated the whole cell molecular weight average ($M_{w,app}^0$) for each of the scans at a set speed using the program MSTARA [16, 17]. If there is no interaction occurring between 7AWSinR and SinI, then the whole cell molecular weight average will simply be that of the tetramer. Any deviation from this will be due to interaction occurring between 7AWSinR and SinI. Table 2 summarises these results. It can be seen clearly that the value of ($M_{w,app}^0$) falls with increasing SinI concentration. This shows that there is an interaction between SinI and 7AWSinR under these conditions. Significantly, the value of ($M_{w,app}^0$) falls from that of the tetramer to close to that expected for the heterodimer (20 kDa), indicating that the binding of SinI disrupts the SinR tetramer, causing this observed decrease.

Table 2 Whole cell molecular weight averages ($M_{w,app}^0$) calculated for the data at 315 nm for a fixed concentration of 7AWSinR with varying concentrations of SinI

7AWSinR (μM)	SinI (μM)	$M_{w,app}^0$
60	1.50	50 600
60	5.00	45 300
60	30.0	33 200
60	45.0	26 000
60	150	19 000

There appear to be no stable high molecular weight intermediates formed from multiple copies of SinI binding to a SinR tetramer.

Conclusions

We have shown here and elsewhere [6, 13, 14] that it is possible to produce functionally active proteins biosynthetically that have the tryptophan analogue 7-azatryptophan in place of the naturally occurring tryptophan. These proteins can be used in binding studies in the analytical ultracentrifuge in place of the native protein, and, as a consequence of the unique absorbance signal at 315 nm, molecular weight information can be derived from these distributions about only one of the components in the system. With 7AWSinR and SinI we are able to show how a potentially complicated interaction (a tetramer interacting with a monomer/dimer equilibrium) can be deconvoluted into simpler binding equilibria. We are currently using direct methods of detailed analysis [18, 19] of the data at 280 and 315 nm to further this aim.

Acknowledgements This work has been supported by the BBSRC (UK) and the Wellcome Trust (UK).

References

1. Errington J (1993) Microbiol Rev 57: 1–33
2. Hoch JA (1993) Annu Rev Microbiol 47:441–465
3. Gaur NK, Oppenheim J, Smith I (1991) J Bacteriol 173:678–686
4. Losick R, Stragier P (1992) Nature 355:601–604
5. Lewis RJ, Brannigan JA, Offen WA, Smith I, Wilkinson AJ (1998) J Mol Biol 283:907–912
6. Scott DJ, Ferguson AL, Buck M, Gallegos M-T, Pitt M, Hoggett JG (1999) Prog Colloid Polym Sci 113:212–215
7. Ross JBA, Szabo AG, Hogue CWV (1997) Methods Enzymol 278:151–190
8. Callaci S, Heyduk T (1998) Biochemistry 37:3312–3320
9. Hogue CWV, Szabo AG (1993) Biophys Chem 48:159–169
10. Soumillion P, Jespers L, Vervoort J, Fastrez J (1995) Protein Eng 8: 451–456
11. Rusinova E, Ross JBA, Laue TM, Sowers LC, Senear DF (1997) Biochemistry 36:12 994–13 003
12. Wong C-Y, Eftink MR (1998) Biochemistry 37:8947–8953
13. Scott DJ, Leejeerajumnean S, Brannigan JA, Lewis RJ, Wilkinson AJ, Hoggett JG (1999) (Submitted to J Mol Biol)
14. Scott DJ, Ferguson AL, Buck M, Gallegos M-T, Pitt M, Hoggett JG (1999) (Submitted to Biochemical Journal)
15. Svensson H (1956) Opt Acta 3:164–183
16. Harding SE, Horton JC, Morgan PJ (1992) In: Harding SE, Rowe AJ, Horton JC (eds) Analytical ultracentrifugation in biochemistry and polymer science. Royal Society of Chemistry, Cambridge, pp 275–294
17. Cölfen H, Harding SE (1997) Eur Biophys J 25:333–346
18. Wills PR, Jacobsen MP, Winzor DJ (1996) Biopolymers 38:119–130
19. Winzor DJ, Jacobsen MP, Wills PR (1998) Biochemistry 37:2226–2233

Progr Colloid Polym Sci (1999) 113:212–215
© Springer-Verlag 1999

D.J. Scott
A.L. Ferguson
M. Buck
M-T. Gallegos
M. Pitt
J.G. Hoggett

Use of the tryptophan analogue 7-azatryptophan to study the interaction of σ^N with *Escherichia coli* RNA polymerase core enzyme

D.J. Scott · A.L. Ferguson
J.G. Hoggett (✉)
Department of Biology
University of York, Heslington
York YO10 5DD, UK

M. Buck · M.-T. Gallegos · M. Pitt
Department of Biology
Imperial College of Science
Technology and Medicine
Biomedical Sciences Building
Imperial College Road
London SW7 2AZ, UK

Abstract The minor sigma factor σ^N of *Escherichia coli* RNA polymerase (RNAP) has been labelled with the tryptophan analogue 7-azatryptophan (7AW) by biosynthetic incorporation. This has the effect of shifting the absorbance spectra of the protein so there is appreciable absorbance at 315 nm. Consequently this makes 7AWσ^N an ideal tool to study the protein/protein interaction of σ^N and RNAP using the absorbance optics in the XL-A analytical ultracentrifuge.

Key words RNA polymerase · *Escherichia coli* · 7-Azatryptophan · Sigma factor · Analytical ultracentrifugation

Introduction

Interactions between of proteins of dissimilar size and shape are essential for cellular function, and consequently protein/protein interactions have formed the basis of many studies in molecular biology. Analytical ultracentrifugation provides a rigorous thermodynamic approach to the determination of molecular weight information and to the study of protein/protein interactions. One problem which has to be overcome, however, is that all proteins give similar signal responses under all optical regimes in the ultracentrifuge. This is especially true of absorbance optics: all proteins containing aromatic residues have similar absorbance properties in the 250–300 nm wavelength region and hence are virtually indistinguisable at most measured wavelengths, making decomposition of the signal into the individual contributions of the components difficult. Recently, it has become possible to incorporate analogues of trytophan such as 5-hydroxytryptophan (5OHW) and 7-azatryptophan (7AW) biosynthetically into proteins in place of tryptophan. The incorporation of either 5OHW or 7AW shifts the absorbance spectra of the proteins so that there is significant absorbance at 315 nm, making them detectable at wavelengths where the absorption of tryptophan containing proteins is transparent [1–6]. As the steric differences between these residues and tryptophan are small (see Scheme 1), substitution is not often accompanied by alteration in functional properties. This makes them ideal for the study of protein/protein interactions in the analytical ultracentrifuge using absorbance optics.

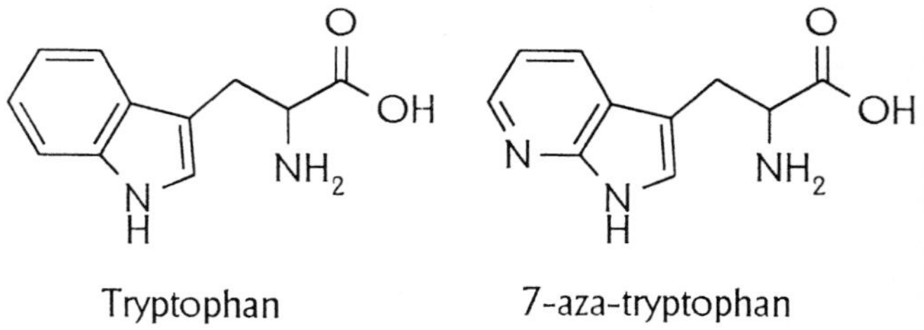

Tryptophan   7-aza-tryptophan

Scheme 1

The methodology has been used to study the binding of the minor sigma factor σ^N (σ^{54}) to *Escherichia coli* core RNA polymerase (RNAP) to form the holoenzyme. Binding of sigma factors to RNAP directs the holoenzyme towards specific classes of genes and therefore is an important feature of selective transcriptional activation in bacteria. σ^N differs from all of the other known sigma factors not only in sequence homology, but in its requirement for an activator protein and the turnover of ATP before being able to open the DNA strands and start transcription. In this it resembles eukaryotic RNA pol II, and as such its mechanism of interaction is a topic of active research as a paradigm of activator protein dependent transcription [7–9]. σ^N is able to bind DNA in the absence of RNAP with a K_d of around 1 μM, as well as being able to bind RNAP on its own in the absence of DNA. However, the detailed characteristics of the binding to RNAP are not known. We have expressed and characterised $7AW\sigma^N$, assayed it functionally with RNAP and activator proteins, and characterised its binding in the analytical ultracentrifuge.

Materials and methods

Incorporation of 7-azatryptophan into σ^N

A high level of incorporation of $7AW\sigma^N$ into the protein is achieved by expression of the protein in the *E. coli* tryptophan auxotroph strain CY15077 (a kind gift of Prof. C. Yanofsky) in Terrific Broth media at 37 °C [1, 11]. Once the cells had grown to $OD_{600} = 0.8$, the media was changed to M9 [10], and after 30–40 min was supplemented with 100 mg/500 ml of 7AW (Sigma-Aldrich, Poole, Dorset, UK). Expression was induced by the addition of IPTG to a final concentration of 1 mM. Protein was purified by refolding from inclusion bodies as detailed previously [11, 12]. Levels of incorporation were estimated to be typically around 70–90%, based on analysis of absorbance spectra (Fig. 1). Purified $7AW\sigma^N$ was assayed for both core RNAP binding and activator protein-dependent transcription and found to act in a wild-type fashion [11], showing that the $7AW\sigma^N$ protein is a suitable substitution for native σ^N in functional assays.

Sedimentation velocity studies

Sedimentation velocity runs were carried out on an XL-A analytical ultracentrifuge (Beckman, Palo Alto, Calif., USA) at 40 000 rpm for σ^N and 30 000 rpm for RNAP using Epon filled centrepieces. Scans were taken every 5 min. Sedimentation coefficients were determined using DCDT [13] and were found to be 3.0 ± 0.2 s and 3.1 ± 0.2 s for native σ^N and $7AW\sigma^N$, respectively. This shows that both proteins have the same hydrodynamic conformation in solution, thus demonstrating that incorporation of 7AW does not cause any gross conformational changes in protein structure.

Sedimentation equilibrium

In order to fully exploit the unique spectroscopic properties of $7AW\sigma^N$, sedimentation equilibrium was employed. By monitoring the signal at 315 nm it is possible to determine the distribution of free and bound $7AW\sigma^N$ by calculating the apparent whole cell weight averaged molecular weight ($M^0_{w,app}$) in the presence of RNAP. To obtain $M^0_{w,app}$, equilibrium traces were analysed using the program MSTARA [14, 15].

The M^* function is defined as:

$$\frac{C(r) - C_a}{M^*(r)} = kC_a(r^2 - a^2) + 2k \int_a^r r[C(r) - C_a]dr \qquad (1)$$

Table 1 Summary of $M^0_{w,app}$ data determined at 315 nm for σ^N:RNAP titrations. The value of $M^0_{w,app}$ for the holoenzyme is 430 kDa. The error in each measurement falls in absolute terms with RNAP concentration. This is because the rotor speed (7000 rpm) is more appropriate for measurements of molecular weights the size of the RNAP complexes, rather than the free $7AW\sigma^N$

$7AW\sigma^N$ (μM)	RNAP (μM)	$M^0_{w,app}$ (kDa)
7.1	0	89 ± 30
7.1	0.52	166 ± 25
7.1	1.57	357 ± 33
7.1	7.84	380 ± 47

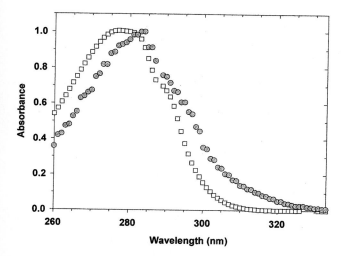

Fig. 1 Ultraviolet absorbance spectra of native σ^N (*open squares*) and $7AW\sigma^N$ (*shaded circles*). It can clearly be seen that the 7AW substituted protein has significant absorbance in the 300–320 nm region. Levels of incorporation of 7AW into the protein, estimated from the absorbance spectra, were typically 70–90%

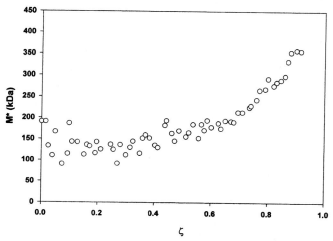

Fig. 2 M^* extrapolation for $7AW\sigma^N$ and RNAP determined from data taken at 315 nm. The concentration of both species was 7.1 μM. As can be seen, the value of M^* tends towards 380–400 kDa. $\zeta = (r^2 - a^2)/(b^2 - a^2)$, where a = radial position of the meniscus and b = radial position of the cell base

It was found that $M^0_{\text{w,app}}$ in the absence of RNAP at 315 nm was around 89 ± 30 kDa (Table 1). The reason for the magnitude of the error was that the rotor speed has to be low enough to see the distribution of core RNAP (378 kDa), thus making the determination of the free $7\text{AW}\sigma^N$ $M^0_{\text{w,app}}$ difficult. However, upon addition of core RNAP, $M^0_{\text{w,app}}$ increased close to a value of 430 kDa, that which would be expected for a 1:1 complex of σ^N and RNAP. A typical M^* plot is shown in Fig. 2.

Discussion

In this paper we show that proteins which have 7AW incorporated in place of tryptophan can be used to study protein/protein interactions. There are several points about the use of these tryptophan analogue containing proteins that merit a mention. Firstly, the expression of the protein must be very tightly controlled so that it only occurs after the growth media has been exchanged for M9 media supplemented with 7AW. Any "leakage" of the promoter will decrease the final level of incorporation. Secondly, we have also found, as originally suggested by Ross et al. [1], that the level of incorporation is increased by starving the cells after changing the media for 1 h (in the case of Ross et al.) or 30–45 min (in our case). This decreases the level of tryptophan in the cells, and therefore increases the final level of 7AW incorporation.

Thirdly, the 7AW protein must be assayed in conjunction with the trytophan containing protein so as determine whether the labelling procedure has affected activity. Fourthly, it is important that the species that is labelled should be chosen so as to maximise the difference in observed molecular weight of the free and complexed forms. In the present case, this means that the smaller protein, σ^N, should be labelled. This gives a better resolution to the data compared with the error associated with each data point owing to the extrapolation procedure in the MSTARA program. Fifthly, it is preferable that each titration should be done at a single rotor speed so as the only quantity that varies in the distribution is the concentration of the tryptophan containing protein (in this case RNAP). This will mean that any higher order complexes present will not sediment with increased rotor speed and therefore

give an erroneous value of $M^0_{\text{w,app}}$. Finally, although the absorption of the tryptophan containing protein is negligible at 315 nm, it will still contribute to the signal if it is present at high concentrations owing to Weiner skewing [16]. This phenomenon is due to the refractive index change that occurs at high protein concentrations. The importance of this phenomenon can be appraised, by scanning at 400 nm as well as at 280 and 315 nm (T. Laue, personal communication). As no absorbance from the proteins should occur at these wavelengths, any change in the apparent absorbance arises solely from skewing of the signal. In the present study it was found that skewing of the signal did not occur appreciably until the RNAP was in excess of 2.2 mg/ml. As the portion of the data that was affected corresponds to the bottom of the cell, this would account for the apparently lower value of $M^0_{\text{w,app}}$ of the final row in Table 1 (380 kDa vs. 430 kDa), where there are 1:1 stoichiometric amounts of both $7\text{AW}\sigma^N$ and RNAP.

Conclusion

7-Azatryptophan can be incorporated biosynthetically to high levels into σ^N to generate a functionally normal analogue protein. The unique absorbance properties of $7\text{AW}\sigma^N$ can be exploited effectively to study its interaction with core RNAP. The analytical ultracentrifuge, in combination with absorbance optics, has provided a method for studying this interaction by calculating the whole cell molecular weight average ($M^0_{\text{w,app}}$) at 315 nm. Provided the certain problems highlighted are avoided, this method provides an excellent way of studying heterogeneous protein/protein interactions in a rigorous manner.

Acknowledgements The authors would like to thank Dr. J.B.A. Ross for his kind donation of the absorbance spectra for tyrosine, tryptophan, 5-hydroxytryptophan and 7-azatryptophan. We would also like to thank Prof. A.J. Rowe and Dr. N. Errington of the NCMH, Nottingham, UK, for their support during the early part of the project. This work has been supported by a grant from the Wellcome Trust, UK.

References

1. Ross JBA, Szabo AG, Hogue CWV (1997) Methods Enzymol 278:151–190
2. Callaci S, Heyduk T (1998) Biochemistry 37:3312–3320
3. Hogue CWV, Szabo AG (1993) Biophys Chem 48:159–169
4. Soumillion P, Jespers L, Vervoort J, Fastrez J (1995) Protein Eng 8:451–456
5. Rusinova E, Ross JBA, Laue TM, Sowers LC, Senear DF (1997) Biochemistry 36:12 994–13 003
6. Wong C-Y, Eftink MR (1998) Biochemistry 37:8947–8953
7. Jishage M, Iwata A, Ueda S, Ishihama A (1996) J Bacteriol 178:5447–5451
8. Ishihama A (1997) In: Eckstein F, Lilley DMJ (eds) Nucleic acids and molecular biology, vol 11. Springer, Berlin Heidelberg New York, pp 51–70
9. Merrick M (1993) Mol Microbiol 10:903–909
10. Maniatis T, Fritsch EF, Sambrook J (1982) Molecular cloning: a laboratory manual. Cold Spring Harbor Laboratory, Cold Spring Harbor, NY

11. Scott DJ, Ferguson AL, Buck M, Gallegos M-T, Pitt M, Hoggett JG (1999) Biochem J (Submitted)
12. Cannon W, Missailidis S, Smith C, Cottier A, Austin S, Moore M, Buck M (1995) J Mol Biol 248:781–803
13. Stafford WF III (1992) In: Harding SE, Rowe AJ, Horton JC (eds) Analytical ultracentrifugation in biochemistry and polymer science. Royal Society of Chemistry, Cambridge, pp 359–393
14. Harding SE, Horton JC, Morgan PJ (1992) In: Harding SE, Rowe AJ, Horton JC (eds) Analytical ultracentrifugation in biochemistry and polymer science. Royal Society of Chemistry, Cambridge, pp 275–294
15. Cölfen H, Harding SE (1997) Eur Biophys J 25:333–346
16. Svensson H (1956) Opt Acta 3:164–183

Progr Colloid Polym Sci (1999) 113:216–220
© Springer-Verlag 1999

C. Walters
M. Cliff
A. Clarke
S.E. Harding

Nonequilibrium self-association of a cpn60 chaperonin induced by tryptophan mutation

C. Walters (✉) · S.E. Harding
NCMH Unit, University of Nottingham
School of Biological Sciences
Sutton Bonington
Leicestershire LE12 5RD, UK
e-mail: scxcw@szn1.nottingham.ac.uk
Tel.: +44-115-9516197
Fax: +44-115-9516142

M. Cliff · A. Clarke
Department of Biochemistry
University of Bristol, Tyndall Avenue
Bristol BS8 1TD, UK

Abstract The tryptophan mutant Y203W of the bacterial GroEL (cpn60) was studied with regard to its hydrodynamic integrity and its oligomeric state. Sedimentation equilibrium using MSTARI gave a weight-average molecular weight of $(905,000 \pm 33,000)$ Da. This is in excellent agreement with results from sedimentation velocity, which revealed three distinct species (19.6S, 26.5S and 38S) in the same proportions by weight for six different loading concentrations, corresponding to 14- and 28-mer subunit compositions with a smaller dissociation product. The relative amounts of each species present, $(23.0 \pm 0.8)\%$, $(61.5 \pm 2.9)\%$ and $(15.5 \pm 3.0)\%$, yielded an estimated weight-average molecular weight of about 870,000 Da. From this we conclude that the tryptophan mutation at the Y203 location causes significant irreversible self-association under the conditions used here, and appears to be yet another example of how sedimentation analysis can be used to probe the effects of a single amino acid substitution in a protein on the conformation and hence the oligomeric state of a protein assembly.

Key words Bacterial GroEL (cpn60) · Hydrodynamic integrity · Sedimentation equilibrium · Sedimentation velocity · Mutation

Introduction

The bacterial chaperonin GroEL (cpn60) has been shown to facilitate the folding and assembly of many unfolded (nonnative) proteins [1–4] and is also known to be part of a class of cpn60s found in mitochondria, chloroplasts and prokaryotic cells [5].

Each GroEL molecule consists of 14 identical monomers of about 57–60 kDa each. These are arranged as two stacked heptameric rings with a sevenfold symmetrical cylindrical complex consisting of apical, intermediate and equatorial domains. Saibil [6] has defined a set of hydrophobic patches in the apical domain area lining the opening of the channel of the GroEL molecule as the binding surface for nonnative proteins. These regions are apparently close to the surface of the cylinder and are facing inwards, which is thought to prevent self-aggregation of the GroEL. Recent studies [7] have shown that replacement of the tyrosine residue (Y203) in the apical region with a nonhydrophobic glutamate (E203) causes a total loss of substrate protein binding to the GroEL molecule and disrupts the binding of the co-chaperonin GroES. This is probably due to a combination of a small conformational change in the GroEL binding regions and reduced hydrophobic interactions with substrate proteins. Similarly, fluorescence studies on a GroEL mutant containing a single tyrosine-to-tryptophan replacement (Y203W) have revealed some increased hydrophobic exposure in the apical domain region of the molecule [3] suggesting that this point mutation causes a conformational change that leads to the increased exposure of the hydrophobic binding regions which are normally less exposed facing the inward binding cavity.

In this study we use sedimentation equilibrium and sedimentation velocity to determine the hydrodynamic stability and possible further self-association interactions of the tryptophan mutant GroEL Y203W of *Escherichia coli* due to possible increased hydrophobic exposure caused by the mutation.

Materials and methods

Preparation of the GroEL mutants was as described by Gibbons et al. [3] and the mutants were stored as precipitates in a 75% saturated ammonium sulphate buffer containing 50 mM tetra-ethylammonium sulphate (pH 7.5), 50 mM KCl and 20 mM MgCl$_2$. Prior to use the protein was resuspended in phosphate buffer of pH 7.0 and ionic strength $I = 0.1$ mol/l [8] containing 4.595 g/l Na$_2$HPO$_4 \cdot$ 12H$_2$O, 1.561 g/l KH$_2$PO$_4$ and 2.923 g/l NaCl (Fisher Scientific, Loughborough, UK), typically at 500 μl for every 1 ml of precipitate originally used, and was dialysed overnight at 4 °C against 1 l of the same buffer.

Sedimentation equilibrium analysis

Sedimentation equilibrium was performed using both an Optima XL-I (Beckman Scientific Inc, Palo Alto, USA) analytical ultracentrifuge, and for the lowest concentration a Beckman Model E analytical ultracentrifuge equipped with a laser light source and interference optics coupled in-house to a purpose-built diode array charge-coupled-device camera linked to a Macintosh computer for on-line analysis.

The low- or intermediate-speed method [9] was used in the Optima XL-I. At this speed the concentration at the meniscus remains finite and can be found from mathematical manipulation ("intercept/slope" method) of the fringe data. Molecular weights [including distribution weight averages, $M_{w,app}$, and point averages, $M_{w,app}(r)$] were obtained using MSTARI [10]. Double-sector cells of 12-mm optical path length were used with solution in one sector and buffer from the dialysis in the other. Sample loading concentrations of 0.5, 0.7, 1, 2, 2.35 and 4 g/l were used with a rotor speed of 6000 rpm and a temperature of 10 °C.

The Beckman Model E ultracentrifuge was employed taking advantage of the longer optical path length ultracentrifuge cells (30 mm) to run the 0.3 g/l sample. The conditions and methods used for analysis with the Model E ultracentrifuge were as described for the XL-I ultracentrifuge.

Sedimentation velocity

All sedimentation velocity experiments were performed in the XL-I ultracentrifuge using interference optics. The run speed and temperature were set at 40,000 rpm and 20 °C, respectively, throughout with a scan interval of 4 min. All sedimentation coefficients were determined using the time derivative, $g^*(s)$ analysis, software (DC/DT) provided by Stafford [11], and were corrected to standard conditions of 20 °C and water as the solvent ($s_{20,w}$) using Eq. (1) [12].

$$s_{20,w} = s_{T,b} \left(\frac{\eta_{T,b}}{\eta_{20,w}} \right) \left(\frac{1 - \bar{v}\rho_{20,w}}{1 - \bar{v}\rho_{T,b}} \right) \quad (1)$$

Results

Sedimentation velocity analysis

Sedimentation velocity data analysis, using DC/DT analysis software, shows the presence of three distinct sedimenting species present throughout all the captured data (Fig. 1) with $s^0_{20,w}$ values of (19.6 ± 0.4) S, (26.0 ± 0.9) S and (38.3 ± 1.9) S for the three $g^*(s)$ "peaks" (sedimentating boundaries), respectively. The total relative amount of each sedimenting species also remains essentially independent of cell loading concentrations giving mean percentage values of (23.0 ± 0.8)% for the 19.6S component, (61.5 ± 2.9)% for the 26.0S component and (15.5 ± 3.0)% for the 38.3S component. The consistency shown throughout the concentration ranges suggests that no concentration dependence of the ratio between the three components and hence no dynamic equilibrium is taking place at the concentration levels used in these experiments.

Sedimentation equilibrium

For sedimentation equilibrium analysis an extrapolated "ideal molecular weight", $(M_w) = M_{w,app}$ $(c \rightarrow 0)$, of (905,000 ± 33,000) Da was obtained from a plot of $M_{w,app}$ versus initial loading concentration (Fig. 2). Figure 3 shows a comparison of the point-average molecular weight distribution $M_{w,app}(r)$ versus local (fringe) concentration $J(r)$ at six different loading concentrations. These point-average molecular weight plots appear to be consistent with the $M_{w,app}$ versus loading concentration plot (Fig. 2) in showing the same trend: an increase in $M_{w,app}$ or $M_{w,app}(r)$ with initial increase in concentration (through the influence of solute heterogeneity) followed by a steady decrease due to nonideality effects.

Discussion

The sedimentation equilibrium results give an unequivocal demonstration of heterogeneity: for a normal 14-mer of GroEL we would expect a molecular weight of 800,000–840,000 Da. It is also clear that a significant proportion of the species of different molecular weight are not in thermodynamic equilibrium: this is because plots of point-average $M_{w,app}(r)$ versus fringe displacement $J(r)$ for different loading concentrations, c, do not overlap (Fig. 3): a purely reversibly associating system should give superposition [12]. This view appears to be strengthened by the results from sedimentation velocity (Fig. 1) where there is clear heterogeneity from the $g^*(s)$ versus s plots with three

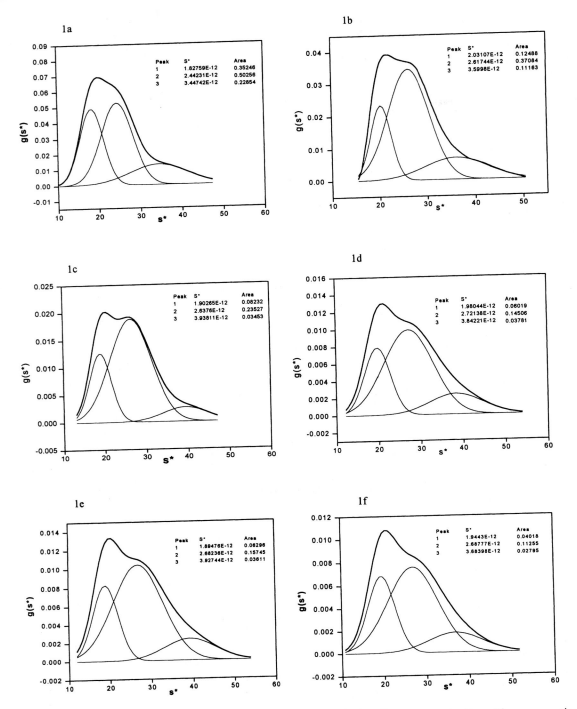

Fig. 1a–f Apparent distribution of sedimentation coefficients $g^*(s)$ for GroEL Y203W (in phosphate buffer pH 7.0, $I = 0.1$) from sedimentation velocity showing three distinct sedimenting species throughout the entire range of concentrations loaded into the ultracentrifuge cell. The routine DC/DT was used. **a** 0.3 g/l, **b** 0.5 g/l, **c** 0.7 g/l, **d** 0.8 g/l, **e** 1 g/l, **f** 1.1 g/l

discrete sedimenting species present, (19.6S, 26.0S and 38.3S), the proportions of which do not change with a change in loading concentration (the proportion of the more slowly moving species should increase with a decrease in c).

Molecular weights can also be estimated from the sedimentation coefficients, assuming the particles of each species are approximately spherical and employing the Mark–Houwink–Kuhn–Sakurada (MHKS) relation: $(s \sim M^b)$ with the MHKS coefficient of 0.667 for spheres

Fig. 2 Plot of weight-average molecular weight $M_{w,app}$, versus loading concentration, c, for GroEL Y203W (in phosphate buffer pH 7.0, $I = 0.1$). This gives an extrapolated M_w ($c \rightarrow 0$), of (905,000 $\pm$ 33,000) Da. The routine MSTARI was used for the analysis

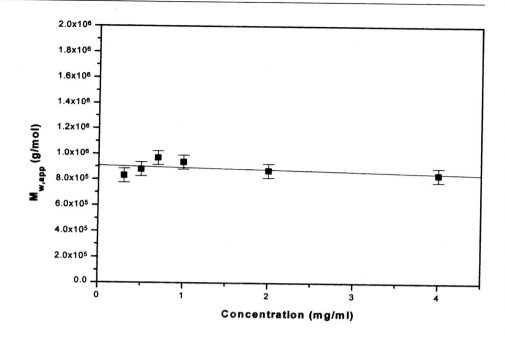

Fig. 3 Point weight-average molecular weights, $M_{w,app}(r)$, versus fringe displacement, $J(r)$, plots for different loading concentrations, c, of GroEL Y203W: 4 g/l ($\square$), 2 g/l ($\triangledown$), 1 g/l ($\bullet$), 0.7 g/l ($\circ$), 0.5 g/l ($\times$), 0.3 g/l ($\blacktriangle$)

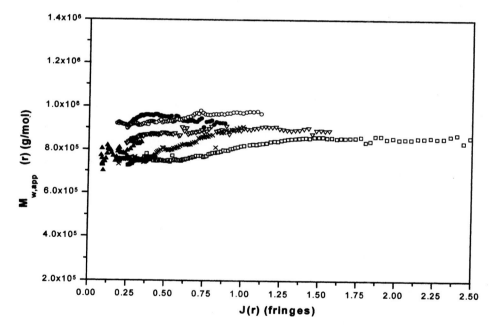

[14]. If we consider the main 26.0S (approximately 62% by weight) component as the 14-mer then the 38.3S species would correspond to a 28-mer (i.e. a dimer of 14-mers) as an aggregation product. However, the mutation appears to have caused some disruption of the GroEL 14-mer, with about 23% by weight of a 19.6S component, which from the MHKS relation would have a molecular weight of about (520,000–550,000) Da. From the relative concentrations of the three species, an estimated weight-average molecular weight of about

875,000 Da is obtained; this is well within the experimental error of the $M_{w,app}$ of (905,000 $\pm$ 33,000) Da determined directly by sedimentation equilibrium.

These observations may be compared with those of Gibbons et al. [3], who showed that the tryptophan mutation Y203W causes conformational change and an increased hydrophobic patch exposure in the GroEL molecule. These workers also observed that the tryptophan mutant Y203W still retained its ability to fold rhodanese but the folding yield obtained was of only

30% efficiency compared to 41% for wild-type GroEL, indicating a nonfatal change (insofar as functionality) in conformation had taken place. The present study appears to confirm that view.

References

1. Gething MJ, Sambrook J (1992) Nature 355:33–45
2. Saibil HR, Zheng D, Roseman AM, Hunter AS, Watson GMF, Chen S, auf der Mauer A, O'Hara BP, Wood SP, Mann NH, Barnett LK, Ellis RJ (1993) Curr Biol 3:265–275
3. Gibbons GL, Hixon JD, Hay N, Lund P, Gorovits BM, Ybarra J, Harowitz PM (1996) J Biol Chem 271:31989–31998
4. Behlke J, Ristau O, Schönfeld HJ (1997) Biochemistry 36:5149–5156
5. Hemmingsen SM, Woolford C, van der Vies SM (1988) Nature 333:330–334
6. Saibil HR (1994) Nature Struct Biol 1:838–842
7. Fenton WA, Kashi Y, Furtak K (1994) Nature 371:614–619
8. Green AA (1933) J Am Chem Soc 55:2331–2336
9. Creeth SE, Harding SE (1982) J Biochem Biophys Methods 7:25–34
10. Cölfen H, Harding SE (1997) Eur Biophys J 25:333–346
11. Stafford WF III (1992) In: Harding SE, Rowe AJ, Horton JC (eds) Analytical ultracentrifugation in biochemistry and polymer science. Royal Society of Chemistry, Cambridge, pp 359–393
12. Tanford C (1961) Physical chemistry of macromolecules. Wiley, New York
13. Roark DE, Yphantis DA (1969) Ann NY Acad Sci 164:245–278
14. Harding SE (1995) Biophys Chem 55:69–93